ADVANCED
ELECTRONIC PRACTICALS

ADVANCED ELECTRONIC PRACTICALS

ADVANCED ELECTRONIC PRACTICALS

[For Honours, B.Tech. and Postgraduate Students]

Dr A B Bhattacharya MSc, PhD, FIETE

Professor and Former Head, Department of Physics
University of Kalyani

New Central Book Agency (P) Ltd

LONDON
HYDERABAD ERNAKULAM BHUBANESWAR
NEW DELHI KOLKATA PUNE GUWAHATI

NCBA

REGD OFFICE

8/1 Chintamoni Das Lane, Kolkata 700009, India
e mail : ncbapvtltd@eth.net

OVERSEAS

NCBA (UK) Ltd. 149 Park Avenue North, Northampton, NN3 2HY, UK
e mail : ncbauk@yahoo.com

EXPORT

NCBA Exports Pvt. Ltd. 212 Shahpur Jat, New Delhi 110049
e mail : ncbaexports@ncbapvtltd.com

BRANCHES

2/27 Ansari Road, 1st Floor
Daryaganj, **New Delhi** 110 00 2
email: ncbadel@ncbapvtltd.com

Shop Nos 3& 4, Vinayak Towers
681/B Budhwar Peth, Appa Balwant Chowk
Pune 411 002
email: ncbapune@ncbapvtltd.com

Shop No -15, Bharati Towers, Ground Floor
Forest Park
Bhubaneswar 751 001
email: ncbabub@ncbapvtltd.com

House No 3-1-315, 1st Floor
Nimboliadda, Kachiguda, **Hyderabad** 500 027
email: ncbahydb@ncbapvtltd.com

GSS Shopping Complex, 1st Floor
Opposite College Post Office, Convent Road
Ernakulam 682 035
0484-4021054/53

Radhamayee Complex (Second Floor)
Opp. SBI, Panbazar, Jaswanta Road
Guwahati - 781 001
email: ncbaguwahati@ncbapvtltd.com

ADVANCED ELECTRONIC PRACTICALS. Dr A B Bhattacharya

First Published: July 2011
Reprinted: March 2015
Reprinted: 2018

PUBLISHER AND TYPESETTER
New Central Book Agency (P) Ltd
8/1, Chintamoni Das Lane, Kolkata 700 009

PRINTER
New Central Book Agency (P) Ltd
Web-Offset Division, Dhulagarh, Sankrail, Howrah

COVER DESIGNER
Soumen Pal

COVER PRINTER
Biswanath Studio, Kolkata

TECHNICAL EDITOR
Dipan Roy

PROJECT TEAM
Somenath Ganguly, Koushik Samanta and Prabhat Jas

ISBN: 978 81 7381 659 8

Contents

Section A

Advanced Electronic Practicals

Section B

Experiments with Brief Procedure

Section C

Advanced Questions for Viva Voce

Preface

Today, it can be very rightly said that we live in the age of electronics. Faced with a rapidly advancing technology causing a need for increasing number of technicians for an expanding industry, educational materials related to these are facing tremendous challenges. In fact, the pressing need for training programs to serve students of various backgrounds and abilities has also become evident owing to an increasing critical shortage of trained technicians.

This laboratory manual on '**Advanced Electronic Practicals**' provides detailed technical materials covering a closely coordinated program of experiments. The book is concerned with the fundamental building blocks in electronics technology—highlighting the students an understanding of the basic control circuits and their applications. It fills the need for basic training in the complex field of industrial instrumentation to produce a technician tailored to the industry's needs.

Digital techniques have been known for years. But they are now used virtually in every area of electronics—in computers, in communications, in industrial controls, and consumer equipments and where not! The aim of this book is to present, in a very simple and practical manner starting from preliminary experiments on vacuum tubes, semiconductors, transistors, FETS, Thyristors to various advanced level and application oriented experiments in a very systematic, lucid and illustrative way. With a view to provide a deeper understanding of the subject, a good number of oral questions with answers and exercises have also been included.

I sincerely acknowledge assistance obtained from my colleagues and students—of my University and also from other universities and colleges—with whom I discussed freely during the long preparation of the manuscript. My special thanks are due to Mr Amitabha Sen, publisher of this book, and to the staff members of New Central Book Agency (P) Ltd, particularly to Mr Prabhat Jas, Mr Himangshu Acharya, Mr Dipan Roy, Mr Soumen Pal and Mr Koushik Samanta for helps in many ways.

For different experiments in this book suitable scale divisions along the X- and Y-axes suggested on the graphs have been drawn for illustration only.

I believe that the users of this book will find the information useful and interesting. Any suggestion for its improvement shall be gratefully accepted and acknowledged.

January 24, 2011 **A B Bhattacharya**

SECTION A

ADVANCED ELECTRONIC PRACTICALS

CHAPTER 1

ELECTRONIC PRELIMINARIES

VACUUM TUBES

1.1 Electron Emission

Electrons perform a certain amount of work to escape from the surface of emitter. To do this work energy is imparted to them from some external sources such as heat energy, kinetic energy of electric charges, energy stored in electric or magnetic field and the light energy. Accordingly, the emission of electrons from a metal surface is classified as follows :

(i) *Thermionic (Primary) emission*: If electrons are emitted from a metal by supplying thermal energy, the process is called *thermionic emission.* The number of electrons released depends on temperature. At a given temperature the thermionic emission current density is obtained by an equation known as Richardson's law.

(ii) *Secondary emission*: When electrons at a high speed suddenly strike a metallic surface, some will collide directly with free electrons on the metal surface and project them outward. The electrons freed in this manner are called *secondary emission.*

(iii) *Field emission* : The presence of strong electric field set up by a high positive voltage outside the emitter surface may cause electron emission from it. The stronger the field applied, the greater will be the field emission from the cold emitter surface.

(iv) *Photo-electric emission* : In this process light energy called 'quanta' falling upon the emitter is transferred to the free electrons within the metal and speeds them up sufficiently to eject from the surface. The number of electrons emission will depend on the intensity of the beam falling upon the metallic surface.

Out of the four methods of emission discussed above, the thermionic emission is the most vital one and most commonly used in electron tubes.

1.2 Thermionic Emitters

Thermionic emitters are of two different types : (i) directly heated emitter or filamentary emitter and (ii) indirectly heated emitter or oxide-coated emitter.

In the direct heating, the electric current is applied directly to a wire called filament which itself serves as electron emitter while in indirect

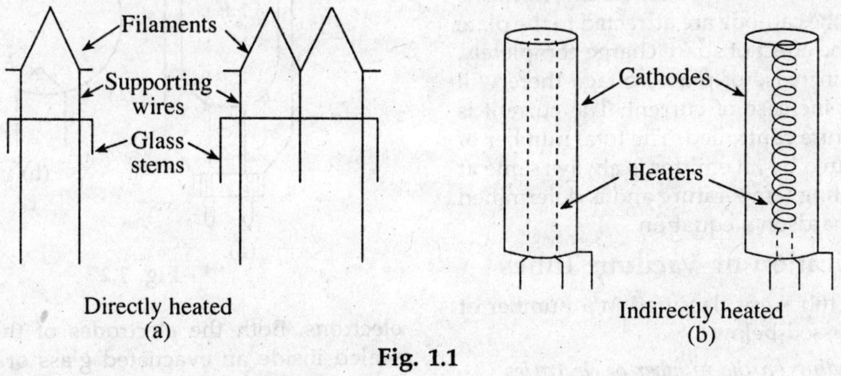

Fig. 1.1

heating, electric current is applied to a separate heater located inside a cylindrical cathode that serves as emitter. The arrangement of these two different types of heating is shown in Fig. 1.1. Both alternating or direct current can be used in either methods of heating. Two important directly heated emitters are (a) tungsten emitter and (b) thoriated tungsten emitter. Indirectly heated cathodes always use oxide-coated emitters. Receiving tubes are, in general, indirectly heated types.

1.3 Space Charge

For simplicity, let us consider a vacuum diode consisting of an emitter called *cathode* and a collector of electrons called *anode*. At low temperatures, there is no emission of electrons from the cathode surface and hence anode current is zero. As the temperature is raised to emit electrons from the cathode, and the anode is disconnected from a source of potential, electrons from the emitter form a cloud in the space. This atmosphere of thermoelectrons that is formed in the interelectrode space between cathode and plate is known as the *space charge*. Since it is made up of electrons, this atmosphere constitutes a negative charge that has a repelling effect on the electrons being emitted from the cathode. So the effect of the negative space charge alone is to force the emitted electrons back into the cathode by a considerable amount and prevent others from reaching the plate.

However, the space charge does not act alone. It is counteracted by the positive plate which attracts electrons through the space charge. When the plate voltage is low, only those electrons which are nearest to the plate are attracted to it and constitute a small plate current. With the increase of plate voltage, a greater number of electrons is attracted towards the plate and eventually a point is reached when all the electrons emitted from the cathode are attracted to the plate overcoming the effect of space charge completely. With further increase of plate voltage there will be no further increase of current. The current is then temperature controlled. The total number of electrons emitted by an emitter is always same at a given operating temperature and is determined by using Richardson's equation.

1.4 Classification of Vacuum Tubes

Vacuum tubes are classified in a number of ways as discussed below :

(i) *Depending on the number of electrodes* :

(a) Diode has two electrodes : an anode and a cathode. (b) Triode has three electrode : anode, cathode, control grid. (c) Tetrode has four electrodes : anode, cathode, control grid and screen grid. (d) Pentode has five electrodes : anode, cathode, control grid, screen grid and suppressor grid. Similarly, hexode has six electrodes, heptode has seven electrodes and so on.

(ii) *Depending on the presence or absence of gas* :

(a) High vacuum tubes : After mounting of the electrodes in the tube envelope, the gas is removed from the electron tube with a vacuum pump of the order of 10^{-8} atmosphere, (b) Gas tubes : After evacuation an inert gas is inserted within the tube envelope.

(iii) *Depending on the type of cathode used* :

(a) Thermionic tubes : In these tubes cathode is heated for emission of electrons. (b) Electron multiplier tubes : Electron emission at a number of electrodes, called dynodes, takes place due to secondary emission. (c) Cold cathode tubes : Electron emission is obtained by high field. (d) Phototubes : Cathode emits electrons when light falls on it.

1.5 Diode Valve

The term 'di-ode' means two electrodes. The valve consists essentially of a cathode which serves as an emitter of electrons surrounded by a plate or anode which acts as a collector of

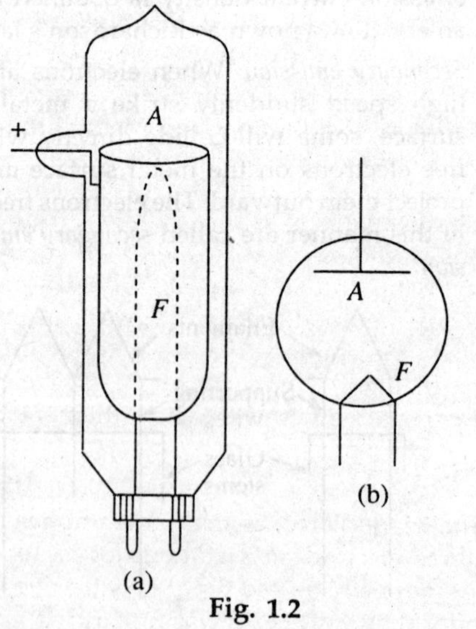

Fig. 1.2

electrons. Both the electrodes of the valve are sealed inside an evacuated glass or metalic envelop. The emitter may be either directly or indirectly heated type. A diagram of the directly heated diode is shown in Fig. 1.2(a) and its schematic symbol in Fig. 1.2 (b). In the figure, *F* is the filament surrounded by a hollow cylinder *A* known as the anode. The anode is usually made of nickel, molybdenum, graphite, iron, etc.

1.6 Operation

From the basic law of electricity we know that like charges repel and unlike charges attract each other. Electrons emitted from the cathode surface may be either attracted to or repelled from the anode depending on whether the anode is charged positively or negatively. When the anode *A* is kept at a positive potential relative to the filament *F* by means of a high tension battery as shown in Fig. 1.3 (a), electrons are able to move from cathode to anode owing to an attractive effect of the electrons and the valve is said to conduct. But if *A* is maintained at a negative potential relative to *F*, as shown in Fig. 1.3 (b), no electrons would reach at *A* due to the repulsive effect on them and so the valve would not conduct. Thus the anode current flows in a diode, only when the anode is made positive with respect to filament. No current flows when the anode is negative with respect to filament.

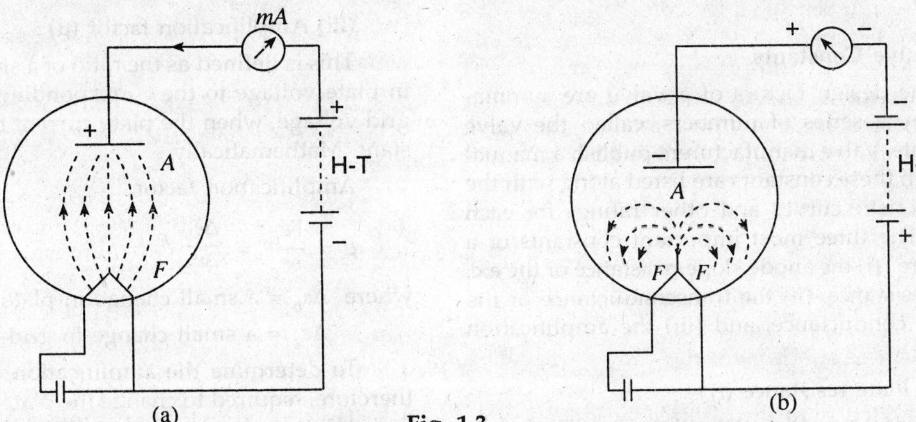

Fig. 1.3

1.7 Triode Valve

The triode valve, as the name implies, contains three electrodes : a cathode of either filament or indirectly heated type, a control grid and an anode. The control grid was first introduced in 1907 by Lee de Forest. It is usually made by a spiral of fine wire placed between the cathode and anode, usually close to the cathode.

The components of a triode valve and its schematic symbol are shown in Fig. 1.4. The filament is made by a fine tungsten wire, the two terminals of which are sealed in a glass base *g* and finally connected to two long pins (*F*, *F*). Around this filament the grid made by molybdenum wire is situated in the form of spiral as shown; the interspaces between the turns of which are sufficiently wide so that electrons can flow easily through them. The grid is sealed to the base and connected to a pin *G* attached to it. Next, the anode which is a nickel cylinder is constructed surrounding the grid. This is supported on the base by two wires, one of which is connected to *A* attached to the base. The whole arrangements viz., the filament, grid and the anode are enclosed in a glass casing which is evacuated to a pressure of about 10^{-6} mm.

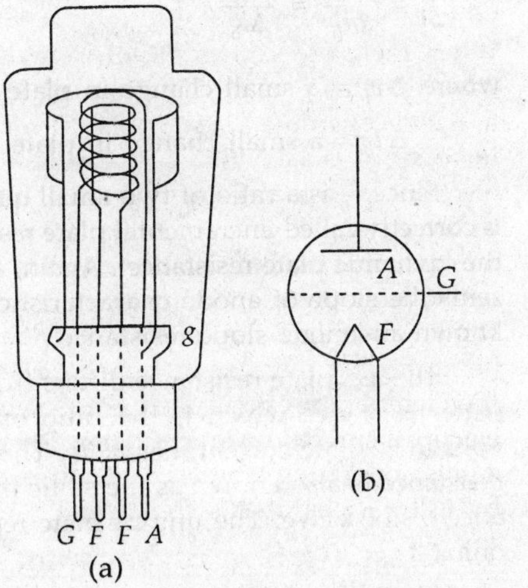

Fig. 1.4

1.7.1 Function of Control Grid

Since the control grid is usually placed close to the cathode, the potential applied to the grid has a greater effect on the electric field inside the tube. The grid, therefore, has a controlling effect on the plate current. The control grid in operation is kept at a negative potential with respect to the cathode. As a result, it does not draw any current and consume power.

As the voltage applied to the grid is made gradually negative, the space current also reduces. Ultimately a certain value is reached when the space current and hence the plate current becomes zero. Such a grid voltage at which plate current vanishes is known as 'cut-off grid voltage'.

1.7.2 Valve Constants

The design factors of a valve are summarized by a series of numbers, called the valve constants. Valve manufacturers publish a manual in which these constants are listed along with the characteristic curves and other ratings for each valve. The three most important constants of a valve are : (i) the anode slope resistance or the a.c. plate resistance, (ii) the transconductance or the mutual conductance, and (iii) the amplification factor.

(i) Plate resistance (r_p) :

This is defined as the ratio of a small change in plate voltage to the corresponding change in plate current, when the grid voltage is kept constant. Mathematically,

a.c. Plate resistance,

$$r_p = \frac{\delta e_b}{\delta i_b} = \frac{\Delta e_b}{\Delta i_b} \Big/_{e_c = \text{constant}} \tag{1.1}$$

where Δe_b = a small change in plate voltage,

Δi_b = a small change in plate current.

Since r_p is a ratio of two small quantities, it is correctly called 'incremental plate resistance' or the 'dynamic plate resistance'. Again, as it represents the slope of anode characteristic, it is also known as 'anode slope resistance'.

The a.c. plate resistance differs from the d.c. plate resistance which is the ratio of the plate voltage to plate current, i.e., e_b/i_b. The d.c. plate resistance is also known as the static plate resistance of the valve. The unit of plate resistance is ohm.

(ii) Transconductance (g_m) :

This is defined as the ratio of a small change in plate current to the corresponding change in control grid voltage, when the plate voltage is kept constant. Mathematically,

Transconductance,

$$g_m = \frac{\delta i_b}{\delta e_c} = \frac{\Delta i_b}{\Delta e_c} \Big/_{e_c = \text{constant}} \tag{1.2}$$

where Δi_b = a small change in plate current,

Δe_c = a small change in grid voltage.

Since the transconductance relates a quantity on the plate side to a quantity on the grid side it is also called 'mutual conductance'. The unit of mutual conductance is expressed in mho (ohm spelled backward).

(iii) Amplification factor (μ) :

This is defined as the ratio of a small change in plate voltage to the corresponding change in grid voltage, when the plate current is kept constant. Mathematically,

Amplification factor,

$$\mu = \frac{\delta e_b}{\delta e_c} = \frac{\Delta e_b}{\Delta e_c} \Big/_{i_b = \text{constant}} \tag{1.3}$$

where Δe_b = a small change in plate voltage,

Δe_c = a small change in grid voltage.

To determine the amplification factor it is, therefore, required to change the plate voltage by a certain amount, record the change in plate current and then change the grid voltage by an amount just to attain the previous value of the plate current.

1.7.3. Relation between r_p, g_m and μ

We can write the mathematical form of the amplification factor as,

$$\mu = \frac{\delta e_b}{\delta e_c} = \frac{\delta e_b}{\delta i_b} \times \frac{\delta i_b}{\delta e_c}$$

$$= r_p \times g_m. \tag{1.4}$$

That is, the amplification factor is equal to the product of plate resistance and mutual conductance. The three valve constants are thus interrelated in accordance with their definition.

1.8 Tetrode

A tetrode valve has two grids in addition to cathode and anode. The first one, as mentioned before, is called the control grid while the second grid is known as the screen grid. In the construction, the screen grid is placed between the control grid and the plate, and thus avoids the undesirable grid-to-plate capacitance. The schematic symbol of a tetrode valve is shown in Fig. 1.5.

1.8.1 Functions of Screen Grid

The screen grid of a tetrode valve is usually maintained at a high potential value, even sometimes greater than that of the plate. The screen grid of a tetrode valve has two important functions. First, it being maintained at a positive potential, helps to overcome the space charge

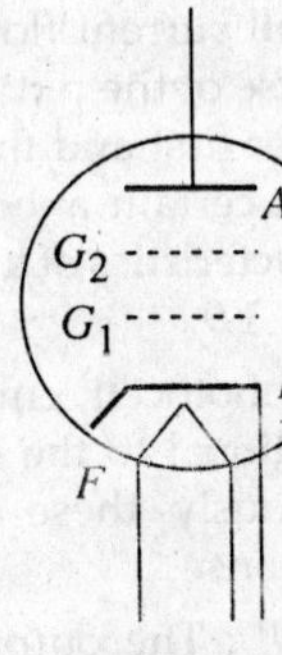

Fig. 1.5

effect and hence increases the plate current and mutual conductance. Secondly, when the valve is used as an amplifier, the screen grid eliminates the evil effect of interelectrode capacitance between the control grid and the anode.

1.9 Pentode

A pentode valve has five electrodes. These are : (i) cathode, (ii) control grid, (iii) screen grid,

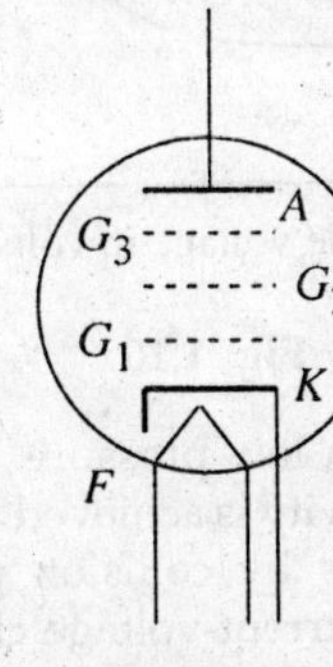

Fig. 1.6

(iv) suppressor grid, and (v) plate. In the construction of a pentode the suppressor grid is placed between the screen grid and the plate as shown in the schematic symbol of Fig. 1.6. Since the performance of a pentode is much superior to that of a tetrode, the former is widely used in preference to tetrode.

1.9.1 Functions of Suppressor Grid

In a pentode, the suppressor grid is connected directly to the cathode and thus maintains a negative potential with respect to anode. The secondary electrons emitted by the anode are, therefore, repelled by the negative suppressor grid and are driven back to the plate again. In this fashion the grid eliminates the effect of secondary electrons and hence the plate current rises smoothly with the increase of plate voltage.

The addition of suppressor grid also increase the shielding action between the plate and the control grid and thus reduces the grid to plate capacitance. Because of high screening the plate current is little affected by the plate voltage and so both the plate resistance and the amplification factor of a pentode become extremely high.

1.10 Beam-power Tube

Beam-power tube is a very useful hybrid between a tetrode and a pentode. Like a tetrode

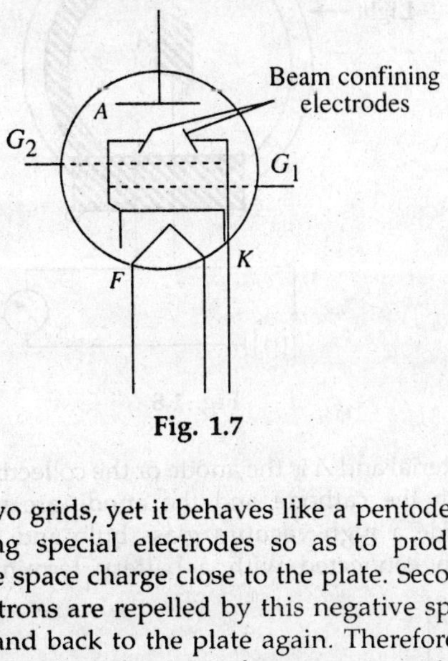

Fig. 1.7

it has two grids, yet it behaves like a pentode by arranging special electrodes so as to produce negative space charge close to the plate. Secondary electrons are repelled by this negative space charge and back to the plate again. Therefore, it is similar to the behaviour of a suppressor grid in a pentode. The schematic symbol of a beam-power tube is shown in Fig. 1.7.

Beam-power tube is superior to that of a pentode as it gives higher plate efficiency and higher power output for a given harmonic distortion. Moreover, it gives a small ratio of screen grid current to the plate current than in a pentode.

1.11 Photo-electric Cells

Photo-electric cells are devices with the help of which light energy can be converted into electrical energy and wherein the photoelectrons are the main agents for such conversion. Three different types of photo-electric cells are there, viz., (i) *Photo-emissive cells,* (ii) *Photo-voltaic cells,* and (iii) *Photo-conductive cells.*

1.11.1 Photo-emissive Cells

The cells consist of a photo-sensitive surface from which electrons are emitted when light is made to fall on it. As a result a current will flow between this surface (called cathode) and a collecting plate (called anode) which is kept at a considerably high positive potential. Photo-emissive cells are either of vacuum types or gas-filled types.

(a) *Vacuum Cells* : A vacuum type of emission cell is shown in Fig 1.8. Here *K* is the cathode of large surface area coated with photo-sensitive

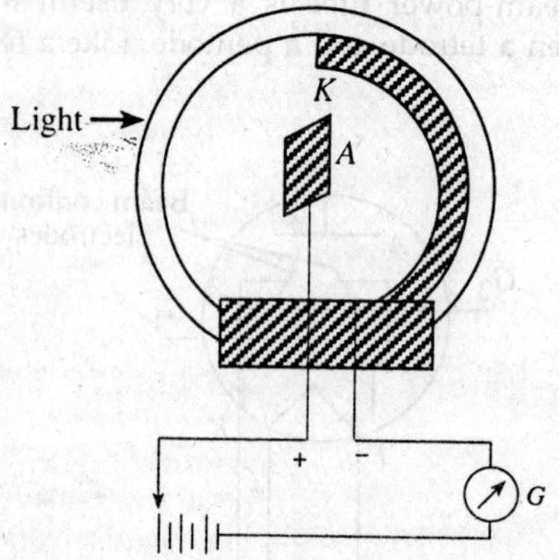

Fig. 1.8

material and *A* is the anode or the collecting plate. Both the cathode and the anode are mounted inside a high vacuum glass bulb and they are then connected with a battery terminals. The

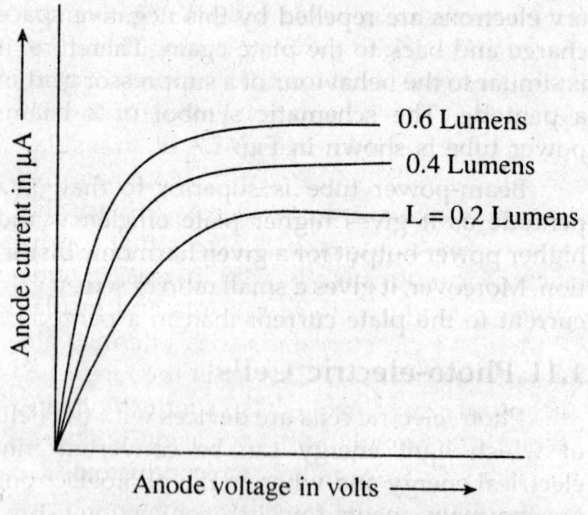

Fig. 1.9

anode *A* is connected to the higher potential point while the cathode *K* is connected through a galvanometer *G* to the lower potential point of the battery. When light is allowed to fall on the cathode surface, photo-electrons are liberated, and are repelled on to *A* from the cathode. As a matter of fact, a small current flows through the tube. With the increase of the p.d. between *A* and *K*, current increases at first and then a saturation stage is reached at a certain anode voltage. The current-voltage characteristics of a vacuum photo-cell is shown in Fig. 1.9.

In this type of photocell, emission is found to be strictly proportional to the intensity of the incident light. Obviously, these are very quick and accurate in response.

(b) *Gas-filled Cell* : The output current of a gas-filled photocells is considerably increased than in a vacuum photocell by inserting an inert gas,

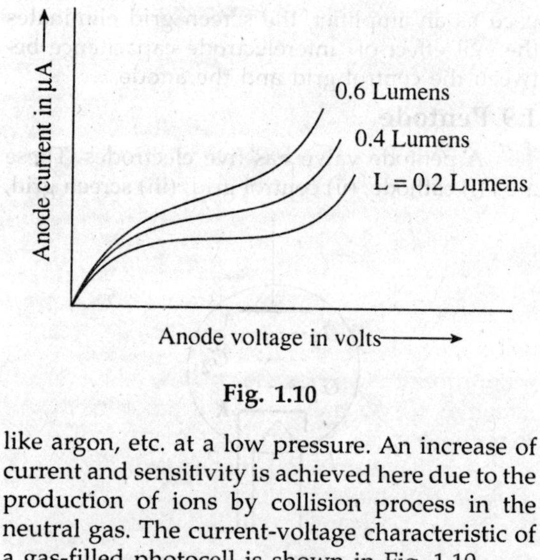

Fig. 1.10

like argon, etc. at a low pressure. An increase of current and sensitivity is achieved here due to the production of ions by collision process in the neutral gas. The current-voltage characteristic of a gas-filled photocell is shown in Fig. 1.10.

Photocell of this type is usually used for commercial purposes because of high sensitivity. But since there is a fluctuation of current even for a constant intensity beam, the gas-filled cells are seldom used for precision work.

1.11.2 Photo-voltaic Cells

Becquerel in 1839 discovered that when a pair of electrodes were immersed in an electrolyte and light was made to fall on one of the electrodes, a potential difference was created between them. This phenomenon is known as photo-voltaic effect. Devices which exhibit this effect are called photo-voltaic cells. Substances like selenium and silicon are mostly used for preparing photo-voltaic cells. The recent forms of such cells are commonly known as barrier layer cells. Barrier layer cells are two different types : (i) front-effect cell and (ii) back-effect cell.

In front-effect cell a thin metallic film F of silver, gold or platinum is deposited on a semiconducting layer like cuprous oxide or on selenide. The whole arrangement is then attached to a metal base plate as shown in Fig. 1.11(a). When light falls on F, it penetrates easily and at the barrier layer between the metallic film and the semiconductor, photo-electric emission occurs. Light further travels and at the barrier layer between the semiconductor and the metal base cause photo-electric emission which is much less than in the previous case. As a result, the conventional current moves from the base to the metal film in the external circuit which is recognized by a meter M.

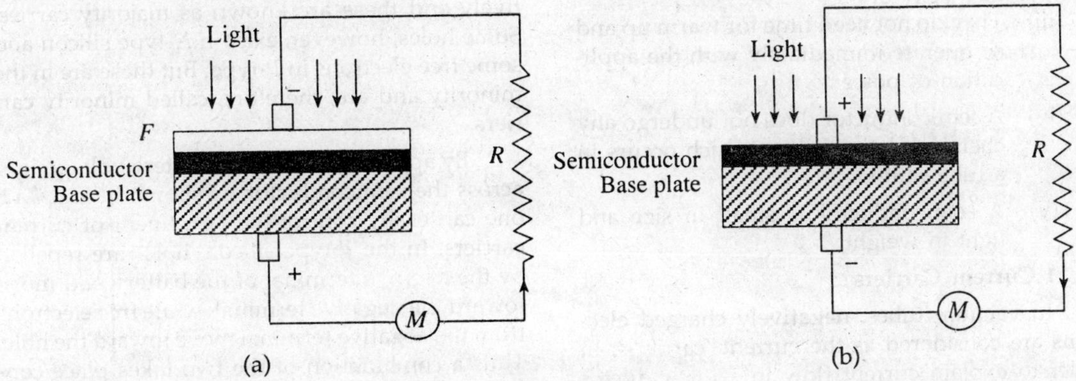

Fig. 1.11

In the back-effect cell the light passes directly through the semiconductor as shown in Fig. 1.11(b). Here electrons flow across the barrier layer from the semiconductor to base plate and so conventional current flows in the opposite direction.

Photo-voltaic cells are used in photographic exposure meter and for operation of relays. Silicon cells, because of their better response in the infra-red region, are widely used in artificial satellites for conversion of solar energy into electrical energy. No separate battery is required to run photo-voltaic cell.

1.11.3 Photo-conductive Cells

Substances like selenium, cadmium sulphide, cadmium selenide, etc. decrease their resistance when light falls on them. The amount of this decrease, of course, depends on the intensity and colour of the incident light. So insertion of such a substance in the circuit and allowing light to fall on it will cause a change of electrical resistance. Hence there will be a variation of current in the circuit in steps with the variation of the intensity of the incident beam of light. This property is utilized in constructing photo-conductive cells.

Photo-conductive cells have a high sensitivity, low cost and long life. They can be used directly with a power supply of low voltage to operate a relay. These are used extensively in street lighting controls, door-openers, burglar alarm systems and also in modern television camera tubes.

SEMICONDUCTORS

1.12 Some Preliminaries

Semiconductor is a solid whose resistivity lies between conductor and insulator. A conductor and an insulator are first distinguished by their extreme values of electrical conductivity. A conductor has a conductivity of 10^4 to 10^6 ohm^{-1} cm^{-1} while the conductivity of an insulator is less than 10^{-6} ohm^{-1} cm^{-1}. Moreover, the conductivity of a conductor decreases with the increase of its temperature whereas that for an insulator increases slightly with increasing temperature. In addition to conductors and insulators, there is a class of solids with intermediate values of electrical conductivity in the range 10^6 to 10^8 ohm^{-1} cm^{-1} and whose conductivity increases more strongly with increasing temperature. Such solids are called semiconductors.

Transistors, metallic rectifiers, integrated circuits, junction diodes, zener diodes, tunnel diodes, etc. are examples of semiconductors. They are used widely in radio and TV receivers, in computers and in various electronic instruments.

The controlling function of a vacuum tube is similar to that of a semiconductor but it is rapidly replacing vacuum tubes because of added advantages, some of which are mentioned below :

(i) Semiconductor is solid. Therefore, there is a little chance of vibration as in vacuum tube.

(ii) Little power is required for a semiconductor and also it radiates less heat than in a tube.

(iii) They do not need time for warm up and so operate immediately with the application of power.

(iv) A semiconductor does not undergo any chemical deterioration which occurs in a tube cathode.

(v) A semiconductor is small in size and light in weight.

1.12.1 Current Carriers

In vacuum tubes, negatively charged electrons are considered as the current 'carriers'. In order to explain current flow in semiconductor diodes and transistors this concept must be modified by the addition of positive charge carriers known as 'holes', which have mass mobility and velocity. Current flows in semiconductors is thus carried on by the flow of both negative charges (free electrons) and positive charges (holes). Holes are usually attracted by free electrons and when they combine, the free electron 'fills' the hole neutralizing its charge. When this happens, the hole and free electron both are lost as current carriers and form new current carrier at other points in the semiconductor.

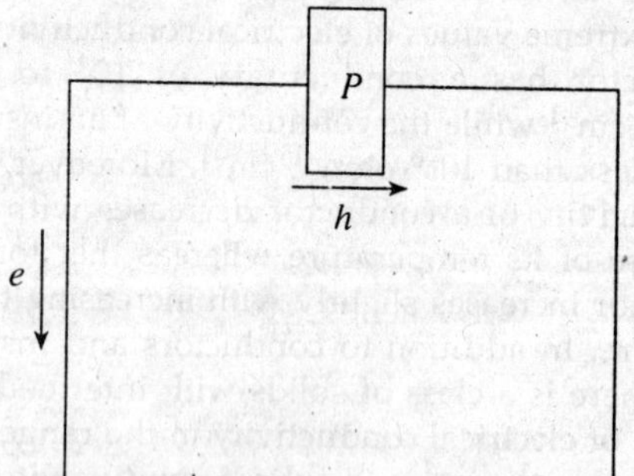

Fig. 1.12

Impurity like arsenic or antimony (pentavalent atoms) increases the conductivity of silicon, which has four valence electrons in its outer shell, by increasing the number of negative charge carriers. Silicon which has been doped here with arsenic or antimony is designated as '*N*-type'. Similarly, impurity like indium or gallium (trivalent atoms) increases the conductivity of silicon by increasing the number of positive charge carriers. For this reason, silicon which has been doped with indium or gallium is called '*P*-type'. Current flow in *N*-type and *P*-type silicon is carried on by 'free electrons' and 'holes' respectively and these are known as majority carries. Some holes, however, exist in *N*-type silicon and some free electrons in *P*-type. But these are in the minority and are, therefore, called minority carriers.

By applying an external battery voltage V_{AA} across the semiconductor, as shown in Fig. 1.12, one can easily control the movement of current carriers. In the *P*-type silicon, holes are repelled by the positive terminal of the battery and move toward the negative terminal, while free electrons from the negative terminal move toward the hole. Thus a combination of the two takes place continuously and the flow of current is maintained in the external circuit.

1.12.2. Operation of Junction Diode

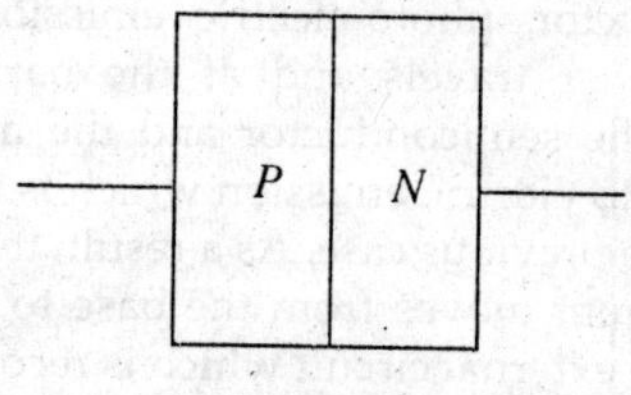

Fig. 1.13

When *P*-type and *N*-type silicon are joined together a junction diode is formed. The diode

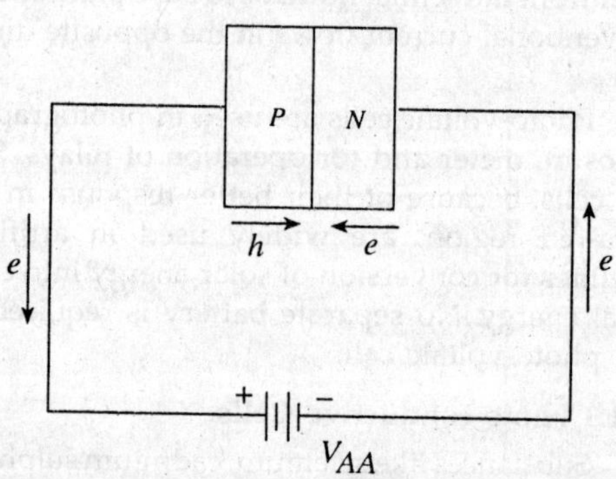

Fig. 1.14

thus created (Fig. 1.13) has a characteristic to pass current in one direction readily but not in the other. This unique characteristic may be explained by considering the theory of negative and positive carriers.

In Fig. 1.14, the *P-N* junction has been connected with the battery V_{AA} as shown. Free electrons from the negative terminal of V_{AA} enter the *N*-type silicon and repel the free electrons therein. Similarly, the holes in the *P*-type silicon are repelled by the positive battery terminal and thereby move toward the *P-N* junction where a combination of free electrons and holes takes place. Although there is usually potential barrier in the diode that prevents electrons and holes to combine, under the influence of the electric field

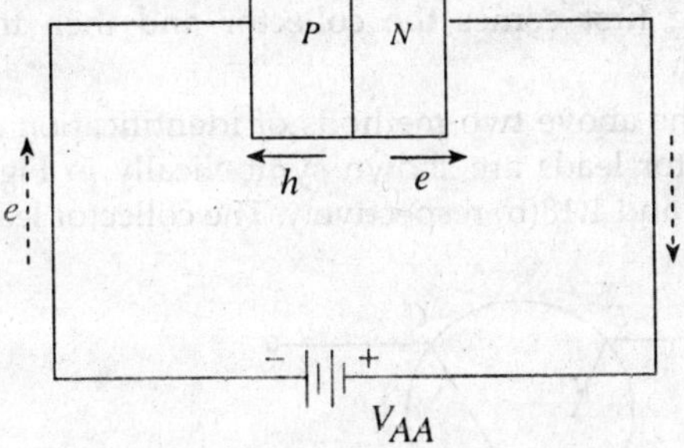

Fig. 1.15

of the battery the holes move to the right and the electrons to the left. At such times a covalent bond near the positive terminal of the battery breaks down causing an electron to liberate and enter the positive terminal. As a consequence, a new hole is further created which moves to the right toward the *P-N* junction. The process is a continuous one and so a current flow is maintained in the circuit. If the value of V_{AA} is increased, the current flow in the diode also increases. A diode connected in such a fashion with positive battery terminal to the *P*-type and negative battery terminal to the *N*-type is called 'forward bias' and the diode is said to have a low forward resistance.

If the polarities of the battery V_{AA} are reversed, as shown in Fig. 1.15, no combination of free electrons and holes takes place. As a matter of fact, majority carriers in the diode do not support any current flow. Only due to minority carriers, in this reverse-bias condition, there is a minute flow of current in the diode.

1.13 Transistors

Scientific interest in semiconductor devices led to the discovery of the transistor. In 1948 Bardeen and Brattain while working at the Bell Telephone Laboratories announced the development of the first transistor. One of the great advantages of a transistor over a vacuum tube is that it is small in size and it does not require any filament power. The supply voltage of a transistor is also very low and it operates as soon as power is applied. In addition, the circuit connection of a transistor is very simple than that of a tube.

1.13.1 Classification

Transistors are of many types which may be broadly classified as follows :

(i) *Depending on the basic materials from which they are formed* : In this category we have silicon and germanium transistors. Nowadays most transistors are of silicon type.

(ii) *Depending on the number of elements used :* Here we have triode or three-element transistor and the tetrode or four-element transistor.

(iii) *Depending on the ability to dissipate power :* Here we have a wide range from low power (below 50 mW) to high power (2 W and above) type.

(iv) *Depending on the process of construction* : In this category we have various junction transistors such as grown junction, alloy junction, mesa, epitaxial mesa, planar transistors, drift-field and point-contact transistors.

Transistors are of different in sizes and shapes. There are variations also in the manner of mounting and in the basing arrangements. Some transistors are socket mounted while others have flexible leads for directly soldering in a circuit.

1.13.2 Junction Transistor

The junction transistor, a radically new type of transistor, was first developed by Shockely in

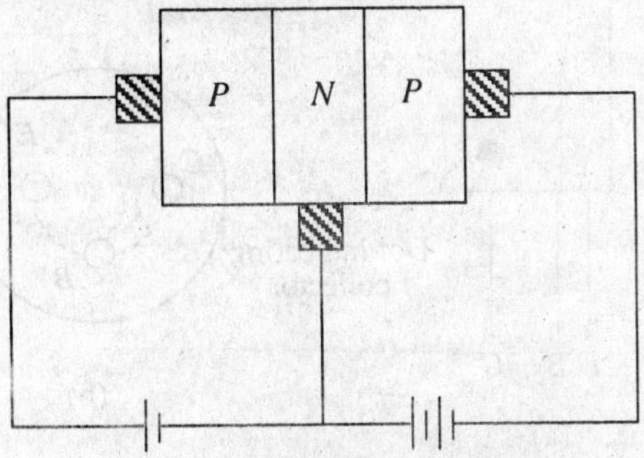

Fig. 1.16

1949 just a year later of the discovery of the point contact transistor. A junction transistor, now widely used is nothing but a combination of two *P-N* junctions. A typical junction transistor formed by sandwiching a thin *N*-type layer between two *P*-type layers is illustrated in Fig. 1.16. Such a transistor is called a *P-N-P* junction transistor or, in short, a *P-N-P* transistor. If, alternately, there are two *N*-type layers on either side of a *P*-type layer it is called *N-P-N* transistor.

Of the two *P-N* junctions one is forward biased while the other is reverse biased, i.e., the *P*-layer is connected respectively with the positive and negative terminals. There are three leads from individual metallic plates which make contact with the respective crystals. Referring to Fig. 1.16, the *N*-layer in the middle is called the 'base', the forward biased *P*-layer on the left is the 'emitter' while the reverse biased *P*-layer on the right is the 'collector'.

1.13.3. Transistor Symbols

Internationally used symbols for *P-N-P* and *N-P-N* transistors are shown in Fig. 1.17(a) and 1.17 (b) respectively. These symbols are required for representing transistors in electronic circuits. In the figure, the element with the arrowhead is the emitter and its symmetrical counterpart is the collector. Of the three leads, the third one is called base. In a *P-N-P* transistor, the emitter arrow points to the base, while in the *N-P-N* type, the arrow points away from the base. The significance of the arrow is that it indicates the direction of emitter current flow, i.e., opposite to the direction of electron flow, under forward biased conditions.

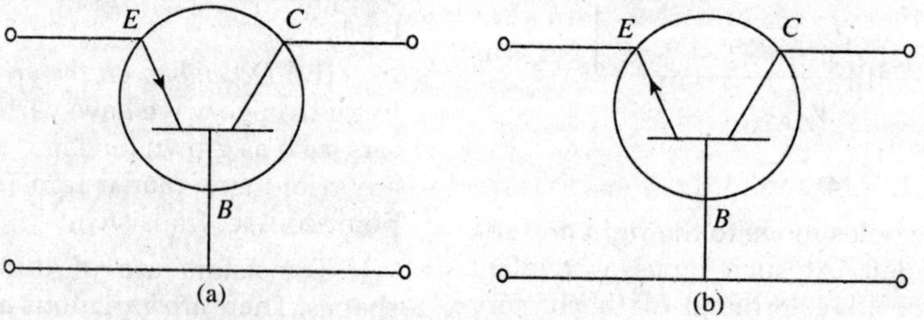

Fig. 1.17

The emitter, base and collector leads of a transistor can be conveniently identified by the following two ways :

(i) The central lead corresponds to the base. The spacing between the emitter lead and the base lead is less than the spacing between the base lead and the collector lead.

(ii) If the leads are along the circumference of a circle then the centre one is called base. The emitter and the collector leads are then identified by reading clockwise from the base. In such reading first comes the collector and then the emitter.

The above two methods of identification of transistor leads are shown symbolically in Figs. 1.18(a) and 1.18(b) respectively. The collector lead can also be identified by a dot which is pointed on the body of the transistor close to the collector.

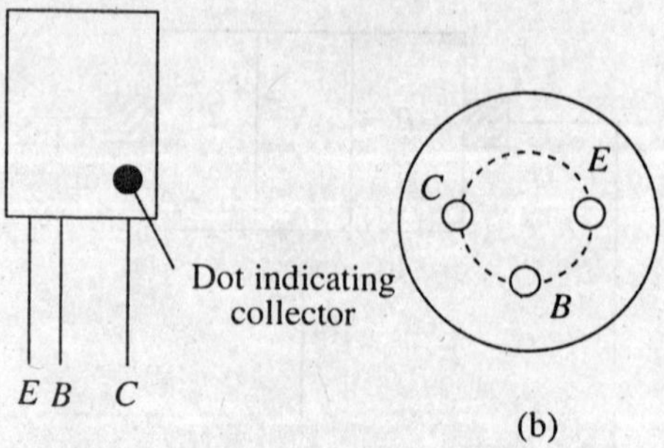

Fig. 1.18

According to the sign convensions, a current flowing into a transistor through any terminal is considered positive while a current coming out of it is considered negative. Thus in a *P-N-P* transistor, the emitter voltage and the current are positive while the collector voltage and the current are negative. In an *N-P-N* transistor the signs are obviously just opposite.

1.13.4 Circuit Configurations

As discussed earlier, a transistor has three internal electrodes known as emitter, base and collector. If it is to be used to study as a four-terminal device, two for the input and two for the output, then one electrode must be kept common. Accordingly, there are three possible circuit arrangements. These are : (i) the common emitter, (ii) the common base and (iii) the common collector circuits. Instead of the term 'common', sometimes the word 'grounded' is also used, such as grounded emitter, etc. The three circuit configurations are illustrated in Fig. 1.19(a), (b) and (c) respectively. When a transistor is used in common emitter configuration, the input signal is applied between base and emitter while the output is obtained from the collector to emitter. The emitter in such a case is thus common to both the input and the output circuits, and hence the name common-emitter. Similarly, in a common-base or in a common-collector circuit the base or the collector serves as a common reference for both the input and output signals. The common emitter is, however, most frequently used as it provides suitable voltage, current and power gains in different electronic circuits.

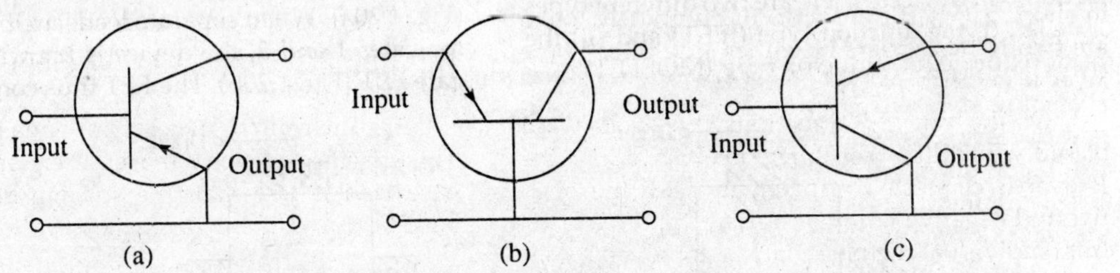

Fig. 1.19

1.13.5 Current Gain

In a common-base circuit the collector current is controlled by the variations of emitter current. This gives one important control characteristic of a transistor designated by α, where α is defined as the ratio of the change in collector current (ΔI_C) to the change in emitter current (ΔI_E), when the collector voltage (V_{CB}) is maintained at a constant value. Mathematically, this can be expressed as

$$\alpha = \left(\frac{\Delta I_C}{\Delta I_E}\right)_{V_{CB} = \text{constant}} \tag{1.5}$$

α is also sometimes designated by the letters h_{fb} and is called the forward-current-transfer ratio. In a common-base arrangement, since the collector current is slightly less than the emitter current, α will be slightly less than unity.

In a common-emitter circuit, the input signal is applied to the base. This gives another important control characteristic designated by β, where β is defined as the ratio of the change in collector current (ΔI_C) to the change in base current (ΔI_B), when the collector voltage (V_{CE}) is kept constant. Mathematically, this can be expressed as

$$\beta = \left(\frac{\Delta I_C}{\Delta I_B}\right)_{V_{CE} = \text{constant}} \tag{1.6}$$

β is also sometimes designated by the letters h_{fe} and it is the current amplification factor in a common-emitter amplifier. Since small changes in base current produce large changes in collector current, the value of β will be always greater than unity.

1.13.6 Relation between α and β

The relationship between currents in the elements of a transistor is expressed as

$$I_E = I_B + I_C. \tag{1.7}$$

Now we know that for a small change in emitter current ΔI_E, there is a corresponding change in base current ΔI_B and in collector current ΔI_C. Thus,

$$\Delta I_E = \Delta I_B + \Delta I_C$$

$$\text{or, } \Delta I_B = \Delta I_E - \Delta I_C.$$

Therefore,

$$\frac{\Delta I_C}{\Delta I_B} = \frac{\Delta I_C}{\Delta I_E - \Delta I_C}$$

$$= \frac{\Delta I_C / \Delta I_E}{1 - \dfrac{\Delta I_C}{\Delta I_E}}.$$

As defined, when V_{CB} is constant the ratio $\Delta I_C / \Delta I_E$ equals to α and when V_{CE} is constant the ratio $\Delta I_C / \Delta I_B$ equals to β. So for constant V_{CB} and constant V_{CE}, we may write

$$\beta = \frac{\alpha}{1 - \alpha}$$

$$\text{or, } \alpha = \frac{\beta}{1 + \beta}. \tag{1.8}$$

It is clear that as α approaches towards unity, β becomes increasingly larger. For a transistor whose α is 0.98, β becomes 49 while an α of 0.99 gives β of 99. So during experiment α must be measured very accurately to avoid formula error in β.

1.14 Field Effect Transistors

Field effect transistors, abbreviated as FETs are unipolar devices. There are two different types of FETs : (i) the junction type (JFET) and (ii) the metal oxide semiconductor type (MOSFET).

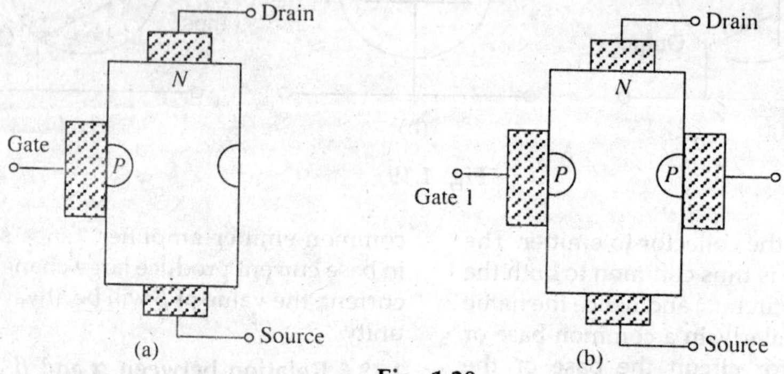

Fig. 1.20

1.14.1 JFET

An N-channel JFET is illustrated in Fig. 1.20.

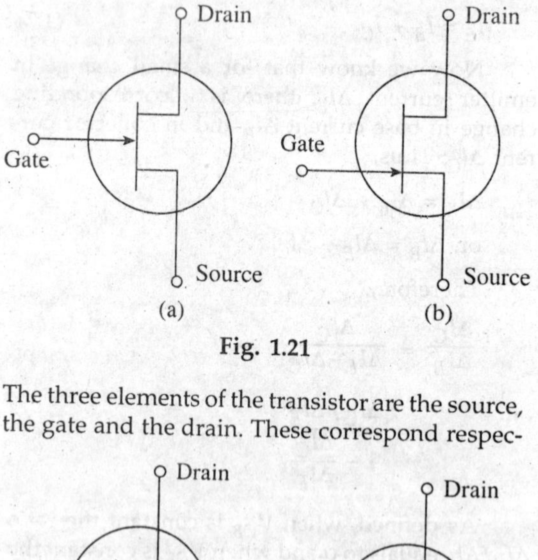

Fig. 1.21

The three elements of the transistor are the source, the gate and the drain. These correspond respectively the emitter, the base and the collector of a bipolar transistor. In the figure, the body or channel of the transistor is an N-type semiconductor while the gate embeded on both sides of channel, is the P-type material and hence the name junction FET. When gates 1 and 2 are internally connected, it is called a single gate FET (Fig. 1.20a). When separate leads are brought out for gates 1 and 2, the device is known as a dual gate FET (Fig. 1.20b). The FET thus constructed is an N-channel, P-gate device, the symbol of which is shown in Fig. 1.21. Similarly, it is also possible to make a P-channel. For an N-channel device the gate arrow points towards the channel while for a P-channel the arrow points away from the channel (Fig. 1.22). For both the types the gate arrow may be centred on the channel or close to the source.

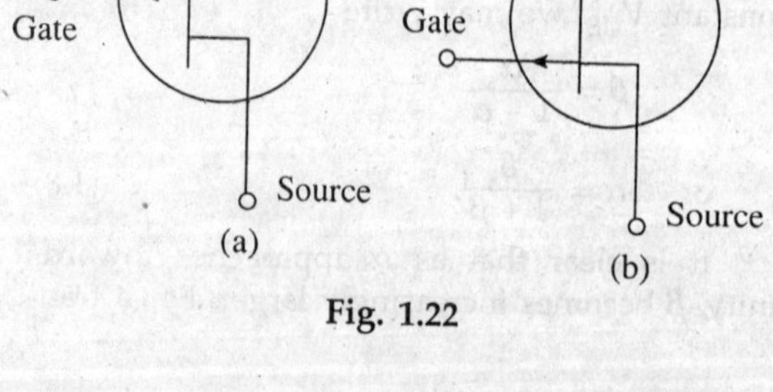

Fig. 1.22

1.14.2 MOSFET

Like JFET, just discussed, the MOSFET is also a field effect transistor of much higher input impedance than the JFET. Depending on the manner of construction MOSFETs are mainly classified into two different types : (i) Enhancement type and (ii) Depletion type.

MOSFETs have some added advantages. These are : (a) lower noise (2 to 6 dB typical), (b) reduced cross modulation effects, (c) low feedback capacitance (0.02 to 0.17 pF typical), (d) high transconductance (up to 15,000 micromhos typical), and (e) high maximum operating frequency (up to 500 MHz, depending upon the model).

The only pronounced disadvantage of the MOSFET is the delicateness of its gate insulation.

Enhancement-type MOSFET

The construction of an enhancement-type MOSFET is shown in Fig. 1.23. In the enhancement type, there is no channel between the source and the drain but separating the *N*-type source and drain is a *P*-type substrate. A very thin, fragile layer of silicon dioxide (SiO_2) is deposited

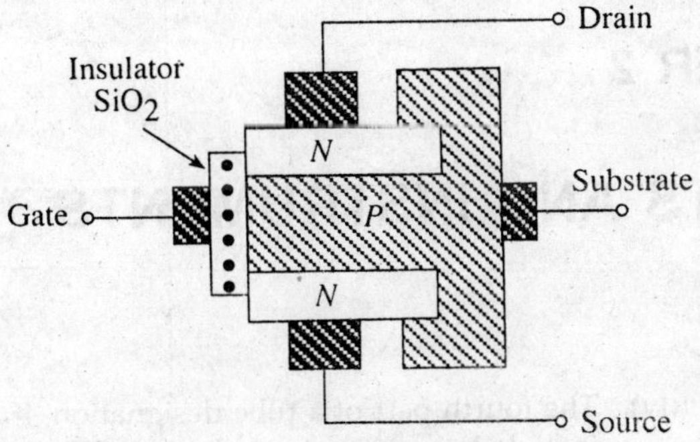

Fig. 1.23

over the substrate. On the silicon dioxide insulator a metallic film is deposited which serves the purpose of the gate. Usually there are four ohmic contacts, one each for the source, drain, substrate and the gate. But sometimes the substrate is internally connected to the source and so no substrate lead is required then.

It may be noted that a *P*-channel normally of MOSFET may be constructed by using *P*-type source and drain and *N* substrate. The symbols for both *N* and *P*-types are shown in Fig. 1.24.

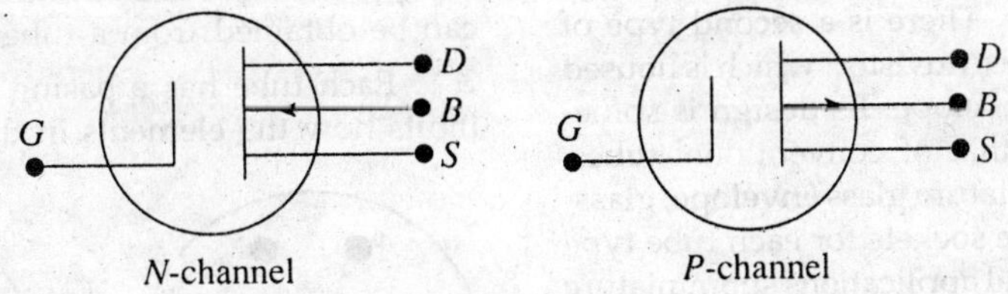

Fig. 1.24

Depletion-type MOSFET

Since the gate is insulated from the body of the FET, the MOSFET is also called an insulated gate FET (IGFET). Although no physical channel is there, the MOSFET is sometimes termed as *N*-channel enhancement-type FET.

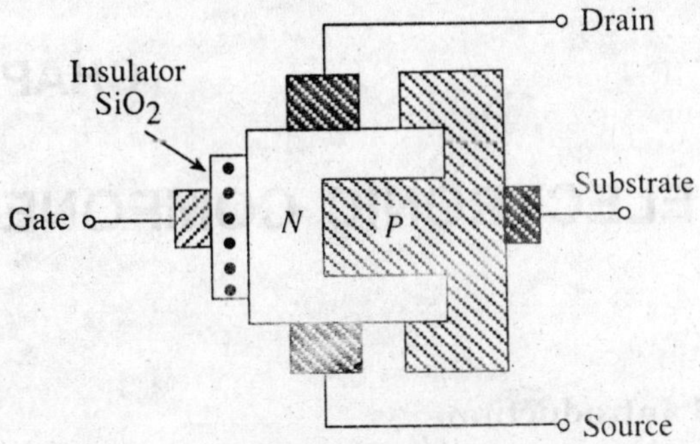

Fig. 1.25

The construction of a depletion type MOSFET is shown in Fig. 1.25. The manner of construction of depletion MOSFET is similar to that of an enhancement type with one exception. The depletion MOSFET, as evident from the figure, has an *N*-channel while the enhancement type has not.

The depletion MOSFET may also be constructed using a *P*-channel and *N* substrate. The symbols for both *N*- and *P*-channel depletion MOSFETs are shown in Fig. 1.26. As the depletion MOSFET is normally on, the vertical channel

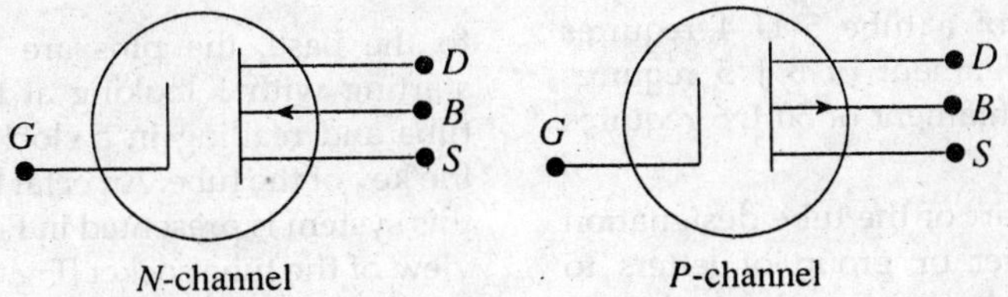

Fig. 1.26

line is not broken here as in the enhancement type.

CHAPTER 2

ELECTRONIC COMPONENTS AND INSTRUMENTS

2.1 Introduction

Various electronic components used in the instruments play a very vital role for precise measurement of any quantity in the laboratory. In this chapter we have briefly focussed on this important aspect.

2.2 Physical Structure of Vacuum Tube

Nowadays vacuum tubes employ glass bases which are nothing but an extension of their glass envelope. The 7-pin and 9-pin miniature tubes are examples of this type. There is a second type of vacuum tube, known as 'nuvistor' which is housed in a ceramic metal envelope. Its design is somewhat different from that of conventional tubes. There are also subminiature glass envelope, glass-based tubes. There are sockets for each tube type though for some special applications subminiature tubes are not socket-mounted.

Generally, tubes are identified by a fairly standard numbering system. However, some exceptions are there as noted below :

(i) The first part of a tube designation is a number or numbers which provide us the approximate voltage necessary for the filament of the tube. As for example, the filament of a tube 5 U 4 requires 5 volts, the filament of 6 J 5 requires 6.3 volts, the filament of 50 L 6 requires 50 volts, etc.

(ii) The second part of the tube designation is a code letter or group of letters to identify the tube types and functions.

(iii) The third part of a tube designation is a number which indicates the number of useful elements brought out to the base terminals. As for example, the tube 6 S N 7 has seven elements, if the filaments are considered as one element.

(iv) The fourth part of a tube designation, if included any, consists of a letter or letters giving some particular characteristic of the tube. As for example, the tube 5 Y 3-G has a glass (G) envelope. The tube 5 Y 3-GT is same as the 5 Y 3-G, except that the glass envelope is smaller (GT) than that of the G.

It may, however, be noted that the above description is only general and there are many exceptions to it. An exact information of the tube can be obtained from a tube manual.

Each tube has a basing diagram which exhibits how the elements in the tube are brought

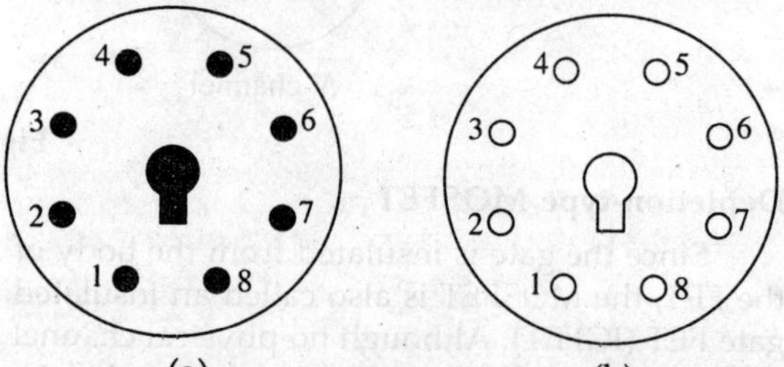

(a) (b)

Fig. 2.1

to the base, the pins are numbered in order, starting with 1 looking at the base pins of the tube and reading in a clockwise direction from the key of the tube. An octal tube base illustrating this system is presented in Fig. 2.1(a). The bottom view of the tube socket [Fig. 2.1(b)] is identically numbered. Tubes which employ an integral glass base such as the 7-pin and 9-pin miniatures do not have a separate key way.

Extra spacing between the first and the last pins serves as the key to seating the tube in its socket. It also serves as the guide to the identification of pin numbers.

2.3 Transistor Manual

Manufacturers of electron devices make a paperback handbook or the so-called manual which lists the characteristics of each device. Presently quite a good number of manuals are available to the electronic engineer. These manuals are obtained usually from bookstores at a nominal cost or those can be obtained free upon request to the manufacturing company.

2.3.1 Manual Organisation

Transistor manuals can be classified into two parts: one dealing with the basic theory and applications while the other devoted to transistor data. Out of these two, the transistor data portion is of primary interest. In this section the description of the transistor is divided into six categories. The description of the device is followed by information on absolute ratings, electrical thermal characteristic, mechanical data and general applications.

2.3.2 Data Sheets

In a data sheet most of the currents and voltages are d.c. values at an ambient temperature of 25 °C. The transistor symbols are not standardized by industry; common usage has established the familiar notation.

Just by going through the transistor data sheet a technician can realize whether the transistor is constructed by silicon or germanium, *NPN* or *PNP*, designed for audio or high-frequency applications, etc. In order to select a transistor for a particular circuit design, one must consider the above points in addition to its electrical ratings and characteristics.

The system of maximum ratings of a transistor has been defined by the Joint Electron Device Engineering Council (JEDEC). It has been standardized by the National Electrical Manufacturers Association (NEMA) and the Electronic Industries Association (EIA).

Dissipation ratings for transistors consider the heat dissipated at the collector. The power-dissipation factor gives the difference between the power supplied to the collector and that dissipated by the load. Since semiconductor material is extremly temperature-sensitive, so its temperature ratings depend on the temperature. Some of the manuals provide a power dissipation chart for different methods of transistor installation. Transistor voltage ratings are always established with reference to a specified point on the device, e.g., collector-to-base, emitter-to-collector. The maximum voltage ratings of a transistor indicate the maximum potential that can be placed across two leads while the ratings for leakage current are taken with the third lead open.

(i) ***Ratings versus characteristics***

Rating is a value of the transistor given by its manufacturer. If it exceeds, then the component can be damaged. Characteristics of a transistor, on the other hand, are measurable parameters under specific conditions and these illustrate regions of working performance.

In most of the manuals the maximum ratings are given at different temperatures. Design considerations should be less than the maximum ratings. In Fig. 2.2, an example of both characteristics and maximum power ratings is given.

(ii) ***Collector-to-base cut-off current*** (I_{CBO})

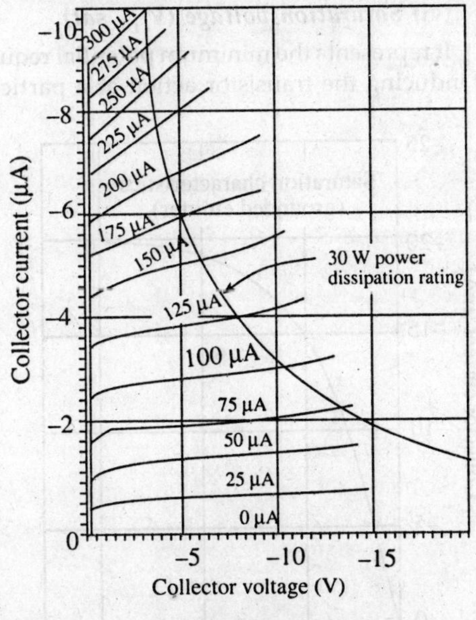

Fig. 2.2

It signifies the quality of the transistor. This current I_{CBO} flows between collector and base with an open emitter circuit. As for example, the transistor 2*N* 139 has an I_{CBO} of 6 μA and the collector to base voltage is 12 V. I_{CBO} is given with three different collector voltages—low, medium and high. The lower is the I_{CBO}, the better is the transistor as a useful device.

(iii) ***Emitter-to-base cut-off current*** (I_{EBO})

This represents the d.c. value for the emitter when the emitter function is reverse-biased and the collector is open-circuited. This parameter is required for considering a typical design region. The transistor 2*N* 139 has an I_{EBO} 40 μA.

(iv) ***Collector-to-emitter leakage current*** (I_{CBO})

This is the d.c. value for the collector with the collector function reverse-biased and a specified base-emitter circuit connection. This rating is very important in considering the switching-circuit applications.

(v) ***Collector-to-base breakdown voltage*** (BV_{CBO})

This rating is considered occasionally when the collector-to-base voltage in the circuit to be designed. For the common-emitter configuration, the same transistor would have a rating of BVCEO. The transistor 2*N* 384 has a breakdown voltage of 80 V when I_C is 50 μA and the emitter current equals zero.

(vi) ***Saturation voltage*** (***V_{CE}, sat***)

It represents the minimum potential required for inducing the transistor action at a particular

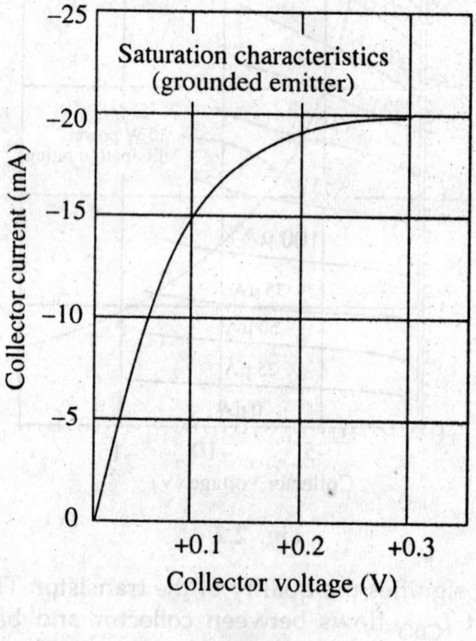

Fig. 2.3

collector current. From Fig. 2.3, it can be seen that the saturation takes place at a collector voltage of about 0.2 V when the flow of collector current is about 20 μA. Saturation voltage increases with the increase of the collector current and varies inversely as the gain of the transistor.

(vii) ***Current gain*** (h_{fe})

It is a function of output current to input current with a constant output voltage. This is illustrated in Fig. 2.4. This is an arbitrary figure which depends on a specific collector-to-emitter voltage and collector current. The figure also

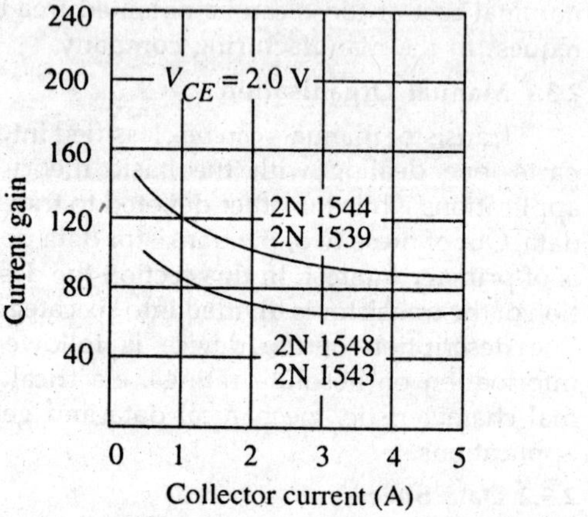

Fig. 2.4

shows that with the increase of the collector current the gain decreases.

(viii) ***Frequency cut-off***

This characteristic defines the frequency at which α or β drops to 0.707 times its value at 1000 cycles. This has been illustrated in Fig. 2.5. For most of the cases the transistors data provide a figure for gain-bandwidth product or the fre-

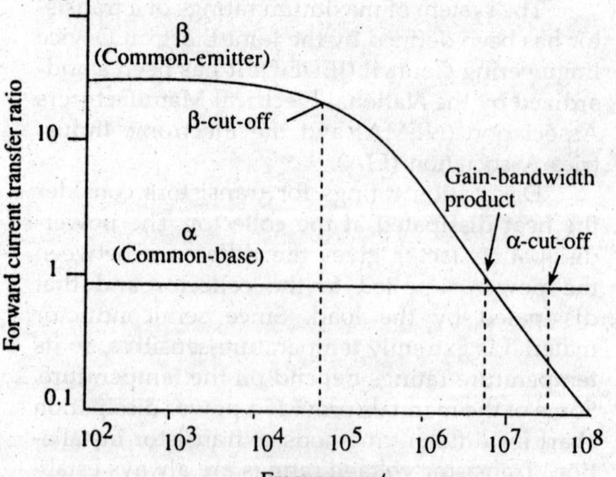

Fig. 2.5

quency at which β becomes equal to unity. From both characteristics, the useful frequency range can be achieved.

2.3.3 Testing of a Semiconductor Diode by an Ohmmeter

As a very rough test, a resistance check may be used in a semiconductor diode's operation. An ohmmeter test of a diode operating normally will show that the diode has low forward resistance and a high reverse resistance.

As shown in Fig. 2.6, lead *A* is positive and lead *B* is negative. If the positive ohmmeter lead

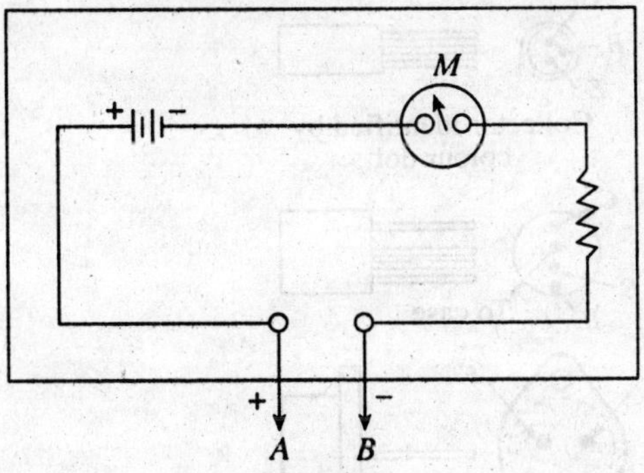

Fig. 2.6

A is connected to the anode of a diode and the negative lead *B* to the cathode, the diode will be forward biased. As a result, there will be a flow of current and the diode will measure low resistance. If now the ohmmeter leads are interchanged, then the diode will be reverse biased causing a very little flow of current. The diode will now measure a very high resistance.

In this experiment, if a semiconductor diode reveals a very *low forward and low reverse* resistance, then one may conclude that it is probably damaged (i.e., fused). Again, an open diode indicates that the forward resistance is infinite or usually high.

2.4 Transistor Basing and Mounting

In general two methods are used for bringing transistor connections out of the envelope. The first one employs long and flexible leads. An example of this is the transistor as shown in Fig. 2.7(a). All three leads are in line and may be identified as emitter, base and collector as shown. The collector spacing is greater than the emitter spacing from the base. The collector can sometimes be identified by a red dot painted close to it [Fig. 2.7(b)]. The method of connecting this type

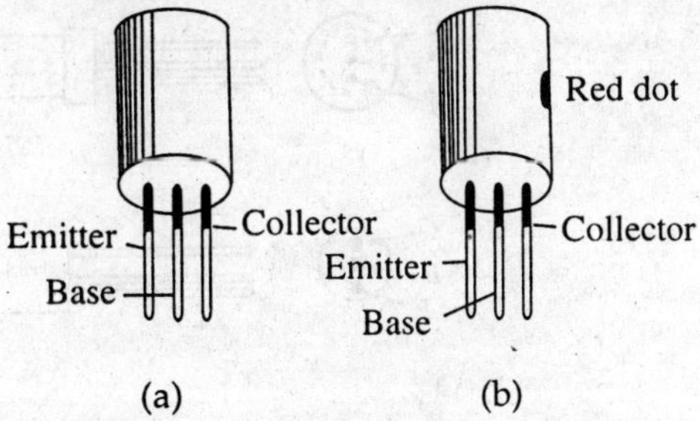

Fig. 2.7

of transistors into the circuit is to solder the flexible leads to the proper terminals in the circuit.

The second method uses rigid pins to bring transistor connections out of the envelope. This is shown in Fig. 2.8(a). This is a plug-in type of transistor and it requires a matching socket to hold it. Flexible leads are used if a transistor is to be soldered in the circuit but if a socket is to be

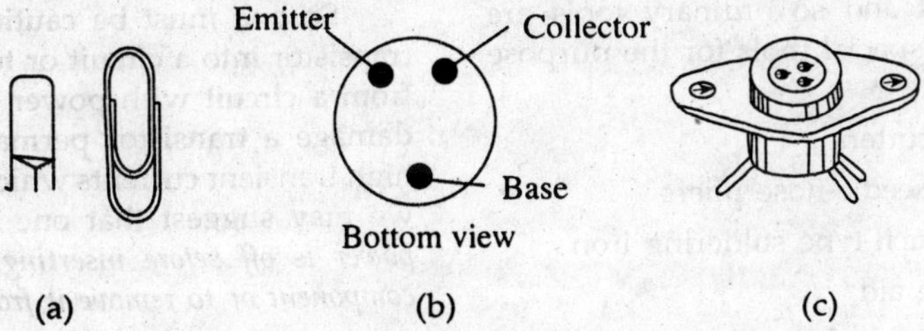

Fig. 2.8

used, then the flexible leads can be easily removed with a pair of cutters. As shown in Figs. 2.8(b) and (c), a triangular type of socket and basing arrangement used on other socket-mounted transistors has also wide application.

Some other typical arrangements of transistor basing and socket arrangements are presented in Fig. 2.9.

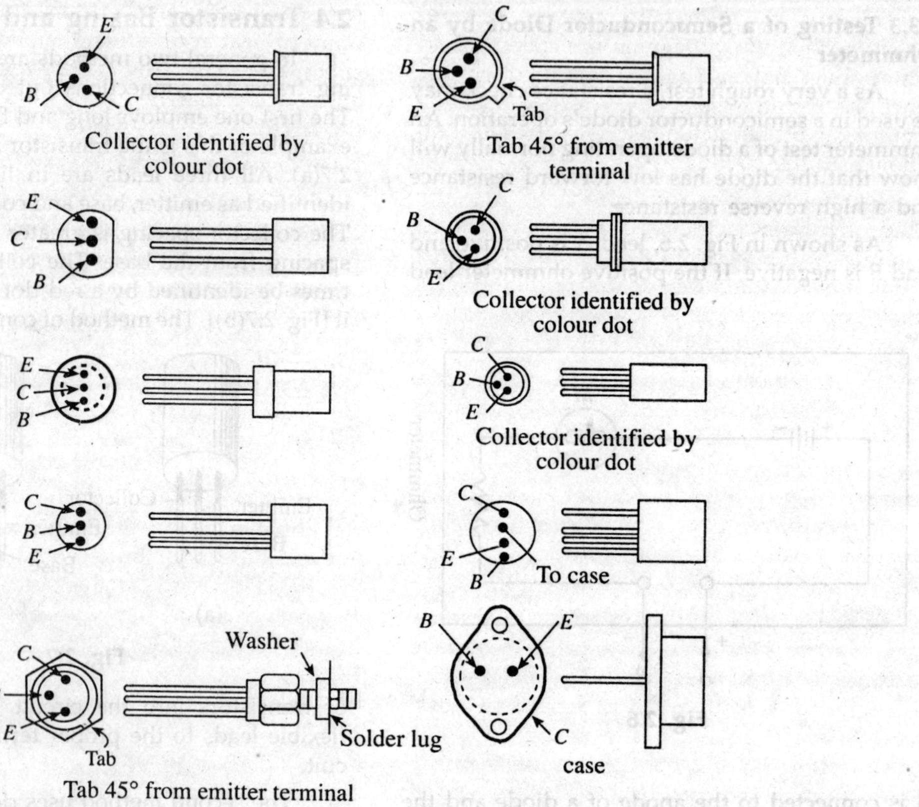

Fig. 2.9

2.4.1 Special Tools for use in Transistors

Due to the small size of the transistors and their associated component some special tools are required to use. Transistorized equipment is miniaturized in general and so ordinary tools are inapplicable. Some special tools for the purpose are listed below :

(i) Miniature cutters

(ii) Miniature needle-nose pliers

(iii) A 25 W pencil-type soldering iron

(iv) A soldering aid

(v) An assortment of service-type tweezers

(vi) A magnifying lens or a jeweler's eye-piece.

If improperly handled, transistors are damaged and so one should exercise great care to work with low-power transistors employing flexible leads which are fragile and can be easily broken. The same precautions should also be applied when working with transistor components.

In contrast to vacuum tubes, transistors can be destroyed almost instantly. In an operating circuit, a short circuit from base to collector will almost always destroy a transistor.

So one must be cautious against inserting transistor into a circuit or to remove a transistor from a circuit with power on. In fact, this may damage a transistor permanently owing to the high transient currents which may develop. Thus we may suggest that one must be certain that *power is off before inserting a transistor or other component or to remove it from a circuit.*

2.4.2 Some Additional Precautions

(i) ***Checking of circuit connections*** : Before the application of power all connections should be checked carefully with the circuit diagram. Transistor should never be connected to a voltage source without limiting resistance in the circuit.

(ii) ***Collector bias and voltage values*** : Before the application of power be sure that the polarity of bias voltage on the collector is correct. The collector in the circuit must be reverse-biased. Further, the voltage on the collector and emitter must not exceed the specified values. Hence one should measure the voltage and adjust them for the proper value before any application of power to the circuit.

(iii) ***Voltage measurements*** : To measure voltage in transistor circuits, care should be taken to minimize the possibility of accidental short circuits between the closely spaced terminals since it may apply excessive voltage to the transistor elements causing a destroy of the transistor.

(iv) ***Resistance measurements*** : Transistors are damaged sometimes during the measurement of resistance. If a shunt-type ohmmeter is employed on the low resistance range, it may supply excessive current to the transistor and destroy it. So one should *check the ohmmeter leads potential and polarity* before checking the transistor junctions.

In order to check the resistance of components in a transistor circuit, it is better to remove the transistor first if it is a plug-in type. Reversing the ohmmeter leads two sets of resistance measurements are to be made. Out of the two the higher reading is more correct because the transistor was reverse-biased on the higher reading.

(v) ***Use of signal generators as a signal source in transistor circuits*** : It has been seen that an excessive signal-generator output may destroy a transistor. So at the start of signal injection, the generator output should be set at minimum. Moreover, the generator should not be coupled directly to the circuit.

2.4.3 Transistor Soldering

For the soldering of transistor, low-wattage irons (25 W) are recommended. Heat sink should be used for soldering transistor leads in a circuit. If long-nose pliers are used for grasping the transistor pigtail between the transistor body and the point of heat application, then an effective heat-sink may be created. Fig. 2.10 reveals how pliers act as heat-sink when transistors are soldered in the circuit.

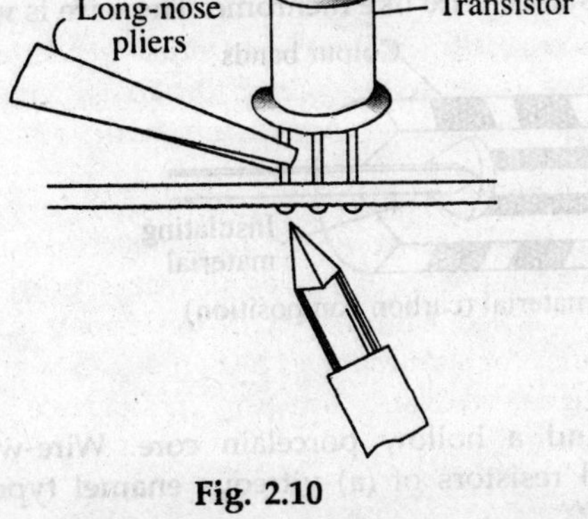

Fig. 2.10

2.4.4 Testing of Transistors with an Ohmmeter

A defective transistor can be tested with the help of an ohmmeter. Resistance tests are made between (i) emitter and base, (ii) base and collector and (iii) emitter and collector.

To test the resistance between two terminals, the ohmmeter leads are first connected in one direction and then they are reversed. For one of the lead positions, the resistance between any two transistor terminals should be very high (10 kΩ or more) while in the other lead direction, the resistance between two terminals is relatively small. Thus the forward resistance between emitter and base, and collector and base is found to be 100 Ω or less. On the other hand, the low resistance reading between emitter and collector may be around 1000 Ω.

Special Note :

When a transistor fails this test, it is certainly defective. If it passes this test by an ohmmeter it may still be defective. While a transistor is in a line circuit, it is better to test its performance by a transistor tester or a curve tracer in comparison to an ohmmeter.

It may also be noted that the ohmmeter terminal voltage must not be greater than the maximum voltage rating between any two terminals of the transistor. Moreover, the current delivered by the ohmmeter must not be higher than the maximum current rating of any junction.

2.4.5. Checking Diodes and Transistors with a Tester

A transistor tester can be suitably used for checking solid state diodes and transistors. Three different types of testers are there :

(i) Quick-check in-circuit tester

(ii) Service-type tester

and (iii) Laboratory-standard tester.

Moreover, the signature-pattern checkers *also called* curve tracers are good indicators of transistor performance.

In-circuit tester is used for checking whether a transistor performing previously is still operational. The amplification ability of a transistor is considered as a rough index of its performance. This tester indicates whether a transistor is dead or still operative. The transistor for this test, is not required to remove from the circuit. *Service-type testers* perform three check, viz., (a) forward current gain of a transistor, (b) base to collector leakage current with emitter open and (c) short circuits from collector to emitter and base. Some of the testers also include a go–no go feature. Some testers are also used to identify transistor elements. Measurements with this tester are relative rather than absolute. **Laboratory-standard testers** are employed to measure the transistor parameters dynamically under operating conditions. The readings thus obtained are absolute.

A transistor tester has the necessary controls and switches to make proper voltage, current and signal settings. A meter with a calibrated scale is there on the panel.

2.5 Electronic Components

All electronic circuits contain a few basic components. In general, there are only five components—three passive and two active. This we have shown in Table 2.1. An integrated circuit (e.g., a microprocessor) contains a large number of transistors and resistors within a very small chip, yet each component will be one of the above five components.

Table 2.1

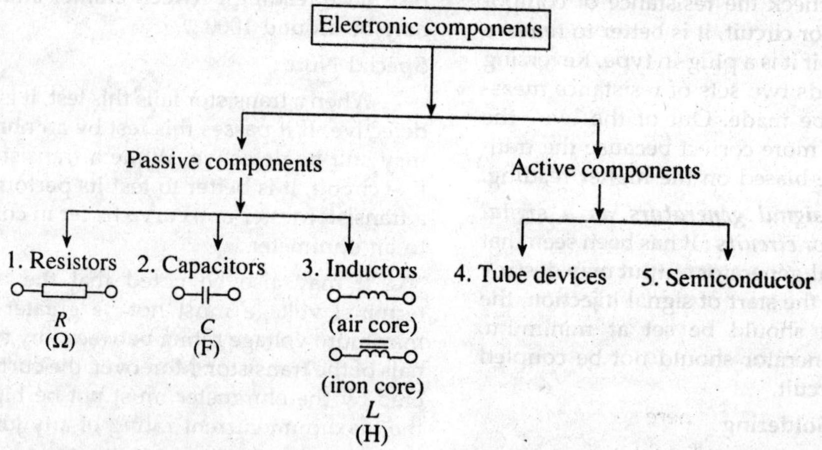

2.6. Passive Components

Resistors, capacitors and inductors are known as passive components.

Resistors

Commonly used low wattage, fixed resistor is the moulded-carbon composition resistor. The basic construction of a fixed composition resistor is shown in Fig. 2.11. Resistors of this type are available in the range from a few ohms to about 22 MΩ. The relative size of resistors increases with the increase of wattage rating. The relative sizes of moulded-carbon resistors for different power ratings are available. An alternative variety of carbon composition resistor is the metallized type whose construction is shown in Fig. 2.12. It is constructed by depositing a homogeneous film of pure carbon or some metal over a glass or ceramic. Film resistors are also called as precision type, since these are obtained with an accuracy of ±1%. A wire-wound resistor uses a length of resistance wire like nichrome. The wire is wound around a hollow porcelain core. Wire-wound fixed resistors of (a) vitreous enamel type and

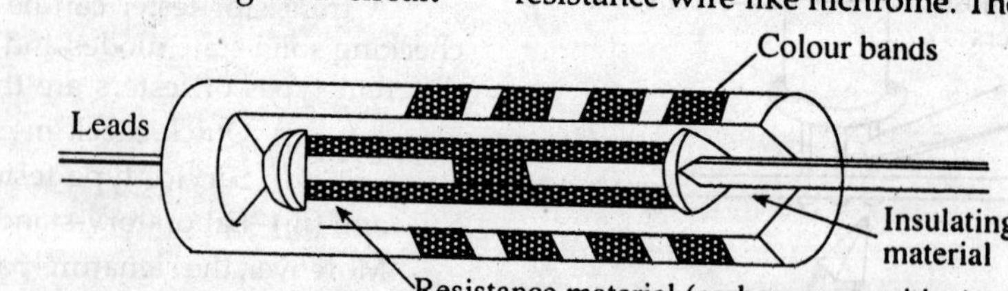

Fig. 2.11

(b) ceramic type are shown in Fig. 2.13. This type of resistors have a range from 1 Ω to 100 kΩ and wattage ratings up to about 200 W.

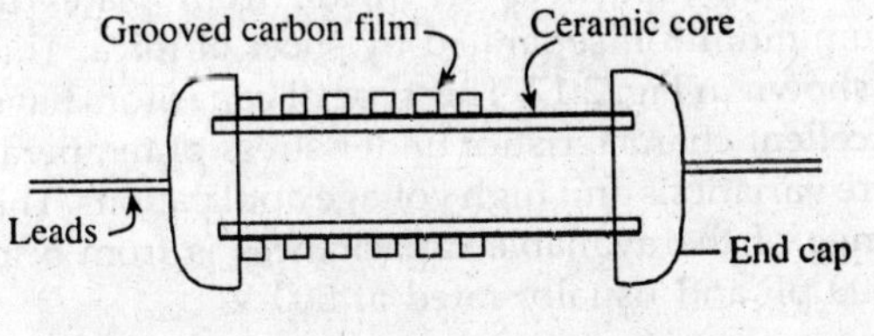

Fig. 2.12

The variable resistors also known as rheostats are used in electronic circuits to adjust the values of currents and voltages. The smaller variable

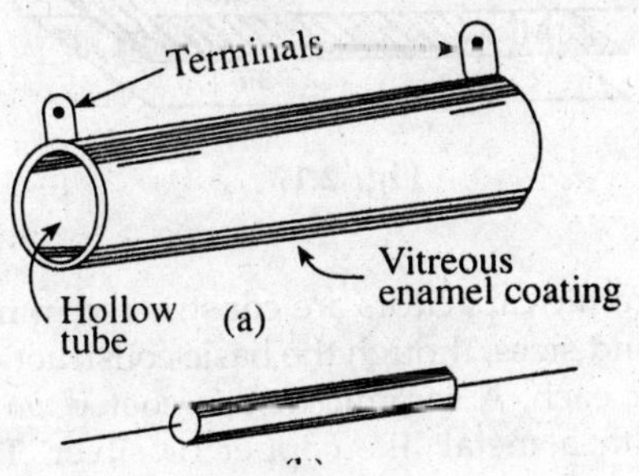

Fig. 2.13

resistors are called potentiometers usually abbreviated as 'pots'. The basic construction of a wire-wound potentiometer is shown in Fig. 2.14. As shown in the figure, there is a rotating shaft at the centre of the core which moves an arm and a contact point from end to end of the resistance element. In a potentiometer there are three terminals : the outer two are the end points of the resistance element while the middle leads to the rotating contact. A potentiometer may be either linear or nonlinear type. Fig. 2.15(a) shows the construction of a linear type potentiometer while Fig. 2.15(b) shows a nonlinear (tapered type) potentiometer. The part over which the wire is wound is called 'former'. In a linear wire-wound potentiometer the 'former' is of uniform height but it is not uniform for nonlinear type. For constructing such potentiometer, a tapered strip is taken and the wire is wound over it after which the strip is bent into a round shape.

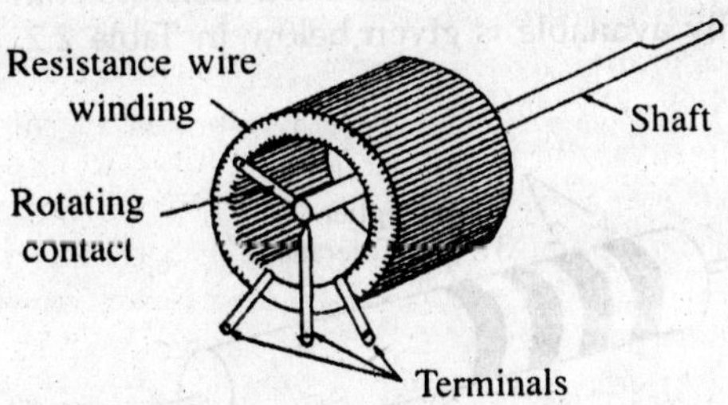

Fig. 2.14

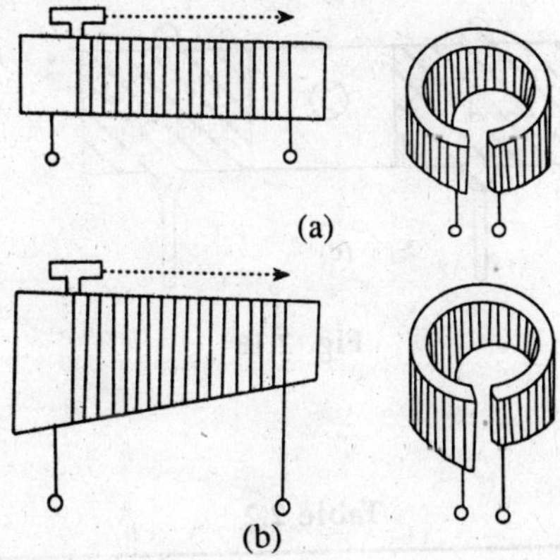

Fig. 2.15

Colour Codes and Standard Resistors

In order to indicate the values of the resistances a system of colour coding is used. For a fixed resistor four bands are printed on one end of the outer casing in a manner as shown in Fig. 2.16(a). The colour bands are read left to right from the end that has the bands closest to it. The first and second bands represent the first and second significant digits respectively while the third band is for the number of zeros that follow the second digit. If the third band is gold or silver, then it represents a multiplying factor of 1 or 0.01. The fourth band gives the tolerance of the resistance. If there is no fourth band, then the tolerance is assumed to be ±20%. The colour coding for wire-wound resistor and moulded composition resistor with radial leads is shown in Fig. 2.16(b) and (c) respectively. The system of colour coding used for moulded resistor is called as body-end dot system. A list of the standard resistors commercially available is given below in Table 2.2.

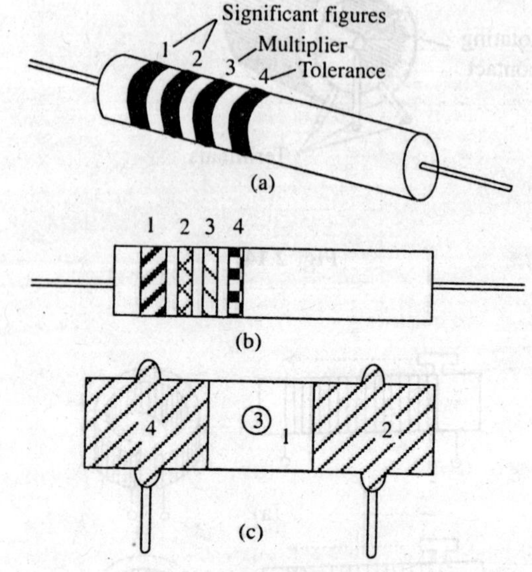

Fig. 2.16

Table 2.2

ohms (Ω)			kiloohms (kΩ)			Megohms (MΩ)	
1.0	10	100	1.0	10	100	1.0	10
1.2	12	120	1.2	12	120	1.2	12
1.5	15	150	1.5	15	150	1.5	15
1.8	18	180	1.8	18	180	1.8	18
2.2	22	220	2.2	22	220	2.2	22
2.7	27	270	2.7	27	270	2.7	
3.3	33	330	3.3	33	330	3.3	
3.9	39	390	3.9	39	390	3.9	
4.7	47	470	4.7	47	470	4.7	
5.6	56	560	5.6	56	560	5.6	
6.8	68	680	6.8	68	680	6.8	
8.2	82	820	8.2	82	820	8.2	

Capacitors

Almost in every electronic circuits capacitors of different types are used. A capacitor has two conducting plates separated by a dielectric. Capacitance is a measure of the capacitor's ability to store charge and it is measured in farad (F) or in microfarad (μF) or picofarad ((pF). Like resistors, capacitors can either be fixed or variable.

Some commonly used fixed capacitors are mica, ceramic, paper, electrolytic, etc. Variable capacitors are mostly air gang capacitors.

Mica capacitors are made from plates of aluminium foil separated by sheet of mica. This is shown in Fig. 2.17. This type of capacitors have excellent characteristics under stress of temperature variations and high voltage applications. The range of the available capacitances is from 5 to 1000 pF and usually rated at 500 V.

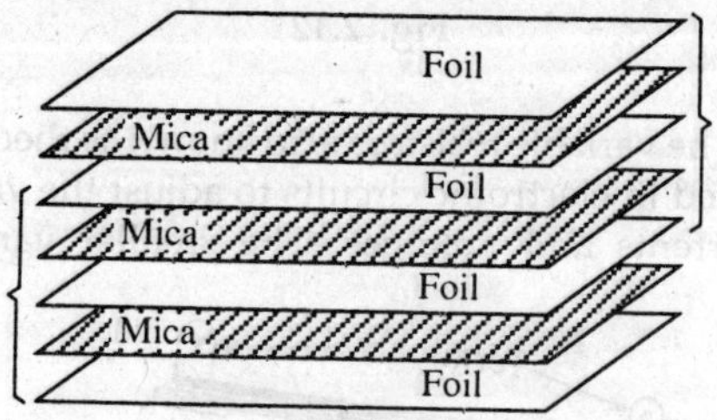

Fig. 2.17

Ceramic capacitors are constructed in many shapes and sizes, though the basic construction is same for each. A ceramic disc is coated on two sides with a metal like copper or silver. These coatings behave as the two plates (Fig. 2.18). During construction tinned wire leads are attached to each plate and then the whole unit is coated with plastic. Fig. 2.19 explains the colour code used for resistors and capacitors. The range of the capacitor is 3 pF to 2 μF. It can be used in both d.c. and a.c. circuits.

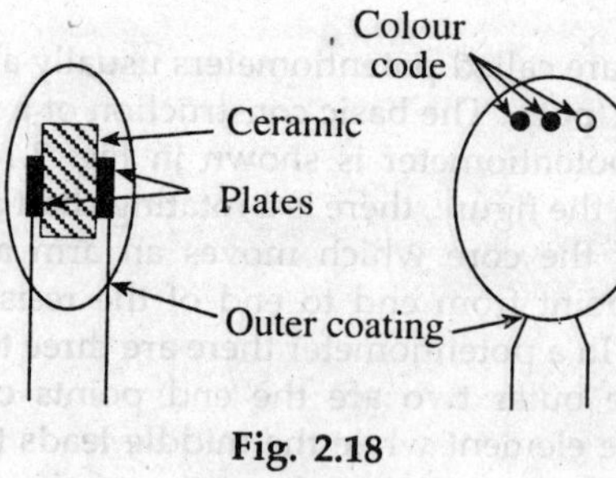

Fig. 2.18

Paper capacitors can be rolled between two metal foils and so it is possible to concentrate a large plate area within a small volume. The construction of paper capacitor is shown in Fig. 2.20. It consists of two metal foils separated by strips of paper. The range of paper capacitors lies

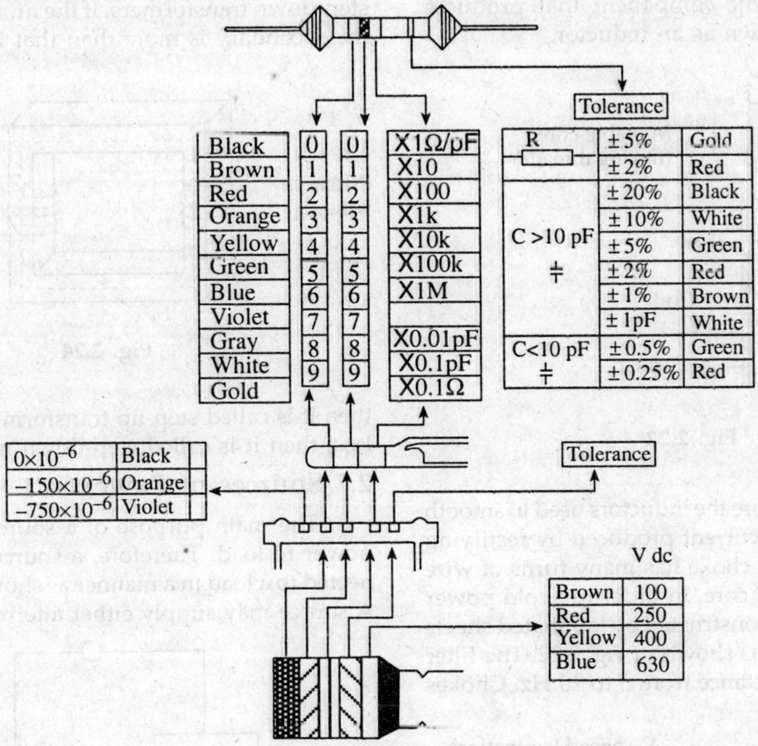

Fig. 2.19

within 0.0005 μF to several μF and are rated from 100 volts to a few thousand volts.

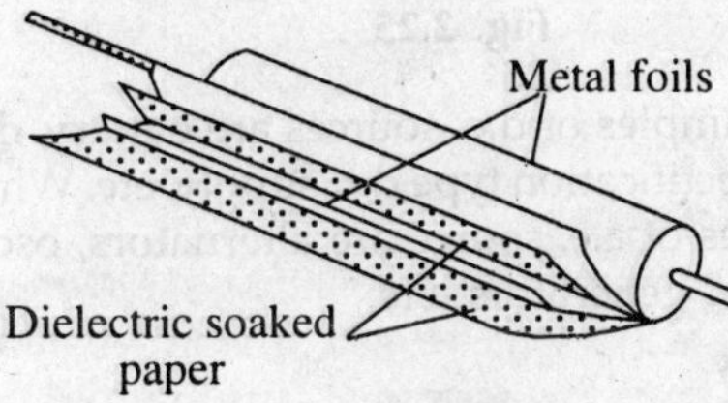

Fig. 2.20

The electrolytic capacitor has an aluminium foil electrode with an aluminium oxide film covering on one side. The aluminium plate behaves as a positive plate and the aluminium oxide film as the dielectric. The oxide is in contact with a paper or gauze saturated with an electrolyte. The electrolyte forms the negative plate of the capacitor. The capacitance value may range from 1 μF to a few thousand μF and are rated from 1 volt to 500 volts or more.

The value of the capacitance becomes necessary to vary in some circuits, like tuning circuit, etc. This is made by means of variable capacitor. The commonly used variable capacitor is the air-gang capacitor. The dielectric for this capacitor is

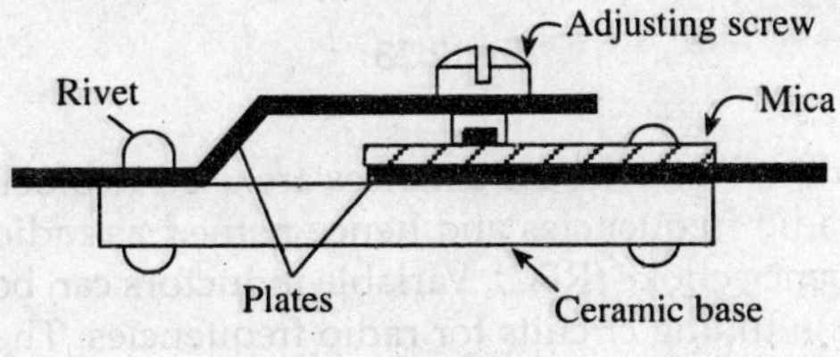

Fig. 2.21

the air. The construction of variable capacitor is shown in Fig. 2.21.

Inductors

If a current is made to pass through a coiled wire, it generates a magnetic field which reacts so as to oppose any change in the current. The relation of the magnetic field, trying to keep the current flowing at a steady rate is called inductance. The electronic component that produces inductance is known as an inductor.

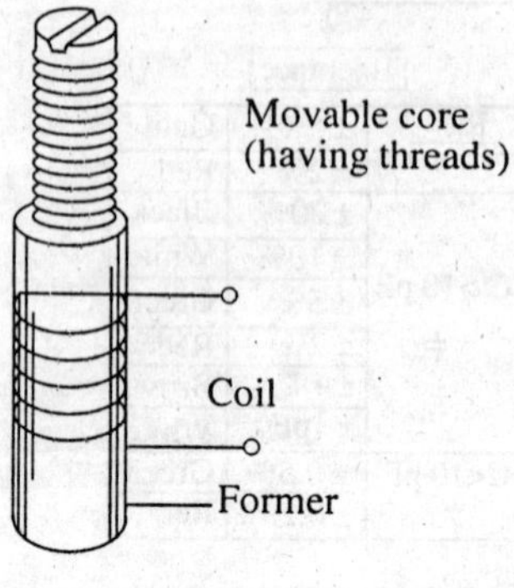

Fig. 2.22

Filter chokes are the inductors used to smooth out the pulsating current produced by rectifying ac into dc. A filter choke has many turns of wire wound on an iron core. In order to avoid power losses, the core is constructed by laminated sheets of E and I shapes as shown in Fig. 2.22. The filter chokes have inductance from 5 to 20 Hz. Chokes

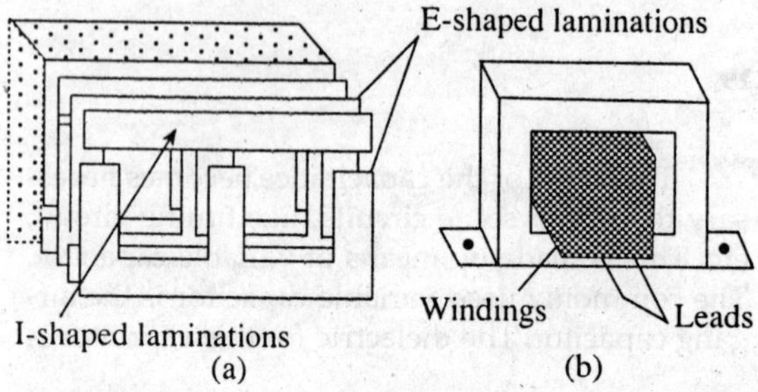

Fig. 2.23

having still smaller inductances are used to block the radio frequencies and hence named as radio frequency choke (RFC). Variable inductors can be used in tuning circuits for radio frequencies. The permeability tuned variable coil has a ferromagnetic shaft which can be moved within the coil to vary the value of the inductance. This is shown in Fig. 2.23.

Transformer

In appearance a transformer is similar to an inductor. Basically, it consists of two inductors having the same core. One of the inductors or windings is called primary while the other is called secondary. This is shown in Fig. 2.24. Transformers are of two types, viz., step-up and step-down transformers. If the number of turns in the secondary is more than that in the primary,

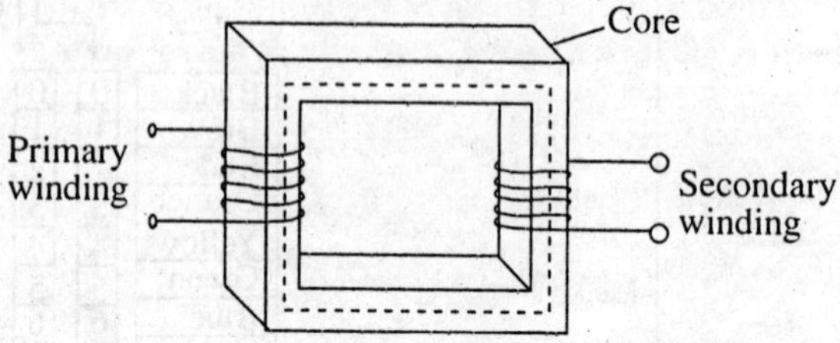

Fig. 2.24

then it is called step-up transformer while if it is less, then it is called step-down transformer.

2.7 Sources of Power

The main purpose of a source is to supply power to load. Therefore, a source is to be connected to a load in a manner as shown in Fig. 2.25. A source may supply either alternating or direct

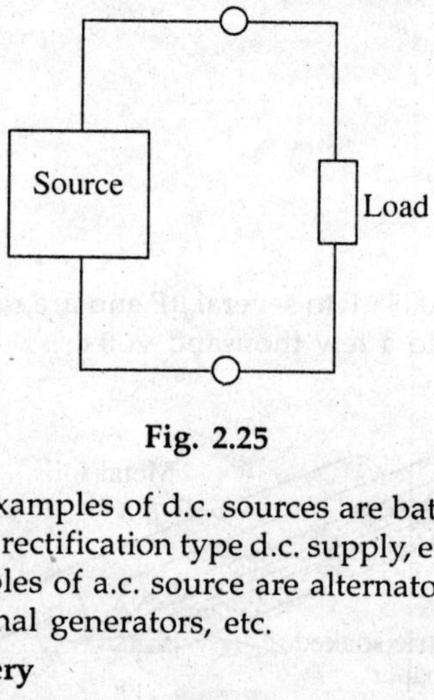

Fig. 2.25

current. Examples of d.c. sources are battery, d.c. generator, rectification type d.c. supply, etc. While the examples of a.c. source are alternators, oscillators, signal generators, etc.

2.7.1 Battery

It is a common d.c. voltage source. It consists of a series or parallel combination of two or more similar cells. A cell is, in fact, the basic source of electrical energy which can be classified into primary and secondary types. The secondary cell is rechargeable while the primary is not. A battery or a cell produces electrical power at the expense of chemical energy. The constructions of both the types are similar. Each has one positive and one negative electrodes which are immersed in an electrolyte. Electrolytes are chemical compounds and when dissolved in a solution, they decompose into positive and negative ions, which carry the charge inside the cell from one electrode to other.

2.7.2 DC Generator

A d.c. generator is quite different from the battery. It has a rotating shaft which when rotated

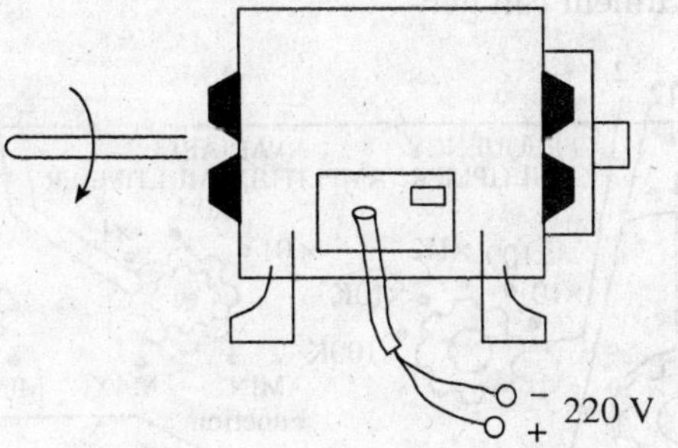

Fig. 2.26

at a specified speed by means of some external agency like a steam turbine or water turbine, a voltage of rated value shows across its terminals (Fig. 2.26). In general, a generator provides higher voltage and power than a battery.

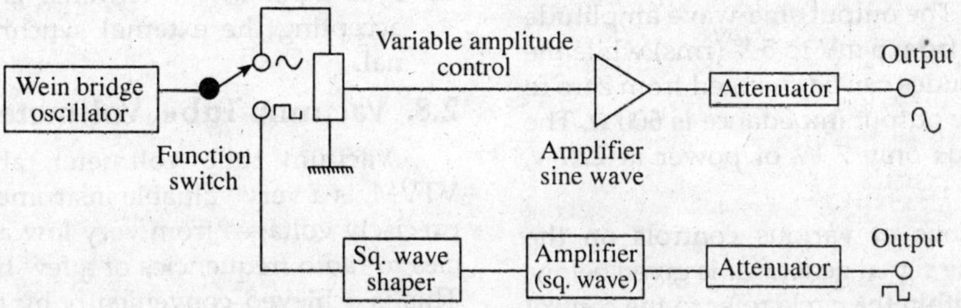

Fig. 2.27

2.7.3 DC Supply

A rectification type supply also known as d.c. supply is frequently used in electronic laboratory. It contains a rectifier which converts an alter-nating voltage to a voltage of fixed value, i.e., d.c.

2.7.4 Alternator

An alternator is similar to the dc generator both in construction and mode of operation. With the rotating of its shaft a sinusoidal voltage is generated across its terminals.

2.7.5 Oscillator or Signal Generator

The oscillator is an equipment which supplies ac voltages. This voltage can be used as a signal to test the working of various electronic circuits, such as amplifier, etc. The frequency of the ac signal as obtained from this instrument can be altered according to requirements. Some signal generators provide triangular, square-shaped waveforms also in addition to the sinusoidal wave.

2.7.6 Audio Signal Generator

The signal generator can generate an electrical signal in either the audio or radio frequency range. An audio signal generator produces audio frequencies in the shape of sine waves and/or square waves. This instrument is extensively used to test amplifiers.

The block diagram of a sine-square audio oscillator is shown in Fig. 2.27. The figure exhibits that the heart of the generator is the Wien bridge oscillator which can provide a better result in the audio frequency range. The frequency of oscillation can be conveniently changed by varying capacitors in the oscillator. The frequency can also be changed in steps by switching in resistors of various values. The output of the Wien bridge oscillator is connected to the 'function switch' which directs the oscillator output either to the square-wave shape or to the sine-wave amplifier. Thus at the output we get either a square wave or a sine wave. The output is then changed with the help of an attenuator.

Front Panel Controls

The front panel controls of a portable sine wave generator (model 51 : 6-I) as manufactured by Ruttonsha Simpson Pvt. Ltd. is shown for illustration in Fig. 2.28. The instrument can generate frequency in the range 10 Hz to 1 MHz. It is continuously variable in 5 decades with overlapping ranges. The output sine-wave amplitude can be changed from 5 mV to 5 V (rms) while the sq. wave amplitudes can be changed from zero to 20 V (peak). The output impedance is 600 Ω. The instrument needs only 7 W of power at 220 V, 50 Hz.

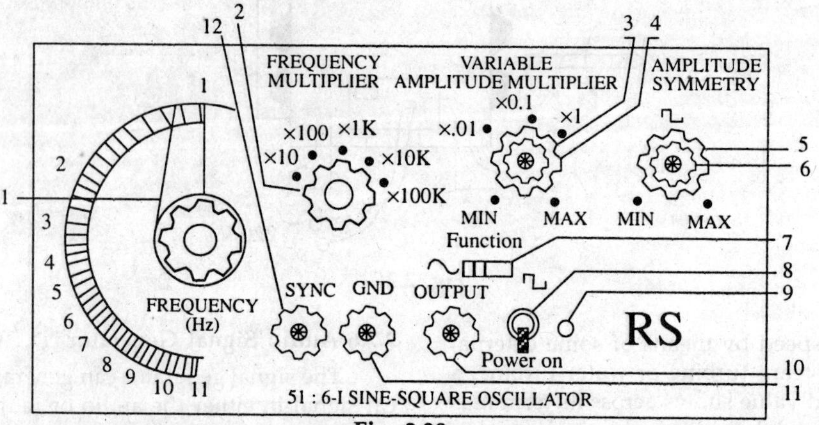

Fig. 2.28

The functions of various controls on the front panel of the signal generator is given below. The numbers within the circle refer to the control number as revealed in Fig. 2.28.

1. *Frequency Selector* : It selects the frequencies in different ranges. The frequency can be varied continuously with the help of it in a ratio 1: 11.
2. *Frequency Multiplier* : It selects the frequency range over 5 decades from 10 Hz to 1 MHz.
3. *Amplitude Multiplier* : It attenuates sine wave in 3 decades × 1, × 0.1 and × 0.01.
4. *Variable Amplitude* : It continuously attenuates the sine-wave amplitude.
5. *Symmetry (SQ)* : It varies the symmetry of square wave from 30% to 70%.
6. *Amplitude (SQ)* : It continuously attenuates the square-wave output.
7. *Function* : It selects either square-wave or sine-wave output.
8. *On Power* : It energizes the instrument in on-position.
9. *Neon Lamp* : It glows when the instrument is on.
10. *Output* : This provides square-wave or sine-wave output.
11. *Gnd* : It provides the ground reference for output (10) and also for Sync. input (12).
12. *Sync Input* : This terminal is provided for accepting the external synchronization signal.

2.8. Vacuum Tube Voltmeters

Vacuum tube voltmeter, abbreviated as VTVM, is a very valuable instrument to measure precisely voltages from very low audio frequencies to radio frequencies of a few hundred MHz. This is achieved conveniently by using vacuum tube detectors, and depending on the type of detectors used they can be broadly classified as : (a) *diode VTVM* and (b) *plate detection type VTVM*.

(a) *Diode* VTVM—Diode VTVMs are usually available in two forms: (i) *peak indicating type* and (ii) *amplifier-rectifier average reading type*.

In the peak indicating type VTVM, similar to the diode dector shown in Fig. 2.29, the rectified d.c. voltage across C–R_L combination is a measure of the peak amplitude of the input signal voltage. The average d.c. flowing through R_L may be measured by connecting a sensitive current-meter in series with it. The sensitivity of the VTVM is increased by amplifying the d.c. voltage developed across R_L. The sensitivity of the VTVM corresponds to a full-scale deflection with only a fraction input voltage. Signals up to several hundred MHz can be measured by such instrument. However, the only drawback is that it is very inconvenient to construct amplifiers with good gain and stability.

In the average reading type VTVM the a.c. signal to be measured is first applied and then it is rectified. The output thus obtained becomes proportional to the average amplitude of the signal and hence it is so named. The sensitivity of the VTVM depends on the gain of the amplifier which is stabilized by making negative feedback. The frequency range of the instrument is rather limited.

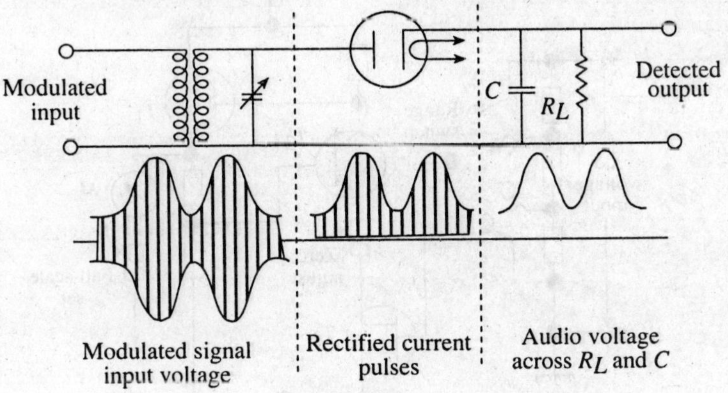

Fig. 2.29

(b) *Plate detection type VTVM*—The principle of this VTVM is similar to that of anode bend detectors. If the tube in Fig. 2.30 is biased to near cut-off value, there will be a small plate current in absence of any input signal. Under such condition if the voltage to be measured is applied at the input, the average plate current increases as recorded by a current-meter connected in the plate circuit. Signals of radio frequency ranges can be measured accurately by such instruments.

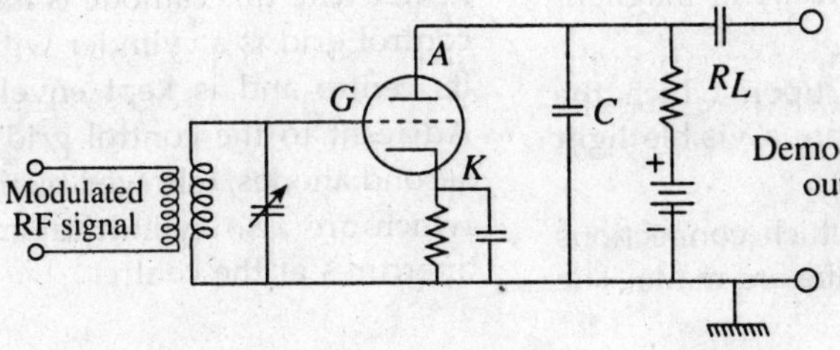

Fig. 2.30

2.9 Solid State Multimeters

Basically, a solid state multimeter is a d.c. voltmeter as shown in Fig. 2.31. The range switch selects the desired range which attenuates the input voltage to a level that can be accommodated by d.c. amplifier. As shown the input stage of the amplifier has an FET which is preferred due to its high input impedance. The two transistors T_1 and T_2 form a d.c. amplifier. Solid state multimeters are available in both analogue and digital type. In analogue type the reading is given on a continuous scale while in a digital multimeter the result of the measurement is displayed in discrete intervals.

2.10 Cathode Ray Oscillograph

Nowadays perhaps all of us are familiar with at least one type of cathode-ray tube used in the television picture tube known as kinescope. However, cathode-ray tubes are of many types, sizes and shapes. The modern cathode-ray tube is, in fact, an important modification of the Crookes tube discovered in 1879.

The CR tube is now widely used in laboratory to study the electric circuit phenomena. It is also extensively used as a viewing device for the reception of radar signals and television images. The CR tube is, in general, used in an assemblage of apparatus called a cathode-ray oscillograph, the essential components of which are : (i) *cathode-ray tube*, (ii) *deflection voltage amplifiers*, (iii) *power supplies* and (iv) *time base circuit*.

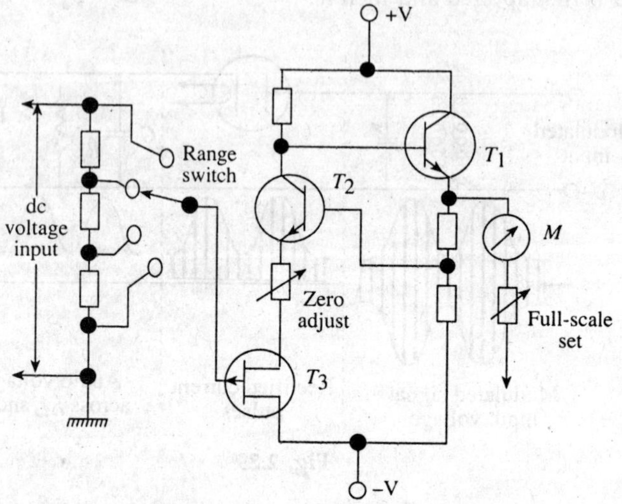

Fig. 2.31

2.10.1 Construction

A cathode-ray tube, as shown in Fig. 2.32, has the following main components :

(i) the electron gun for producing and focussing the emitted electrons in a narrow beam,

(ii) a deflection system, for deflecting the electron beam either electrostatically or magnetically, and

(iii) a fluorescent screen, upon which the electron beam impinges to create a visible light spot.

A tube base is there to which connections from the elements inside the tube are made. The

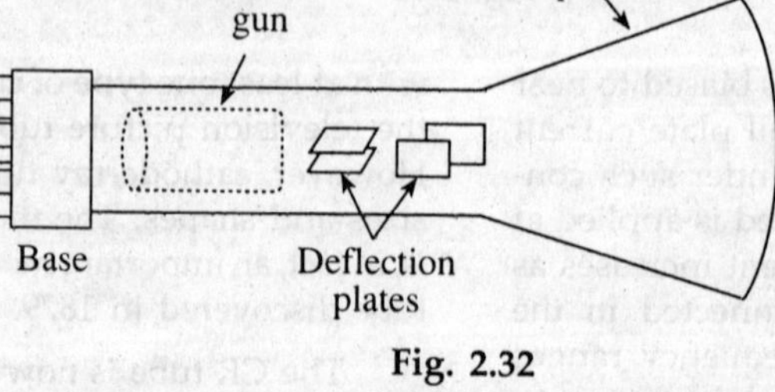

Fig. 2.32

components of a CR tube are mounted carefully inside a highly evacuated glass envelope.

2.10.2 Electron Gun

In an electrostatically focussed CR tube, as illustrated in Fig. 2.33, the electron gun consists of (a) an *indirectly heated cathode*, (b) a *control grid*, (c) an *accelerating electrode*, (d) a *first focussing anode* and (e) a *second*, or *final accelerating anode*.

The cylindrical cathode is usually coated with barium and strontium oxides for providing a plentiful supply of electrons. By a tungsten heater wire this cathode is indirectly heated. The control grid is a cylinder with a tiny aperture at its centre and is kept enveloping the cathode. Adjacent to the control grid there are first and second anodes, followed by the accelerating grid, which are also cylindrical in shape with small apertures at the centre.

The indirectly heated cathode emits a stream of electrons whose density is controlled by the bias voltage applied between the control grid and the cathode. In order to focus the beam of electrons properly, the voltage applied on the first anode is utilized. The second anode which is kept at a high positive d.c. voltage relative to the cathode, accelerates the electron beam.

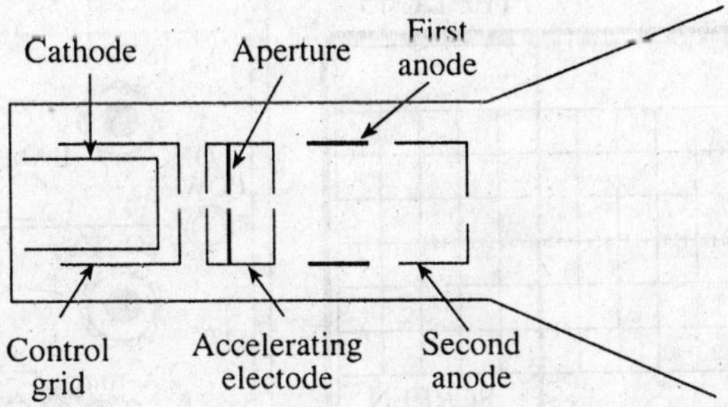

Fig. 2.33

The anode potentials in a CR tube are adjusted by considering the required beam power. The first anode is always kept at an appreciably lower potential than the second one. In television camera tube of iconscope type the first anode potential is of the order of 300 volts while that of the second anode potential is 1000 volts.

2.10.3 Fluorescent Screen

If the beam of electrons coming from the electron gun is not deflected it would form a luminous spot of light at the centre of the fluorescent screen. Though the control grid is usually used to control the intensity of this spot, it is also dependent on the fluorescent material coated inside the tube.

2.10.4 Front Panel Controls of a CRO

The front panel controls of an Aplab general purpose oscilloscope manufactured by Applied Electronics Limited, Thana is shown in Fig. 2.34. The bandwidth of its vertical amplifier ranges from dc to 15 MHz. The function of all the front panel controls are explained below. The numbers in the figure refer to the control number.

General

1. *On Power* : It is a toggle switch used for switching on power. In on position, power is supplied to the instrument and the near lamp 3 glows.

2. *Inten* : It controls the trace intensity from zero to maximum.

3. *Focus* : It controls the sharpness of the trace. A slight readjustment of this control may be required after changing the intensity of the trace.

4. *X-Mag* : It expands length of time base from 1 to 5 times continuously and makes maximum time base to 40 ns/cm.

5. *Sq. Wave* : It provides a sq. wave of 2 V *(p–p)* amplitude to enable one for checking the Y-calibration of the CRO.

6. *Sawtooth Wave* : It provides a sawtooth waveform output coincident to sweep-speed switch with an output of 5 V*(p–p)*. The load resistance must not be less than 10 kΩ.

Vertical Section

7. *Cal Switch* : When pressed, a d.c. signal of 15 mV or 150 mV is applied to a vertical amplifier depending on the position of ×1–×0.1 switch (9) position.

8. **× 1–× 0.1** *Switch* : When switched in ×0.1 position, it magnifies the basic sensitivity.

9. *Y Pos* : It enables the movement of the display along the y-axis.

10. *DC Bal* : This is a preset control and so adjusted for no movement of the trace when either × 1–×0.1 switch is pressed or the position of AC–DC–GND coupling switch (14) is changed.

11. *Volt/cm (Attenuator)* : It is a 10-position attenuator switch and is used to adjust sensitivity of vertical amplifier from 50 mV/cm to 50 V/cm in 1, 2, 5, 10 sequence. The accuracy of the attenuator is ±3%.

12. *Y-input* : It connects input signal to vertical amplifier through AC–DC–GND coupling switch 14.

13. *AC–DC–GND Coupling Switch* : This is used to select coupling to the vertical amplifier.

In d.c. mode, it directly couples the signal to the input while in a.c. mode it couples the signal to

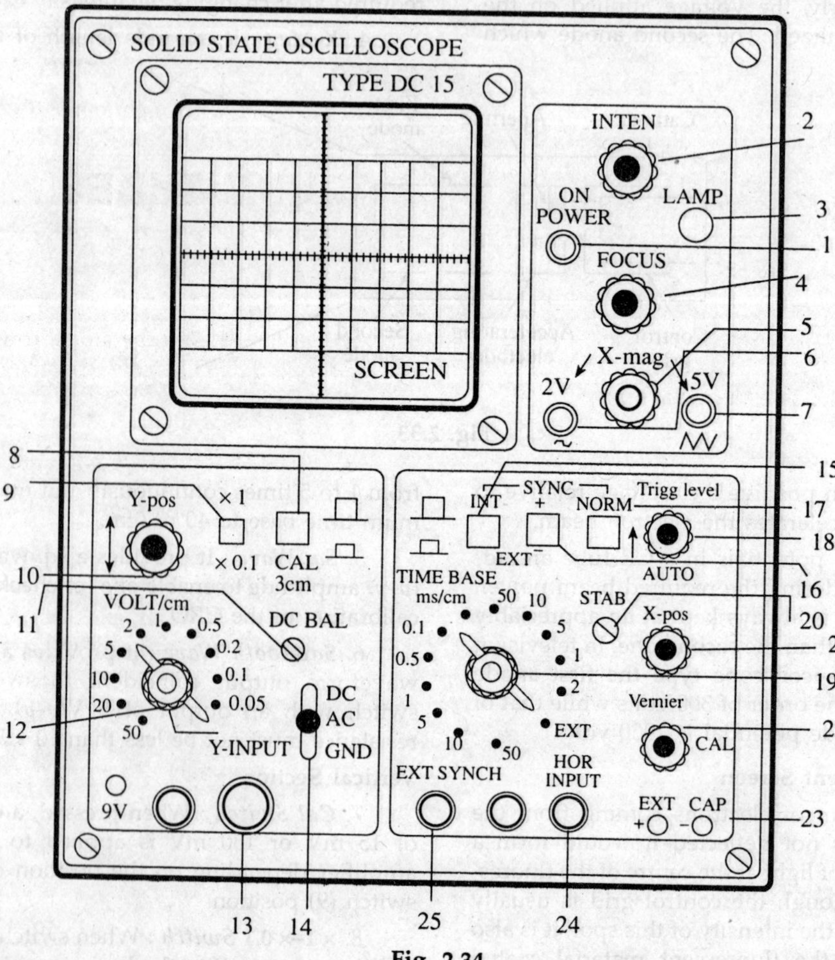

Fig. 2.34

the input through a 0.1 μF, 400 V capacitor. In *GND* position the input to the attenuator 12 is grounded while the Y-input is isolated.

Horizontal Section

14, 15 and 16. Sync : These controls are used for sync. As evident from the figure they are for 'Int.', '±' and 'Norm.' respectively.

17. ***Trigger Level*** : This is used to select the mode of triggering. In auto position, the time base line is displayed when the input signal is absent. But when it is present, the display is automatically triggered.

18. ***Time Base*** : It selects sweep speeds from 50 μs/cm to 0.2 μs/cm in 11 steps. The position marked *EXT* is used when an external signal is to be applied to the horizontal input 24.

19. ***Stab*** : This is a preset control on the panel and is adjusted to get the base line in outs position of Trigger Level Control 18.

20. ***X-pos*** : It enables the movement of display along x-axis.

21. ***Vernier*** : This control is a fine adjustment associated with the Time Base Sweep Selector Switch (9). It extends the range of sweep by a factor 5. It should be clockwise turned to the *CAL* position for calibrated sweep speeds.

22. ***Ext Cap*** : This pair of connectors enables the time base range to be extended beyond 50 m/cm by connecting a capacitor at these connectors.

23. ***Hor Input*** : It connects the external signal to the Horizontal Amplifier.

24. ***Ext Sync.*** : It connects external signal to trigger circuit for synchronization.

CHAPTER 3

EXPERIMENTS ON VACUUM TUBES

3.1 To Draw the Characteristic Curves of a Vacuum Diode

Circuit Diagram

The circuit diagram for studying the characteristics of a diode valve is shown in Fig. 3.1. The indices are given below :

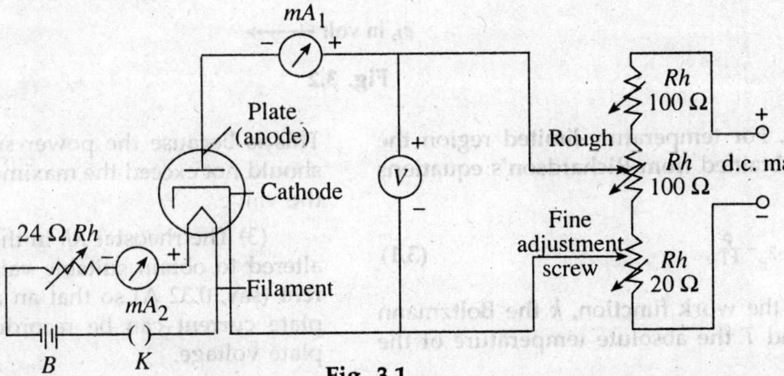

Fig. 3.1

- $V \rightarrow$ Voltmeter [range 0-25 volt]
- $Rh \rightarrow$ Rheostats [24 Ω for filament circuit and 100 Ω, 100 Ω, 20 Ω in series for plate circuit]
- $B \rightarrow$ Battery [6 volts]
- $mA_1 \rightarrow$ Milliammeter, whose negative terminal is connected to anode [range 0-50 A.]
- $mA_2 \rightarrow$ Milliammeter, whose positive terminal is connected to the filament [range 0-1 A.]

Working Formula

The curves obtained on plotting anode current along the Y-axis against anode voltage along the X-axis is called the current-voltage characteristics of a diode. Since current and voltage involved pertain to the tube alone, irrespective of the associated circuits, the characteristics thus obtained usually referred to as 'static characteristics'.

Static characteristics may be subdivided into two categories : (i) Anode current vs. anode voltage characteristics for a fixed filament current and (ii) Anode current vs. filament temperature (which is proportional to filament current) for a fixed anode voltage.

A typical characteristic is drawn in Fig. 3.2 for expt. 3.1 above. Considering Fig. 3.2 we may come to the following results :

Filament current in A.	e_b for	
	space charge limited region	temp. limited region
$I_{f_3} = 0.34$	0–18 volts	>18 volts
$I_{f_2} = 0.33$	0–14 volts	>14 volts
$I_{f_1} = 0.32$	0–11 volts	>11 volts

For space charge region, the anode current is obtained from the Child-Langmuir's equation given by $i_b = ke_b^{3/2}$, where i_b and e_b are the plate current and plate voltage respectively, k is a constant whose value depends on the dimensions of the electrode spacing and on certain other

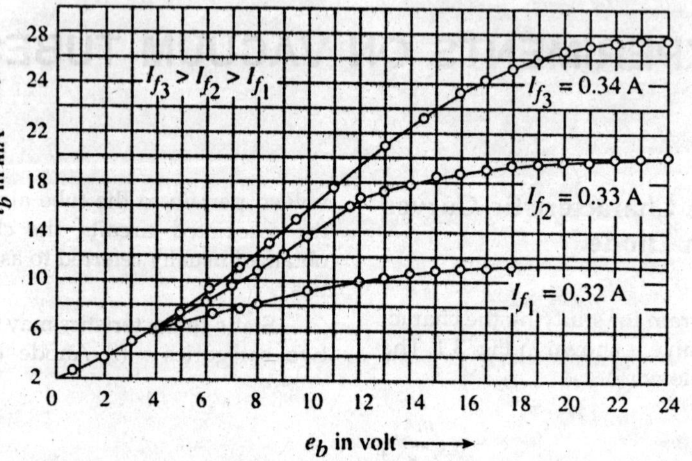

Fig. 3.2

parameters. For temperature-limited region the current is obtained from Richardson's equations given by

$$I = AT^2 e^{-\frac{\phi}{kT}}, \tag{3.1}$$

where ϕ is the work function, k the Boltzmann constant and T the absolute temperature of the filament.

The a.c. anode resistance is obtained from,

$$(r_p)_{ac} = \frac{\delta e_b}{\delta i_b}, \tag{3.2}$$

where δe_b is the change in plate resistance and δi_b is the change in plate current. From $(i_b - e_b)$ characteristics δe_b and δi_b can be chosen for getting r_p.

For d.c. anode resistance we can use the relation,

$$(r_p)_{dc} = \frac{e_b}{i_b} \tag{3.3}$$

Experimental procedure (using basic principle of the circuit shown in Fig. 3.1).

(1) The basic principle of the circuit diagram is drawn in the laboratory notebook and next the connections are made as shown in Fig. 3.1.

The polarities of the milliammeters mA_1, mA_2 and the voltmeter V are properly considered during the connections.

(2) The maximum power ratings of the valve supplied for the experiment are to be seen from the valve manual or as directed by the instructor. This is because the power supply to the diode should not exceed the maximum power rating of the valve.

(3) The rheostat Rh in the filament circuit is altered to obtain suitable value of filament current (say, 0.32 A) so that an appreciable data of plate current can be recorded with change of plate voltage.

In this case the plate voltages are gradually varied by means of the rheostat used for approximate adjustment and the corresponding plate currents are noted. If the plate current gradually increases with plate voltage and attains a *nearly* saturated value, then that filament current may be taken for drawing the characteristic curve.

After this *rough checking*, the filament current is kept to the previously adjusted value. The plate voltage is adjusted so that the plate current becomes nearly zero. The plate voltages and the corresponding plate currents are noted twice and the mean value is calculated. The plate voltage is increased to higher value at regular intervals depending on the power rating of the diode and the smallest division of the plate voltmeter (V) and in each case the plate voltage and the corresponding plate current are noted as before. This is done until the temperature-limited region is attained.

N.B. : For diodes of indirectly heated type, it is very difficult to get the temperature-limited region properly. If a directly heated type is selected for the purpose, then the above region can be more easily obtained. The filament may be short-circuited to the valve base with the cathodes for making it a directly heated type for lack of a directly heated valve. However, in this case the lifetime of the tube may be shorter.

(4) The filament current is altered to a higher value and the above procedure is repeated. If the filament current is now increased from I_{f_1} to I_{f_2} (say), then there will be an increase of anode current.

(5) The above operation is repeated for a third set of filament current.

(6) The plate voltage and the filament current are slowly reduced to minimum and the circuit is made off.

Works with Power Supply

Nowadays a regulated power supply is generally used for studying the characteristics. Some binding tags are provided with the supply which are used for obtaining variable voltages across the plate and across the filament. There is no hard and fast rule for the design of a supply and in fact it varies from one manufacturer to other.

(i) The potential divider regulating the plate supply voltage is identified and it is slowly rotated to the anticlockwise direction to keep the plate voltage minimum.

(ii) The regulator used for changing the filament current from the supply is adjusted to a suitable value as before.

(iii) The regulator for plate supply voltage is slowly rotated in clockwise direction to adjust a suitable plate voltage and the corresponding mean plate current is noted as discussed earlier.

Experimental Results :

[A] *To find the range of the meters used*

Table 3.1

Name of the meters used	*Range*	*Value of this smallest division*	*Zero (or initial) error, if any*
Ammeter to measure filament current	...ampere	...ampere	...ampere
Voltmeter to measure plate voltage	...volt	...volt	...volt
Milliammeter to measure plate current	...mA	...mA	...mA

[B] *To find the plate currents for different values of plate voltage at constant filament current*

Table 3.2

Valve number = ...

Set	Filament current (I_f) in ampere	Plate current (I_b) in mA	Plate voltage (e_b) in volt
I	I_{f_1} = ...	...... (min. : ...... (max)	...... (min) : ......(max)
II	I_{f_2} = ...	etc.	etc.
III	I_{f_3} = ...	etc.	etc.

Drawing of Graphs

For getting the static characteristic curves, the anode current (in milliampere) is plotted against the anode voltage (in volt) for a constant filament current. The nature of the curves obtained is shown in Fig. 3.2. From this the values of $(r_p)_{ac}$ and $(r_p)_{dc}$ can be calculated at three operating points.

Discussion

(i) At very high anode voltages oxide-coated cathodes may become damaged because of abnormally large emission. In general, electron tubes are operated in the space charge-limited (low e_b) regions.

(ii) The saturation current in the characteristic is obtained due to the space charge effect.

(iii) Note that the function is not a linear one. The actual equation was established by Child and Langmuir and is known as three-halves power equation which can be expressed as $i_b = Ke_b^{3/2}$, where K is a constant determined by the physical construction of the tube.

(iv) Two resistance values can be achieved from the curve, the dc or static plate resistance R_p and the ac or dynamic plate resistance r_b.

(v) Instead of taking the observations by increasing and decreasing the plate voltage, the reading should be taken twice under *identical operating conditions*. If the readings are noted (i) for voltage increasing and (ii) for voltage decreasing, and the mean is taken, the final result will be *erratic*. This is because the current during the decreasing value of the plate voltage will be the same if sufficient time is allowed for attaining the thermal equilibrium condition.

3.2 To use the Vacuum Diode as a Half-wave Rectifier

Circuit Diagram

The circuit diagram for using the vacuum diode as a half-wave rectifier is shown in Fig. 3.3.

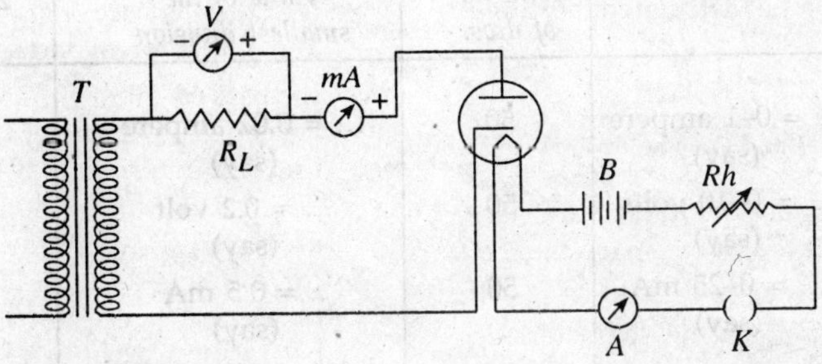

Fig. 3.3

The indices are given below :

T→Transformer

R_L→Load resistance

V→Voltmeter

mA→Milliammeter

A→Ammeter

K→Key

Rh→Rheostat

B→Battery

Working Formula

For the diode valve to act as a rectifier the input voltage should not exceed the space charge control region of the anode voltage and anode current. The cut-in voltage of the diode is taken to be zero. The variation of the dc output voltage as a function of the dc load current is called the regulation. We may write,

Percentage regulation

$$= \frac{V_{no\ load} - V_{rated\ load}}{V_{rated\ load}} \times 100\%, \quad (3.4)$$

where no load refers to zero load current and rated load refers to the current at which the regulation is desired.

The above expression can be further written as

$$\text{Percentage regulation} = \left[\frac{V_{NL}}{V_{RL}} - 1\right] \times 100\%. \quad (3.5)$$

When $V_{NL} \simeq V_{RL}$, then percentage becomes nearly equal to zero. In fact, for a perfect voltage regulator with drawing of current there should not be any change of supply voltage for a desired current range. Hence the V_{NL} should be equal to V_{RL} for a perfect voltage regulator. So when the percentage regulation approaches to zero, then $V_{NL} = V_{RL}$. For a poor voltage regulator the value of percentage regulation is high.

For a half-wave rectifier the variation of V_{dc} with I_{dc} is given by $\frac{V_m}{\pi} - I_{dc} R_f$, where R_f includes the resistance of the secondary transformer.

Experimental Procedure

(1) The circuit diagram showing the basic principle of the half-wave rectifier (Fig. 3.3) is drawn in the laboratory notebook and connections are made.

(2) The frequency and the voltage of the input ac signal are adjusted to suitable value as directed by the instructor. This is because the input signal to be rectified depends on the power ratings of the valve and other circuit parameters.

(3) The range of the meters used and the value of one smallest division are noted. Zero errors of the meters used (if any) are also recorded.

(4) Keeping the filament current constant, the load current I_L is increased in regular steps by changing R_L and at each step the load voltage V_L is noted.

(5) The filament current is next changed to a second value and the above observation is repeated. A third set of readings is also taken for another constant filament current.

Experimental Results :

[A] To find the range of the meters used.

(Data given only for illustration)

Resistance of the rheostat used = ...Ω, its current tolerance = ... ampere.

Table 3.3

Name of the meters used	*Range*	*Total no. of divs.*	*Value of the smallest division*	*Zero (or initial) error, if any*
Ammeter to measure filament current (I_f)	.. = 0–1 ampere (say)	50	... = 0.02 ampere (say)	... ampere
Voltmeter to measure load voltage (V)	... = 0–10 volts (say)	50	... = 0.2 volt (say)	... volt
Milliammeter to measure load current (mA)	... = 0–25 mA (say)	50	...= 0.5 mA (say)	...mA

[B] *To find the load currents for different values of load voltages when the filament current is kept constant.*

Table 3.4

Set No.	*Filament current in ampere*	*Resistance applied in the resistance box (R_L) in ohm*	*Load current (I_L) in mA*	*Load voltage (V_L) in volt*
		...	...	...
		...	...	...
		...	...	...
I	I_{f1} = ...	...	...	...
		...	...	...
		etc.	etc.	etc.
II	I_{f2} = ...	etc.	etc.	etc.
III	I_{f3} = ...	etc.	etc.	etc.

Drawing of Graphs

The nature of the regulation curve is as shown in Fig. 3.4 (for a pure resistance filter).

Similar curves are to be drawn for other sets of observations.

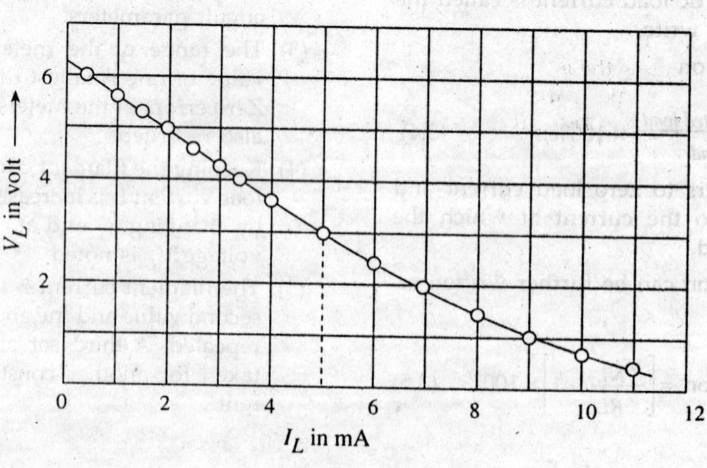

Fig. 3.4

[C] *To calculate the percentage regulation for different load currents*

For this we consider a V_L versus I_L curve for a particular filament current [say at I_{f_1} ampere] as shown in Fig. 3.4. An operating load current is selected at high load current region, say, between 9 to 10 mA. The corresponding value of V_L is recorded from the load regulation curve (i.e., V_L versus I_L curve). *It is reasonable to term this voltage as the rated-load voltage (instead of full-load voltage) at the rated load current* I_L. The load current is selected to another value (say, between 8 to 9 mA) and the corresponding rated-load voltage is again recorded from the regulation curve. In this way various rated voltage are recorded for various load currents. *This procedure is repeated for other values of filament current* (say, I_{f_2}, I_{f_3} ampere, etc.).

Table 3.5

Filament Current (I_f) in ampere	*No load voltage (V_{NL})* in volt	*Rated current (I_L) in* mA	*Corresponding load voltage (for rated voltage)* V_{RL} in volt	*Percentage regulation* $= \frac{V_{NL}-V_{RL}}{V_{RL}} \times 100\%$	*Remarks about the regulation curve*
I_{f_1} = ...	...	... ...	... ...	... ...	
I_{f_2} = ...	etc.	etc.	etc.	etc.	etc.
I_{f_3} = ...	etc.	etc.	etc.	etc.	etc.

Finally, the load is disconnected from the plate circuit and the plate is directly connected to the input of the ac source via the *mA*. The voltage between the cathode and the plate is directly measured by *an* a.c. *voltmeter*. This gives the no-load voltage (V_{NL}) and its value is recorded. Of course, this can be also done by measuring directly the input ac voltage across the secondary terminals of the transformer *T* without disconnecting the load.

[D] *To calculate the percentage regulations at different operating currents.*

N.B. : Instead of a pure resistance load, an L–C or R – L – C filter circuit may be used. In this case, the regulation will be much better than that shown in Fig. 3.4.

[E] *To plot percentage regulation at various operating load currents*

Remarks : Since the percentage regulation changes with change of operating load currents, hence a graph may be plotted with load currents along X-axis and the corresponding percentage regulation along Y-axis (for a particular filament current). The data is taken from Table 3.5 for the filament current I_{f_1} ampere (say) and the *percentage-regulation characteristic* curve is drawn. Similar curves are drawn for other filament currents (say, I_{f_2}, I_{f_3}, etc.).

Discussion : Same as in the expt. no. 3.1.

3.3. To Draw the Plate Characteristic Curves of a Triode

Circuit Diagram

The *basic principle* of the circuit diagram for studying the plate characteristics of a triode valve is shown in Fig. 3.5. The active and passive elements and different electrical accessories required for the purpose are :

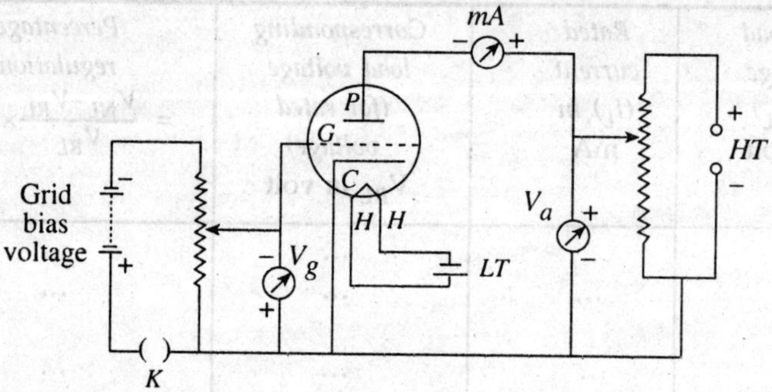

Fig. 3.5

(1) a triod valve, (2) a high-tension supply (0 to 220 volts, d.c.) denoted by *HT*, and a rheostat (of proper power rating) connected in parallel to it, (3) a grid-bias supply with a rheostat in parallel to it, (4) a plate voltmeter (V_a) of suitable range (say, from 0 to 220 volts), (5) a grid-voltmeter (V_g) of suitable range (say, from 0 to –20 volts), (6) a milliammeter (*mA*), (7) a low-tension supply (denoted by *LT*) for heater circuit (*H,H*), (8) a key, *K*, and connecting wires, etc.

N.B. : Instead of using different supply voltages for the plate, grid and the heater circuits, the biasing may be made from the same *HT* supply by using proper potential divider.

Working Formula

It is assumed that e_b, i_b and e_c represent plate voltage, plate current and grid voltage respectively.

The three parameters of a triode valve are defined as follows :

(i) *Anode slope resistance (or the* ac *plate resistance, or the internal resistance, i.e.,* r_p*)*

It is defined as

$$r_p = \frac{\text{change of anode voltage}}{\text{change of anode current for a constant grid voltage}}$$

$$= \left(\frac{\Delta e_b}{\Delta i_b}\right)_{e_c = \text{constant}} \qquad (3.6)$$

where

Δe_b = a small change in plate voltage

Δi_b = a small change in plate current.

N.B. : It is the resistance offered by a triode valve to the flow of alternating plate current, when varying sinusoidal voltages are applied between the electrodes. It is also known as the dynamic resistance. It is measured by the inverse slope of the tangent to a given I_b versus e_b curve at a particular value of e_c. This resistance is not the same as the tube's opposition to dc

(ii) *Transconductance or mutual conductance* (g_m)

The transconductance (g_m) is defined as

$$g_m = \frac{\text{change of anode current}}{\text{change of grid voltage for a constant anode voltage}}$$

$$= \left(\frac{\Delta i_b}{\Delta e_c}\right)_{eb = \text{constant}} \qquad (3.7)$$

where

Δi_b = a small change in plate current

Δe_c = a small change in grid voltage.

N.B. : It signifies the relative effectiveness of the control grid on the flow of the plate current for a fixed plate voltage. It is also frequently termed as the figure of merit for comparing different valves. The value of g_m is a factor used by a tube tester to compare the tube with its original ratings.

(iii) *Amplification factor* (μ)

The amplification factor (μ) is defined as

$$\frac{\text{change of anode voltage}}{\text{change of grid voltage to preserve constant anode current}}$$

$$= -\left(\frac{\Delta e_b}{\Delta e_c}\right)_{i_b = \text{constant}}$$

The –ve sign is due to the fact that the plate voltage (e_b) is to be increased while the grid voltage (e_c) is to be decreased to make the same change in the plate current.

In the above,

Δe_b = a small change in plate voltage

Δe_c = a small change in grid voltage.

Considering, *the magnitude only,* we have

$$\mu = \left(\frac{\Delta e_b}{\Delta e_c}\right)_{i_b = \text{constant}} \tag{3.8}$$

N.B. : The amplification factor gives the relative effectiveness of grid voltage to control the plate current. It expresses the effectiveness of the field of the control grid to overcome the electrostatic field produced by the plate.

(iv) *d.c. plate resistance* (r_{dc})

The d.c. plate resistance is defined as

$$r_{dc} = \left(\frac{e_b}{i_b}\right)_{e_c = \text{constant}} \tag{3.9}$$

N.B. : The d.c. plate resistance of a triode is not a tube parameter, because it can have any value from a few hundred ohms to infinity.

(v) *The relation between* μ, r_p *and* g_m

It can be shown that

$$\mu = r_p g_m \tag{3.10}$$

The values of g_m and r_p vary widely over the operating range of the tube, but the value of μ remains fairly constant.

(vi) It is to be noticed that value of μ for a tube is a good guide to check the negative grid voltage required to stop the anode current. This is because

$$e_c \text{ (cut-off)} = \frac{e_b}{\mu}. \tag{3.11}$$

Actually, e_c (cut-off) tends to be slightly greater than this because of the initial curvature of the i_b–e_b characteristics. Hence, the value of μ calculated from eqn. (3.11) should be slightly less than that of the actual value. This slight decrement of μ in the region of $i_b = 0$ confirms the nonlinearity of the curve at the low plate current region.

Plan of Circuit Connections

(using HT, LT and rheostats)

The circuit diagram using *the basic principle* is shown in Fig. 3.5. The plate *P* is connected to the milliammeter and the voltmeter as shown. The +ve terminal of the milliammeter is connected to the +ve side of the potential divider of HT source. The grid (*G*) is connected to the –ve side of the grid-bias voltage as shown. The voltmeters used in the plate circuit and the grid circuit are properly connected as shown in Fig. 3.5. The heater (*H, H*) is properly heated, either by using transformer heating or by using a separate LT source. The cathode circuit is also properly connected.

Caution : Check the polarities of the HT and LT sources and also the polarities of the respective meters before switching on the circuit.

Experimental Procedure

Procedure (1) : The circuit diagram, shown in Fig. 3.5, is drawn in the laboratory notebook, and finally the connections are made.

N.B. : Before switching on the circuit, the circuit should be approved by the instructor.

Procedure (2) : The value of the one smallest division of the respective meters used, and also the zero error, if any, are calculated and recorded.

Procedure (3) : From the manufacturer's tube manual, the range of the plate voltages for computing the values of μ, r_p and g_m are properly selected. *Otherwise, the plate voltages to be applied during experiment are selected to those values as instructed by the instructor.* The plate voltages and also the grid voltages are to be selected in such a way so as to record the values of μ, r_p and g_m over a considerable region of *the linear portion of the graph.*

N.B. : In the following procedures it is assumed that the plate voltage ranges are from 215 to 5 volts. Of course, the values are selected only for the illustrations.

Procedure (4) : The zero-error in the different meters are at first eliminated. That is, the pointers of different meters are set to the zero-reading by a screw-driver (if, there is any error). This zero-error correction should be done only when the *circuit is made-off. It should never be done when the circuit is made-on.*

Procedure (5) : The plate voltage is *reduced to the minimum value* by adjusting the rheostat (or, the regulator). The grid voltage is adjusted to –1 volt (say) (or –0.5 volt), depending on the smallest graduation of the grid-voltmeter, by adjusting the rheostat (or the regulator). The plate voltage is decreased to such a value that the plate current is zero. This plate voltage is recorded. This recording is done twice (as observation I and observation II) and the mean value of the plate current is recorded. The plate voltage is increased in steps of 10 volts or 5 volts, depending on the smallest graduation of the plate-voltmeter, and the corresponding mean plate current is noted. *This is described elaborately as follows* :

The plate voltage is adjusted to a lower value (say, 35 volts). The corresponding plate current is recorded, *as observation* I. The plate voltage is reduced to a slightly lower value (say, 30 volts) than that as before with the help of the sliding contact of the rheostat (or, the regulator). Next, the plate voltage is again gradually and slowly increased to the previously adjusted value (say, 35 volts), and the plate current is again recorded. This recording of the plate current is recorded as *observation* II. *The mean value of the current from observation* I *and observation* II, gives the value of the plate current for that plate voltage.

Caution :

(I) The grid voltage should be kept constant to the pre-adjusted value (which in the above operation is –1 volt, or –0.5 volt, say). Any fluctuations of the plate voltage and the grid voltage are to be minimised before final recording.

(II) The readings of the pointers of the different meters should be taken by viewing the corresponding pointer normally under the distinct vision condition of the eye. Parallax error should be avoided.

The plate voltage is gradually increased in steps of 10 volts (or 5 volts, depending on the instrument), *and in each case, the value of the mean plate current, from observation* I *and observation* II *are recorded.* The mean value of the plate current for the plate voltage is calculated. Of course, the grid-voltage should be biased to the previously adjusted value (say, at –1 volt). The plate voltage is gradually increased in steps as before up to its *maximum limiting value for that apparatus*, and in each case, the corresponding mean plate current is recorded. *The maximum limiting plate voltage is generally specified by the instructor.* Sometimes, the biasing of the circuit is so adjusted that the plate voltage can only be adjusted up to the maximum value without damaging the valve.]

Special Remarks : (I) The characteristic curves are non-linear in the region of the lower plate currents. Hence, to show the lower bend of the characteristic curves, the plate voltage is to be reduced in such a way that the lower bend of the curves can be shown in the graph. For this, the plate voltage is to be reduced in steps of 5 volts when the plate current attains the value nearer to 3 mA (for 6 *J* 5 or, 12 *AU* 7 valves, etc.) This is because, the bend is appreciable for currents lower than 3 mA for the above two valves.

(II) If the negative grid voltage is reduced, i_b begins to increase until it finally reaches saturation at some positive value of grid voltage. It is not possible to approach saturation with oxide coated cathode without seriously damaging the coating. The grid is rarely kept at a positive potential in practice. Because, as soon as the grid current flows, power is taken from the driving source, and this may cause distortion of the applied controlling voltage.

(III) In practice, grid voltage should not be kept at positive potential. Also the grid should not be kept at zero-potential, specially when the plate potential is high. Because, this will damage the valve in the long run.

Caution :

(I) During observations, the grid voltage is to be kept fixed to its adjusted value. Any fluctuation of the grid bias supply is to be reduced *(if possible) to the minimum value.*

(II) At first, the grid voltage is to be adjusted before applying the plate voltage. For this, the grid voltage is adjusted to its required value keeping the plate supply voltage at the minimum value. Next, the plate voltage is adjusted as required. This is because the excessive plate current may damage the tube and the milliammeter.

(III) The grid voltage should not be kept at zero volt. If it is done, the tube may be damaged, specially when the cathode is indirectly heated oxide-coated filament.

(IV) *For the arguments of taking observation* I *and observation* II *instead of recording the observations,* once for increasing and then for decreasing, student should consult the special remark.

(V) The range of the *mA* is to be adjusted, if necessary. After recording the data for a particular characteristic curve, say, at a particular plate voltage, the circuit should be made-off for 2 to 3 minutes. This is done to avoid heating effect and other stray effects due to the biasing and other circuitry arrangements of the regulated power supply.

Experimental procedure (continued)

Procedure (6) : The grid voltage is adjusted to another value, say, at –2 volt or –1.5 volt (depending on the smallest graduation of the grid voltmeter, by means of the rheostat or the regulator). Proceeding as before, the recordings of the plate voltage and the corresponding plate current are taken until the maximum plate current is attained.

Procedure (7) : The grid voltage is adjusted to –3 volt (or –2.5 volt) and similar recordings are made.

Procedure (8) : This reduction of grid voltage is to be done to such a value, that the variation of the plate voltage gives an *appreciable linear region of the characteristic curve.*

Procedure (9) : The plate voltage is reduced to the minimum value and the grid voltage is also reduced to the *minimum value*, and next the switch *is made-off.*

Procedure (10) : A graph is drawn with plate voltage (e_b in volt) along the X-axis, and the plate current (i_b in milliampere) along Y-axis for each grid-bias voltage. For, all grid bias voltages, such graphs are drawn.

Procedure (11) : The values of μ, r_p and g_m (and also r_{dc}) are calculated using the *approximate linear region of each characteristic curve. The operating points are recorded.* The mean values of the μ, r_p and g_m are taken using approximate linear region only.

N.B. : For details, see the mode of calculations.

Procedure (12) : From the recorded data, the product of r_p and g_m are calculated for a *particular operating point.* Next, it is to be shown that $\mu = r_p g_m$ holds good for linear region only. The validity of $\mu = r_p g_m$ is to be shown for different linear regions of the characteristic curves.

Procedure (13) : Tabulate the value of μ from the manufacturer's tube manual and calculate the percentage difference between the experimental value and the *statistical value (based on the random sampling method)* given in the manufacturer's tube manual.

Percent difference

$$= 100 \times \frac{\textit{Graph value} - \textit{Manufacturer's value}}{\textit{Graph}\text{ value}}$$

Procedure (14) : The values of μ, r_p and g_m may be calculated using non-linear regions also. In this case, different *operating points* should be *clearly mentioned.* From the recorded data, it can be shown that the relation $\mu - r_p g_m$ does not hold for non-linear region.

Procedure (15) : The approximate value may be checked from the calculated data.

Works with a Regulated Power Supply and Closed Box Instrument :

Sometimes a regulated power supply is used to study the characteristic curves of different active devices. A regulated power supply is built with the proper circuitry connections of some active and passive devices. The basic principle of a regulated power supply is to obtain *different dc voltages from ac supply.* For this, the sinusoidal voltage (i.e., the ac voltage is at first step down (or, step up, depending on the parameters by means of a transformer. This diminished (or, increased) ac voltage is rectified by proper semiconductor devices (i.e., semiconductor diodes, or rectifier chips). This rectified output is fed to proper electrolyte condenser (with pre-calculated specified capacitance), and finally proper filter circuits are applied. The final output can be varied by means of a potential divider. Of course, *with an ordinary filter circuit fluctuations of voltage cannot be completely removed owing to the presence of ripple voltages.*

Nowadays, ICs are employed instead of the filler circuit and the rectified output is fed to IC (inside which proper rated operational amplifiers are also designed), and the final output is varied by means of a potential divider. By the use of IC *the fluctuations are completely removed.* For ordinary laboratory purposes (for low power ratings) the use of IC instead of the filter circuit does not increase the cost of the manufacture. Hence, nowadays, ICs are used.

The basic principle of a regulated power supply used for the study of the characteristic curves of a triode is shown in Fig. 3.6. Of course, this block should *not be treated as the standard one, and the design depends on the manufacturer.* The main terminals are the following :

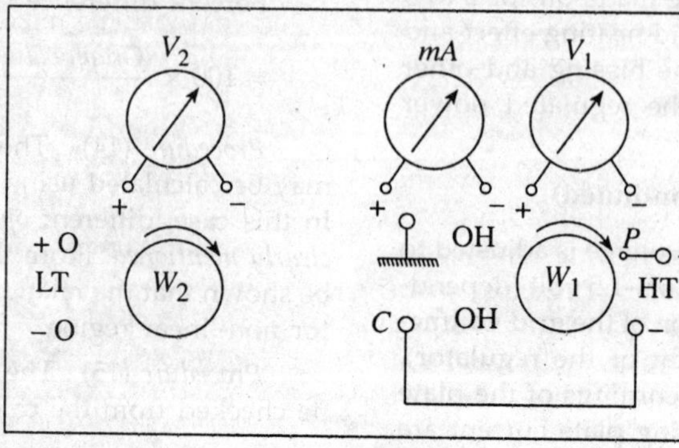

Fig. 3.6

(i) *HT terminals* : The high tension terminals are generally marked as *HT* on the panel board. The +ve terminal is indicated by red colour.

(ii) *The wandering plug* (W_1) : By rotating this plug *to clockwise direction* the potential across the *HT* terminals can be varied from a very low value to a high value (say, up to 220 volts to more depending on the instrument).

(iii) *LT terminals* : The low tension terminals are generally marked as *LT* in the panel board. The positive terminals is marked by the red colour.

(iv) *The wandering plug* (W_2) : The knob is used to vary the potential from 0 to −20 volt (say). By rotating the wandering plug to clockwise direction the grid bias can be changed from a low negative value to a high −ve value.

(v) *The heater terminals* (*H, H*) : The heater terminals of the oxide-coated filament type valve are connected to *H, H.* The heating is done by the use of the transformer heating.

(vi) *Meters* : The voltmeters V_1, V_2 and the milliammeters (*mA*) are *sometimes* fitted with the panel board. In some instruments, the ammeter and the voltmeters are supplied separately.

(vii) *Extra plug points* : Some *extra plug-points* may be fixed on the panel board *for the convenience of electrical* connections. The interconnections of different plug-points are drawn by lines on the panel board.

(viii) *The valve-stand* : The triode valve is mounted on a separate wooden stand. The terminals of the valves are connected properly with plug-points mounted on a wooden panel-board. The terminals are marked as *P,C,G, H, H,* etc.

N.B. : The description of the black-box device is an ideal one. It is given only to understand the working principle with a regulated power supply. If no HT and LT terminals are marked, the students should ask the instructor for proper identification of the terminals.

Plan of Connections

After proper identification of the different terminals, connections are made. Of course, the circuit should be analogous to that shown in Fig. 3.5.

Experimental Procedure (with closed box instrument)

(1) After completing the circuit diagram, connect the *power supply to the ac main supply.* The circuit should be *approved* by the instructor before switching on the circuit.

(2) The wandering plugs are properly identified at first. The wandering knobs W_1 and W_2 are rotated to *the extreme anticlockwise directions.* The switch is then made on. The wandering plug W_1 is rotated to a certain amount to the clockwise direction to adjust the plate voltage to 20 volts, (say).

The pointer of the milliammeter may, at first, move to the left side (i.e., below the zero value). But, after 2 to 3 minutes, it is observed that the plate current slowly increases and attains the steady value. The wander knob W_2 *is slowly rotated to the clockwise direction* to adjust the grid voltage to −1 volt, or to −0.5 volt (say).

As before (i.e., *as described in the procedure no.* 5), the plate voltage is adjusted by the wandering knob W_1, and the mean value of the plate current is noted as before. As described in the *procedure no. 5* for a fixed grid voltage (say –1 volt), the plate currents for various plate voltages are noted.

(3) The procedures as described under the procedure nos. 6 to 15 are the same as before.

Experimental Results

N.B. : The purpose of showing the experimental results in different tables is to help the student to study the trend of the values of different measurable quantities while performing the experiment. The data given below is *for 6J5 triode valve. The value of μ based on the random sampling method as supplied by the manufacturer,* is 20 (say). The variation of the values of the valve parameters may be ± 10% for the valves generally used for common laboratory purposes. Hence, the values given in the following tables are typical and should never be taken as the standard. Also, the value of the valve parameters, given in the valve-manual, should never be taken as the actual correct values, since these values are based on the random sampling method.

[A] *Valve number* : ....

Name of the manufacturer : ....

[B] *To find the range and the smallest division of the meters used*

Special Note :

(1) It is evident from the above table that as the negative grid voltage increases, the range of the linear region for computation of μ, r_p and g_m decreases. Hence, *the negativity of the grid voltage* is to be increased up to such a value that an appreciable linear region can be obtained.

(2) It is reasonable to take the readings by eye-approximation up to one-half of the smallest division of the meter. But, *if the value of the smallest division is large, then one-fourth of the value of the smallest division may be taken for the bent-region only as the approximate value for the common laboratory purpose.* Otherwise, the bent-region of the characteristic curves are difficult to obtain.

(3) The number of significant figures to be recorded in the arithmetic mean (or in the most probable value) is *usually one more than the number given in the individual measurement.*

(4) If the value of the smallest division of the grid voltmeter is 0.5 volt, then the grid voltage should be changed by –0.5 volt instead of –1 volt, *if approved by the instructor.*

Drawing of graphs and hence calculations of Δe_b, Δe_c and Δf_2

Table 3.6

Instrument (i.e., the meter)	Range	Total no. of division	Value of the one smallest division	Zero (or initial) error, if any
Milliammeter	0–50 mA	50	1 mA	...mA
Voltmeter (plate)	0–250 volts	50	5 volts	...volt
Voltmeter (grid)	0–20 volts	40	0.5 volt	...volt

N.B.: If possible, the name of the manufacturer may be mentioned.

[C] *Record of data for drawing (e_b – i_b) curve*

N.B. : If the smallest division of the grid voltmeter is –0.5 volt, then the grid voltage may be set at –0.5 volt, –1 volt, –1.5 volt, –2 volt, etc. for drawing the plate characteristics.

(i) On the same graph paper for each grid voltage, anode voltages are plotted along the X-axis and the corresponding anode currents along the Y-axis. A *typical* plate-characteristic curve for a 6J5 triode valve is shown in Fig. 3.7. From the family of such curves the values of Δe_b, Δe_c and Δi_b can be calculated as shown. *Either the mean operating plate voltage or the mean operating plate current* should be mentioned.

Table 3.7

Applied grid voltage in volt	Applied anode voltage in volt	Corresponding anode current in milliampere for Observation (I) = x	Observation (II) = y	Mean anode current in milliampere $\frac{x+y}{2}$	Region of the characteristic curve	Indexing of the curve
	25	0	0	0		
	35	0.5	0.5	0.5	Non-linear	
	45	1	0.8	0.9		
–1	⋮	⋮	⋮	⋮		curve I
	115	7.5	7.2	7.35		
	125	8.5	8.4	8.45	Approximately linear	
	⋮	⋮	⋮	⋮		
	215	21.5	21.5	21.5		
	45	0	0	0	Non-linear	
	55	0.3	0.3	0.3		
–2	⋮	⋮	⋮	⋮		curve II
	215	18	17.9	17.95	Approximately linear	
	65	0	0	0	etc.	
–3	⋮		⋮	⋮		curve III
	215	15	15	15	etc.	
etc.	etc.	etc.	etc.	etc.	etc.	etc.

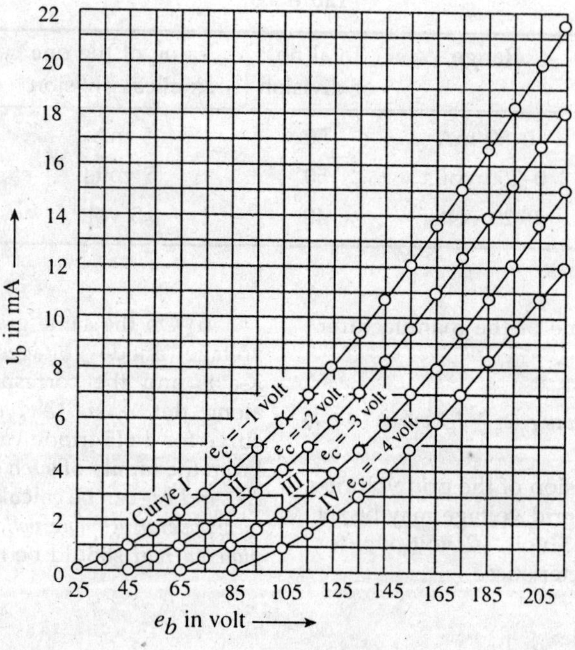

Fig. 3.7

[D] *Recording of* Δe_b, Δe_c and Δi_b *from the graph* (Fig. 3.7)

Table 3.8

Curves employed	Triangles employed	Plate voltage range in volt $= m$ to n	Mean operating plate voltage in volt $= (e_b)_o = \frac{m+n}{2}$	Magnitude of Δe_b in volt	Grid voltage change $\Delta(e_c)$ in volt	Plate current range in mA $= p$ to q	Mean operating plate current in mA $= \frac{p+q}{2}$	Magnitude of Δi_b in mA
I & II	…	146 to 166	156	20	1	11 to 13.8	12.4	2.8
III & IV	…	187 to 206	196.5	19	1	11 to 13.6	12.3	2.6
etc.	etc.	etc.	etc.	etc.	etc.	etc.	etc.	etc.

N.B. : Such triangles are to be drawn considering each characteristic curves. The triangles drawn should be at least 7 to 8 in numbers. The triangles should be drawn in the approximate linear region of the characteristic curves for the calculation of μ, r_p and g_m.

[E] *To calculate* μ, r_p *and* g_m *(data taken from Table 3.8)*

Table 3.9

Curve employed	$\mu = \left(\frac{\Delta e_b}{\Delta e_c}\right)_{i_b}$	$r_p = \left(\frac{\Delta e_b}{\Delta i_b}\right)_{e_c}$ × 1000 in ohm	$g_m = \left(\frac{\Delta i_b}{\Delta e_c}\right)_{e_b}$ × 10^{-3} in ohm^{-1}, i.e., in mho	Mean value of (for linear region) μ	r_p in ohm	g_m in mho
I & II	20	7142	28×10^{-4}			
I & III	…	…	…	…	…	…
II & III	…	…	…			
etc.	etc.	etc.	etc.			

N.B. : Mean value should only be taken for the calculated values of μ, r_p and g_m using approximate linear region of the plate characteristics.

[F] *To calculate $r_p \times g_m$, and to show that the relation $\mu = r_p \times g_m$ is valid for linear region*

Table 3.10

Curves employed	Calculated value of r_p in ohm (from graph)	Calculated value of g_m in mho (from graph)	Calculated value of $r_p \times g_m$	Calculated value of μ (from graph)	Remarks
...	...	...	...	...	Therefore,
...	...	...	...	...	$\mu = r_p \times g_m$
...	...	...	...	...	holds good
etc.	etc.	etc.	etc.	etc.	

Modes of calculation of μ, r_p and g_m

Calculation of 'μ'

$$\mu = \left(\frac{\Delta e_b}{\Delta e_c}\right)_{i_b = \text{constant}}$$

Considering curve I and curve II [Fig. 3.8] we have for the triangle ABC,

$$\mu = \frac{(e_b)_2 - (e_b)_1}{(e_c)_1 - (e_c)_2} = \frac{(e_b)_2 - (e_b)_1}{-1-(-2)} = (e_b)_2 - (e_b)_1.$$

In the above $(i_b)_o = \frac{(i_b)_2 + (i_b)_1}{2}$,

while $(e_b)_o = \frac{(e_b)_2 + (e_b)_1}{2}$.

Calculation of 'g_m'

g_m is defined as

$$g_m = \left(\frac{\Delta i_b}{\Delta e_c}\right)_{e_b = \text{constant}}$$

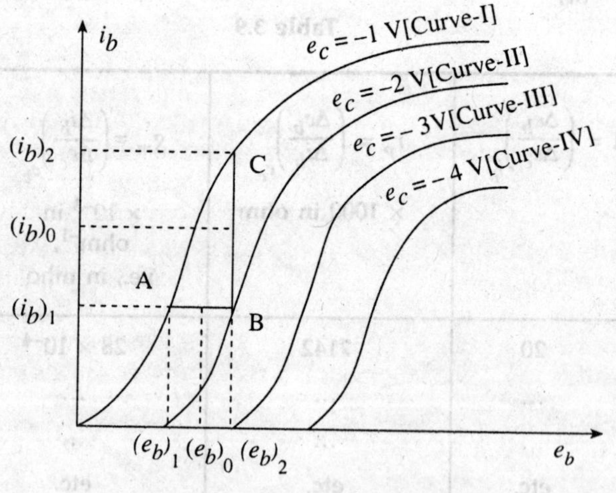

Fig. 3.8

This operating points for this value of μ are given by $(i_b)_0$ and $(e_b)_0$ respectively.

Considering the triangle *ABC* between curves I and II, in Fig. 3.8, the value of g_m is calculated from

$$g_m = \frac{(i_b)_2 - (i_b)_1}{e_{c_1} - e_{c_2}} = \frac{(i_b)_2 - (i_b)_1}{-1-(-2)} \times \frac{1}{1000} \text{ mho},$$

because i_b is expressed in milliampere.

$$\text{Also, } g_m = \frac{(i_b)_2 - (i_b)_1}{-1-(-2)} \times 1000 \text{ micro-mho.}$$

Calculation of 'r_p'

$$r_p = \left(\frac{\Delta e_b}{\Delta i_b}\right)_{e_c = \text{constant}}$$

Considering the triangle ABC between curve I and curve II, the value of r_p is calculated from

$$r_p = \frac{(e_b)_2 - (e_b)_1}{(i_b)_2 - (i_b)_1} \times 1000 \text{ ohm.}$$

Similar triangles are drawn between each curves and the values of μ, r_p and g_m are calculated using all such triangles in the approximate linear regions of the characteristics curves.

Discussion

(i) Since the valve coefficients are constants in the approximate linear region of the curves, hence only linear portions are to be employed to determine these constants.

(ii) In taking readings of voltmeter and ammeter parallax should be avoided as far as possible.

(iii) Grid bias should not be kept positive with respect to cathode.

(iv) In the non-linear region $\mu = p_r \times g_m$ is not valid and the values change rapidly with the change in operating points as the non-linearity increases.

(v) During the observation, the constancy of the grid voltage (i.e., e_c) should be checked from time to time, since a slight change in e_c produces a greater change in i_b (if μ is large).

(vi) In the case of the determination of Y (i.e., coefficient of elasticity) by flexure method, etc., where the system attains its initial equilibrium state, the readings should be taken for load increasing and load decreasing. In this case, the mean value give the nearly accurate depression.

But in this case of experiments on electronics like determination of μ, r_p and g_m for triode, also hybrid parameters, etc. for transistors, the above procedure should not be applied. This is because, the equilibrium condition of the supply voltage and other observational parts (e.g., recording of currents, voltages, etc.) attains the operating condition after some lapse of time. Such effects are notably found with the ageing of the active elements. In fact, experimental results show that if the earlier procedure, as adopted for determination of Y, etc. is imposed for recording for later observations (e.g., plate current, grid voltage, collector current, etc.), then in the second observation a considerable deviation with respect to the previous one occurs. Such difficulties can only be avoided if successive observations (as observation I and observation II) are taken as described in the procedure no. (5).

(vii) The circuit should be made off for 2 to 3 minutes after recording the data to draw each characteristic curve.

(viii) Any fluctuations of the plate voltage, plate current and grid voltage should be minimised during the recordings of the observational data. If the fluctuations cannot be completely removed owing to the fluctuation of the power supply unit, etc., then the mean reading taken from 3 to 4 observations (*for the same set*) minimises the error appreciably.

(ix) The input impedance of the voltmeters used in the plate circuit and grid circuit should be large. The high-sensitivity voltmeter produces a negligible error in the measurement of the parameters.

3.4 To Draw the Dynamic Characteristics of a Triode Valve and to Measure its Voltage Gain for a Specified Plate-supply Voltage for Different Load Resistances

Circuit Diagram

The *basic principle* of the circuit diagram for studying the dynamic characteristic curves of a triode and hence to measure voltage gain is shown in Fig. 3.9.

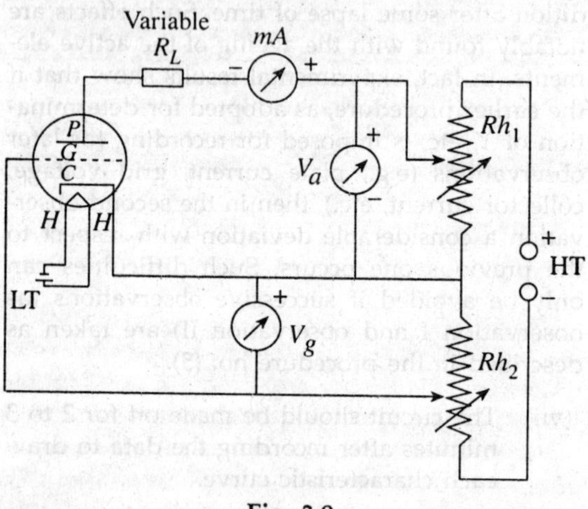

Fig. 3.9

Indexing of different part : (1) *P*, *G* and *C* represent the plate, grid and cathode respectively of the triode valve; (2) Rh_1, Rh_2, are rheostats used for varying the plate and grid voltage respectively; (3) *mA* represents the milliammeter to measure the plate current; (4) V_a and V_g are the high-sensitive voltmeters for plate and grid circuits respectively; (5) R_L is a fixed or a variable load resistance; (6) HT represents the high-tension terminals [say, 220 volts, (dc)]; (7) LT represents the low-tension supply for heater circuit.

Working Formula

If the anode circuit of a triode valve be completed with a load resistance R_L and a milliammeter, then the curve obtained by plotting various negative grid voltages e_c along the negative *X-axis* and their corresponding anode current i_b along the *Y*-axis for a fixed anode supply voltage (e_b) is called the dynamic characteristic of a triode for that given load resistance. Various such curves may be drawn for different load resistances in the anode circuit for that plate supply voltage. Various such curves can be drawn using *different plate supply voltages* for different load resistances.

Let a suitable voltage $-e_c$ is applied to the grid for a given load resistance R_L for a fixed plate supply voltage (e_b, say). If now the grid voltage is changed by an amount Δe_c, then the current change in the anode circuit would be ΔI_b.

Hence, the amplified *output voltage developed* across R_L is given by $\Delta v_L = R_L \times \Delta I_b$, while the *input voltage change* is Δe_c. Therefore, the voltage gain (*considering magnitude only*) is

$$\alpha = \frac{\text{Output voltage}}{\text{Input voltage}} = \left(\frac{\Delta v_L}{\Delta v_c}\right).$$

Actually, input and output voltages are 180° out of phase.

$$\text{Hence, } \alpha = \frac{R_L \times \Delta i_b}{\Delta e_c}. \quad (3.12)$$

In eq. (3.12), R_L is expressed in ohm, Δi_b in ampere and Δe_c in volt.

If R_L be expressed in kilo-ohm, while Δi_b in milliampere, then

$$\alpha = \frac{R_L \times 1000 \times \Delta i_b}{\Delta e_c \times 1000} = R_L \times \frac{\Delta i_b}{\Delta e_c}. \quad (3.13)$$

Also, the voltage gain of a triode amplifier is given by

$$\alpha = \frac{\mu R_L}{r_p + R_L} = \frac{\mu}{1 + \frac{r_p}{R_L}} \quad (3.14)$$

where, r_p is the a.c. plate resistance of the triode valve.

Hence, if dynamic characteristics for various load resistances are drawn from the experimental data, then by finding the slope of the curve for a given R_L, the voltage gain α of the triode valve for a given load resistance and *for a specified plate supply voltage* can be determined by the eq. (3.13).

N.B. : (i) If the value of r_p and μ are known, either from the static characteristics of that valve or from the valve manual, then the value of α can be obtained from eq. (3.14).

(ii) From eq. (3.14), it is evident that the limiting values of α are zero and μ respectively. If R_L approaches to zero, then α tends to zero; also when R_L tends to infinity then $\alpha = \mu$. Hence, the value of the voltage gain varies from zero to μ for load resistances varing from zero to infinity. The curve tends to attain the saturation for very high values of load resistances.

If the values of r_p and R_L are known (either from the experimental data or from the valve manual), then the value of α calculated from eq. (3.13) can be compared with that obtained from eq. 3.14. The variation of R_L versus α are drawn for different specified load resistances.

N.B. : The value of α calculated from eq. (3.13) may not be in close agreement from that calculated from eq. (3.14) if the value of the valve parameters are taken from the valve manual. This is because, the valve parameters given in the valve manual are based on the data obtained from the random sampling of the valves. The variation of the data may be up to ± 10% (according to the international convention). Also, the parameters change with ageing of the valve. But, if the parameters are calculated from the static characteristics of the valve supplied, then the value of α calculated from eqs (3.13) and (3.14) agrees closely.

Experimental Procedure

(With HT *and rheostats, etc. as shown in Fig.* 3.9)

(1) The *basic principle* of the circuit diagram is drawn in the laboratory notebook, and next, the connections are made as shown in Fig 3.9.

(2) Load resistance R_L of minimum value is inserted between the anode terminal and the negative terminal of the milliammeter (*mA*). A suitable value of the *anode supply voltage (as directed by the instructor*) is applied by adjusting the rheostat Rh_1. The gird bias voltage is adjusted by means of the rheostat Rh_2 until the plate current *attains the maximum value.* If possible, adjust the grid bias to zero volt. The plate current is taken twice, once for the observation I and the other for the observation II. The mean of the plate current from the obs. I and obs. II gives the actual plate current. For illustration, it is assumed that the plate supply voltage is 175 volts.

N.B. : (i) The grid voltage should not be kept to zero volt for a long time, specially when the plate voltage is high, because this will damage the triode valve with ageing of the valve. Hence, the reading for the zero grid bias should be taken as quickly as possible.

(ii) The grid voltage should not be made positive value. When the grid voltage becomes positive, electrons are collected by this electrode and hence grid current flows. This condition is *rarely allowed to occur,* because as soon as grid current flows, power is taken from the driving source, and this will cause distortion of the applied controlling voltage. The main importance of the dynamic characteristic curve is to study the behaviour of the valve as an amplifier (e.g., class A, class AB, class B, class C, etc.). For these amplifiers, the range of the grid bias is from the cut-off bias to the zero grid bias.

The grid voltage is made –1 volt (or –0.5 volt depending on the smallest division of the gird-voltmeter supplied). The plate current is noted for that fixed plate supply voltage (say, 175 volts). This plate current is recorded as the *observation* I. The grid voltage is decreased slightly from the previously adjusted value, and then the grid voltage is again adjusted to its previous value (say, –1 volt). Any fluctuation of the plate voltage, if occurs, is removed, and the plate voltage is adjusted to the previous value (say, 175 volts). The plate current is noted *cautiously,* and this current is recorded as *observation* II. *The mean value of the plate current from the obs.* I *and obs.* II *gives the actual plate current under the operating conditions of the instrument.*

The *negativity* of the grid bias voltage is *increased* in steps of –1 volt (or –0.5 volt), and in each case, the value of the mean plate current (as before) is noted. This increment of the negative grid bias is to be done until the plate current attains the zero-value (*if permissible by the instruments supplied).*

(3) The minimum value of the load resistance is replaced by a higher value of the load resistance. Keeping the plate supply voltage fixed as the previously adjusted value (say, 175 volts), all the above operations as described in the procedure no. (2) are performed. This is done *for at least 6 values* of the load resistances (*if supplied).*

N.B. : The curve is steeper if the value of the load resistance is low and the plate supply voltage is high.

(4) The *plate supply voltage is decreased to another value* (say, 160 volts), and all the above operations are done.

(5) All the above operations are to be done for all the plate supply voltages *as directed by the instructor.*

(6) The plate voltage is reduced to the minimum value and next, the grid voltage is reduced to its minimum value. The circuit is then made off.

Works with Regulated Power Supply

For description and different parts of a regulated power supply see the expt. no. 3.3. In this experiment, a variable and is supplied and is supported on a different panel board. Connections are made as shown in Fig. 3.9, showing the basic principle of circuity.

(i) At first the wandering knobs W_1 and W_2 are properly identified.

(ii) The grid voltage is adjusted to –1 volt (or less) by adjusting the wandering plug W_2 (or, by other plug according to the design). The load is adjusted to the minimum value (say, 5 kΩ). The plate supply voltage is adjusted to a suitable value (by the plug W_1) as directed by the instructor. Keeping the plate supply voltage fixed, the negativity of grid voltage is increased in steps of –1 volt (or –0.5 volt) and in each case the value of mean plate current (as before) is noted. This increment of the negative grid bias is to be done until the plate current attains the zero value (*if permissible by the instruments supplied).*

(iii) The load resistance is replaced by a load having higher resistance. This may be done by means of a proper switching device. As before, keeping the plate supply voltage fixed (as adjusted in previous operation), the negative grid voltage is gradually increased and in each case, the mean plate current is noted as before.

Drawing of Graphs : On a graph paper, the grid-voltage is plotted along the X-axis and the anode-current along the Y-axis for each plate-supply voltage. On the same graph paper different curves are drawn for different values of the load resistance (Fig. 3.10).

Another graph is to be drawn by plotting load resistance in kilo-ohm along the X-axis and the corresponding voltage gain along the Y-axis for each plate supply voltage, for *linear region.*

Modes of Calculation

On the dynamic characteristic curves shown in Fig. 3.10, two vertical lines parallel to Y-axis are drawn in the linear region. For triode valves (*generally used for common laboratory purposes*), the dynamic characteristic curves may be treated as linear for grid voltage range from –1 volt to 3 volts. For load resistance R_{L_1}, the voltage gain

$$\alpha_1 = R_{L_1} \times \frac{\Delta i_b}{\Delta e_c}.$$

In the figure, shown in 3.8,

$$\Delta i_b = bc$$

and $\Delta e_c = ab.$

$$\therefore\ \alpha_1 = R_{L_1} \times \frac{bc}{ab}$$

$$= R_{L_1} \times \frac{bc}{2}\ (\because ab = 2 \text{ volts}). \qquad (3.15)$$

Similarly, for load resistance R_{L_2},

$$\alpha_2 = R_{L_2} \times \frac{ef}{2} \qquad (3.16)$$

In a similar way, the voltage gain can be calculated for different values of R_L. In the above example, the operating grid voltage is –2 volt.

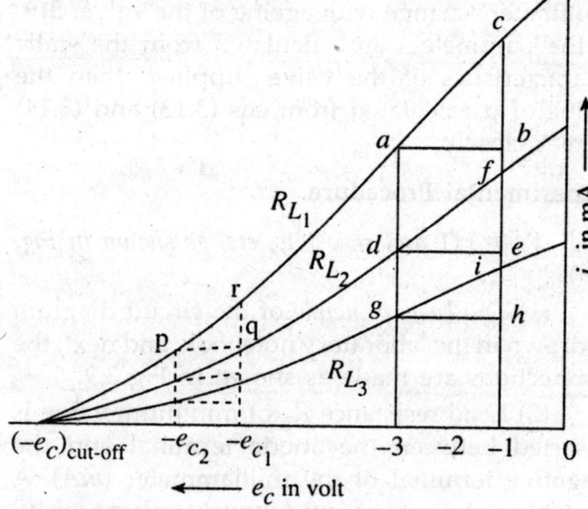

Fig. 3.10

N.B. : In the above procedure, two vertical lines are drawn at a fixed operating point in the linear region. Instead of this, different triangles can be drawn, as before, in different linear regions on the same characteristic to calculate the value of α. In each case, the mean operating point must be mentioned.

Proceeding in a similar way we consider the non-linear region within the grid voltage range $-e_{c_1}$ and $-e_{c_2}$. Hence, the values of α(i.e., voltage gain) for different load resistances can be calculated for *non-linear region.* In this case the operating grid voltage is $\frac{-(e_{c_1} + e_{c_2})}{2}$ volt. Different triangles *pqr*, etc. are drawn, as before, in the non-linear region [as shown in Fig. 3.10].

N.B. : The variation of voltage gain (α) with R_L for linear and non-linear region are shown in Figs. 3.12 and 3.13 respectively for a fixed plate supply voltage (e_b) = 175 volts) for 6 J 5 triode valve. The superimposed variation is shown in Fig. 3.14.

For a class *A* amplifier, the linear region of the dynamic characteristics are taken into consideration, while for a class *AB*, *B* and *C* amplifiers, the operating points lie in the non-linear region. So the variations in the non-linear region has special importance for understanding the principle of the design of different types of power amplifiers with valves. However, nowadays vacuum tubes are not frequently for this purpose.

Experimental Results

N.B. : The data given should not be taken as the standard and are given only for illustration.

[A] To find the smallest division of different measuring instruments.

Table 3.11

Instrument employed	Range	Total no. of scale division	Value of smallest scale division	Zero (or initial -error), if any
Milliammeter	...mA	...	...mA	...
Voltmeter (anode)	...volt	...	...volt	...
Voltmeter (grid)	...volt	...	...volt	...

[B] Record for drawing e_c–i_b curves

Plate supply voltage, e_b = ... volt (200 volts, say)

Load resistance, R_L = ... kΩ

Table 3.12

Applied grid voltage (e_c) in volt	Corresponding anode current (i_b) in mA		
	Observation 1	Observation 2	Mean
0	...	...	...
–1	...	...	...
–2	...	...	...
–3	...	...	...
etc.	etc.	etc.	etc.

[Make similar tables for other fixed load resistances]

[C] Data for drawing e_c–i_b curve for different load resistances.

Plate *supply* voltage = ... volt.

Table 3.13

Applied grid voltage (e_c) in volt	Corresponding mean anode current (i_b) in mA for load resistance			
	11 kΩ (=R_{L_1})	21 kΩ (=R_{L_2})	31 kΩ (=R_{L_3})	41 kΩ (=R_{L_4})
0	...	...	...	...
–1	...	...	...	...
–2	...	...	...	...
–3	...	...	...	...
etc.	etc.	etc.	etc.	etc.

[**Note** : Repeat the above observations for a second plate-supply voltage]

[D] Drawing of the dynamic characteristics and hence to find the voltage gain (α) from the linear region of the curves.

Negative grid voltage in volt is plotted along the X-axis and the corresponding anode current in mA along the Y-axis for different values of load resistances. The nature of typical curve so obtained is as shown in Fig. 3.11, using 6J5 triode valve.

[E] Calculation of voltage gain from linear region R_L = ... kΩ.

[Solid vertical lines *AB* and *CD* are drawn for this purpose as shown in Fig. 3.11.

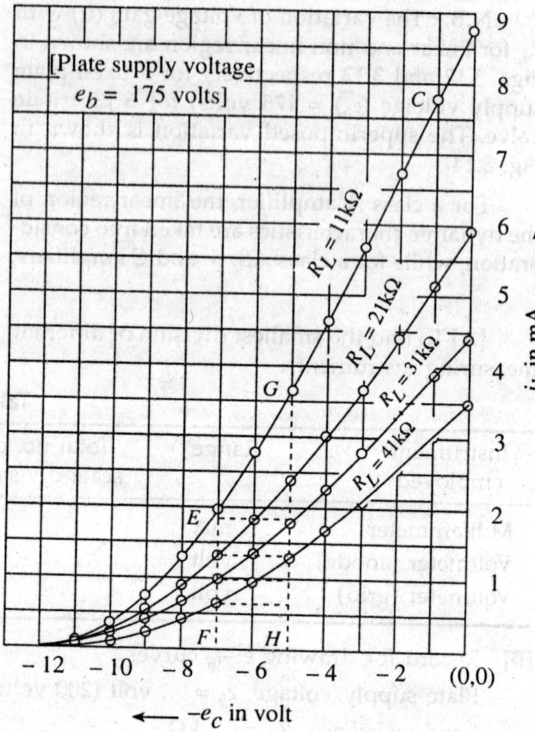

Fig. 3.11

Table 3.14

Load resistance (R_L) in kΩ	Grid voltage range in volt [$-m$ to $-n$]	Mean operating grid voltage in volt = $\left[\frac{(m+n)}{2}\right]$	Change in grid voltage (Δe_c) in volt	Plate current range in *mA* [*p to q*]	Mean operating plate current in mA $\left[=\frac{p+q}{2}\right]$	Change in plate current (Δi_b) in mA	$m = \Delta i_b / \Delta e_c$ mA/volt	Voltage gain α (= $R_L \times m$)
11	–1 to 3	–2	2	5.5 to 7.55	6.5	2.05	1.025	11.3
21	–1 to –3	–2	2	3.85 to 5.2	4.5	1.35	0.675	14.2
31	–1 to –3	–2	2	2.90 to 3.85	3.4	0.95	0.475	147
...	... to ...	...	...	... to ...	...	...	...	
etc.	etc.	etc.	etc.	etc.	etc.	etc.	etc.	etc.

Plate supply voltage = ...volt = 175 volts (say)

N.B. : Instead of two operating points as shown in the table, mention of operating grid voltage alone will serve the purpose.

[E_1] *Drawing of R_L–α curve for plate supply voltage e_b =* 175 volts *(say) using linear region.*

Different values of load resistances R_L is plotted along the X-axis and the voltage gain α along the Y-axis. The nature of the curve obtained in the linear region is shown in Fig. 3.12, for a fixed plate supply voltage.

[F] Calculation of the voltage gain (α) from the non-linear part of the dynamic characteristic curves.

[Dotted vertical lines *EF* and *GH* are drawn as shown in Fig. 3.11 for this purpose.]

Applied plate supply voltage = ...volt.

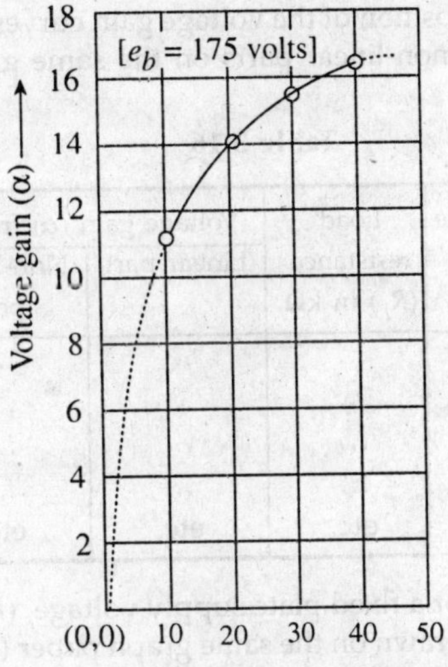

Fig. 3.12

Table 3.15

Load resistance (R_L) in kΩ	Grid voltage range in volt	Mean operating grid voltage in volt	Change in grid voltage in volt $= \Delta e_c$	Plate current range in mA	Mean operating plate current in mA	Change in plate current in mA (Δi_b)	Value of m $\Delta i'/\Delta e_c$ in mA/volt	Voltage gain $\alpha (= m \times R_L)$
...	... to ...	...	...	... to ...	...	...	...	...
...	... to ...	...	...	... to ...	...	...	...	...
...	... to ...	...	...	... to ...	...	...	...	...
...	... to ...	...	...	... to ...	...	...	...	...
etc.	etc.	etc.	etc.	etc.	etc.	etc.	etc.	etc.

N.B. : This table (for non-linear region) should be done if directed by the instructor.

[F_1] *Drawing of* R_L–α curve *(for non-linear region)* :

A curve is drawn by plotting R_L along X-axis and the corresponding values of α along the Y-axis. The nature of the curve obtained from the non-linear part is as shown in Fig. 3.13.

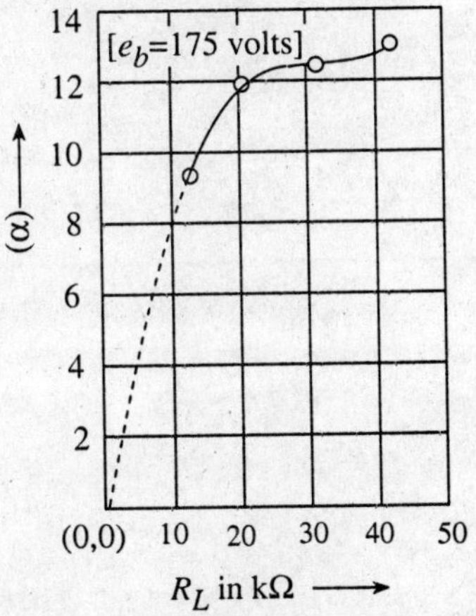

Fig. 3.13

[G] Superposition of the voltage gain curves from linear and non-linear parts on the same graph.

Table 3.16

Applied plate voltage (e_b) in volt	Load resistance (R_L) in kΩ	Voltage gain (α) from Linear part	Non-linear part
	...	...	...
	...	...	...
...	...	...	...
	...	...	...
	etc.	etc.	etc.

[G1] For a fixed plate supply voltage, (R_L–α) curves are drawn on the same graph paper (3.14). The graph shows the nature of variation of the voltage gain α with variation of load resistances (R_L), for both linear and non-linear regions.

Discussion

(i) Since in this experiment, the plate currents are very small, a precise instrument is used to measure them.

(ii) Care is to be taken so that the valve-filament is not allowed to glow continuously.

(iii) In the experiment the grid circuit is cut off from time to time. The voltage applied to the grid is not made positive, because in that case a grid current would flow which is not desirable.

(iv) In non-linear region the tube parameters μ, r_p and g_m vary widely without any systematic variation and hence the graph is not smooth which is evident from the voltage-gain curves.

Also, the relation $\alpha = \dfrac{\mu R_L}{r_p + R_L}$ cannot be strictly

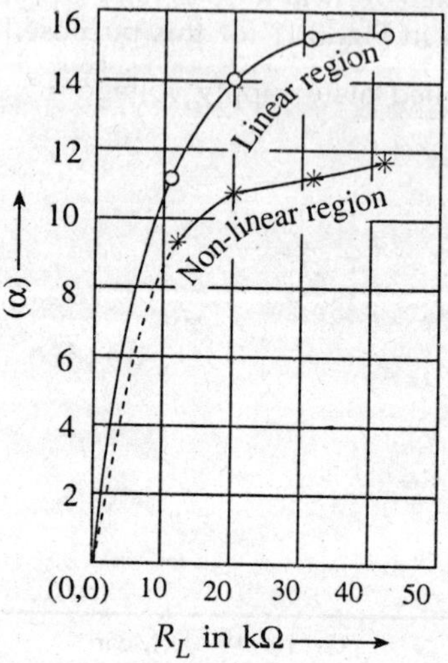

Fig. 3.14

applied in this region, because the above deduction is based on the assumption that $i_b = f(e_b, e_c)$, where f represents a single-valued function of plate voltage e_b and grid voltage e_c.

(v) The variation of α will be uniform only over the linear portion of the dynamic characteristic curve. So, for calculation of the voltage gain, linear region of the characteristics is only required (for class A amplifier only).

3.5 To Draw the Mutual Characteristic Curves of the Triode Valve Supplied and hence to Determine the Valve Parameters

Circuit Diagram

Same as described in the experiment no. (3.3).

Working Formula : *Same as described in the experiment no. (3.3).*

Plan of Circuit Connections : Same as described in the experiment no (3.3).

Experimental Procedure

[Using the basic circuit shown in the Fig. (3.5), using bettery and rheostats, etc.]

Procedure nos. (1), (2), (3), and (4) are the same as described in the experiment no. (3.3).

Procedure No. 5

Note : At first, the grid voltage is to be adjusted before applying the required plate voltage. Hence, before starting the experiment, the grid voltage is adjusted to the required value. During this operation, the plate voltage is set to the minimum value. Next, the plate voltage is adjusted as required. This is because, the excessive plate voltage (at starting) with zero grid bias voltage may damage the tube with ageing. Also, the current sensitivity of the milliammeter (in the plate-circuit) is reduced in the long run, if the above precaution is not adopted.

The plate potential is set to the minimum value before adjusting the required grid bias. The grid potential is next set to –1 volt (say). The plate voltage is then adjusted to the highest permissible value (say, 190 V) *as directed by the instructor (or may be selected from the different parameters and other properties of the valve supplied from the valve manual)*. The corresponding plate current is properly noted. This plate current is recorded as the observation I. The grid voltage is slightly changed from that adjusted before. The grid voltage is again *properly set* to the *previously adjusted value* (say, –1 volt), the plate voltage is kept to the previously adjusted value (say 190 V), and the plate current is again cautiously noted. This reading is recorded as the observation II. The mean value of the plate current from the observation I and observation II gives the nearly accurate value of the plate current for that condition.

The negative grid voltage is gradually increased in steps of –1 volt or, by –0.5 volt (if the instrument permits), and for each case the mean plate current is noted. That is, the grid voltage is successively set to –2, –3, –4, –5, –6 volt, etc. and for each grid voltage the value of the mean plate current, from obs. I and obs. II is calculated.

This increment of negative grid bias voltage is increased until the plate current reaches the zero value. This grid bias is known as *the cut-off bias*. From this cut-off bias, the approximate value of μ may be pre-calculated.

Next, the grid is kept at zero-potential and the value of the plate current is noted as before.

N.B. : (i) The grid-bias should not be kept at zero-voltage for a triode valve having oxide-coated cathode for a long time, because it may damage the valve with ageing (i.e., in the long run).

(ii) For, the *special remarks* and the points for caution, the experiment no. (3.4) should be consulted.

(6) The plate voltage is decreased by 10 volts (or, 15 volts). The negative grid voltage is increased in steps of –1 volt and each case the corresponding mean plate current (as before) is noted for that plate voltage. This increment of negative grid bias is continued until zero plate current is attained.

(7) The above operation is repeated for other plate voltages decreasing in steps of 10 V (or 15 V). The reduction of plate voltage is to be done to such a value that the variation of grid voltages gives an appreciable linear region of the mutual characteristic curve.

(8) The plate voltage is reduced to the minimum value and the grid voltage is also reduced to the minimum value, and after this, the switch is made off.

(9) A graph is drawn with grid voltage (e_c in volt) along the negative X-axis, and the plate current (i_b in milliampere) along the positive Y-axis for each plate voltage.

For all such plate voltages, the mutual characteristic curves are drawn.

(10) The values of μ, r_p and g_m (and also r_{dc}) are calculated.

Works with a Regulated Power Supply.

For description and different parts of regulated power supply, see the previous experiment. Plan of connections are the same as described before.

Experimental procedure :

(1) Same as described in experiment no. 3.3).

(2) The wandering plugs for adjusting the grid bias and the plate bias are at first identified. The wandering knobs are set to the extreme anticlockwise direction (i.e., both the bias are set to their minimum values).

(3) The switch is then made on. After 2 to 3 minutes, the current slightly increases after slight rotations of the adjusting knobs W_1 and W_2.

(4) The wandering plug W_2 is adjusted to set the grid bias at –1V, say, (or –0.5V depending on the meter supplied). The wandering plug W_1 is adjusted to adjust the plate bias to the required value (say, at 190 V). The grid bias is gradually adjusted to the successive increasing negative values for that fixed plate voltage, and in each case, the mean plate current (obtained from obs. I and obs. II) is noted as described.

Experimental Results :

N.B. : The data given should not be taken as the standard and are given only for illustration.

[A] Valve number ......

[B] Scale divisions of the meters supplied.

Table 3.17

Instrument	Tatal value of full scale	Total number of divisions	Value of smallest scale divisions	Initial (or zero) error, if any
Milliammeter	50 mA	50	1mA	...
Voltmeter (Plate)	250 volts	50	5 volts	...
Voltmeter (Grid)	30 volts	60	0.5 volt	...

[C] Record of data for drawing (e_c–i_b) curve.

Table 3.18

Plate voltage in volt	Applied grid voltage (e_c) in volt	Corresponding anode current (i_b) in milliampere		Mean anode current (i_b) im mA	Region of the characteristic curve	Indexing of the curve
		Observation I	Observation II			
190	0	23.0	22.8	22.9	Nearly linear	Curve I
	–1	20.2	20.2	20.2		
	–2	17.5	17.4	17.45		
	⋮	⋮	⋮	⋮		
	–11	1.0	1.0	1.0	Non-linear region	
	–12	0.5	0.5	0.5		
	–13	0.25	0.25	0.25		
	–14	0	0	0		
175	etc.	etc.	etc.	etc.	etc.	Curve II
⋮	⋮	⋮	⋮	⋮		
115	etc.	etc.	etc.	etc.	etc.	etc.

N.B. : It is evident from the above table that as the plate voltage decreases, the range of the linear region for computation of μ, r_p and g_m decreases. Hence the plate voltage is to be decreased to such a value that an appreciable linear region can be obtained.

[D] To draw the mutual characteristics and hence calculation of Δe_b, Δe_c and Δi_b.

For each anode voltage a curve is drawn with the *negative* grid voltage along the *X*-axis and the anode current along the *Y*-axis. The nature of the curves obtained is as shown in Fig. 3.15 for a typical triode valve. From the approximate linear portions of these curves, μ, r_p and g_m are determined.

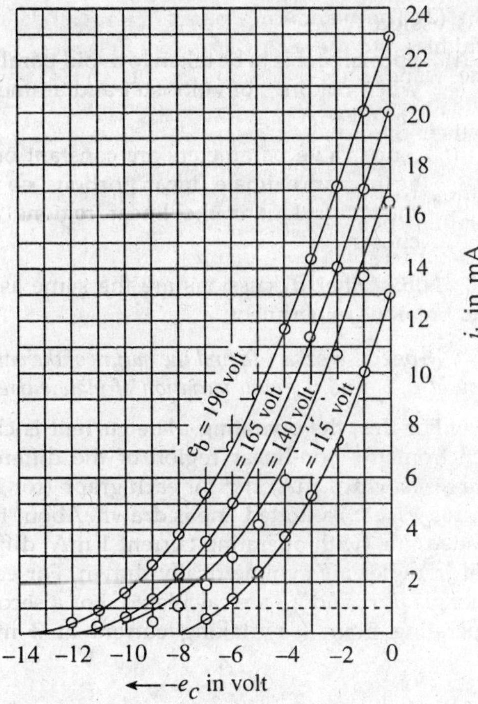

Fig. 3.15

Table 3.19

Curves employed	Triangle employed	Grid voltage range in volt (–ve)	Mean operating grid voltage in volt (–ve)	Magnitude of $-\Delta e_c$ in volt	Plate voltage change Δe_b in volt	Plate current range in mA	Mean operating plate current in mA	Magnitude of Δi_b in mA
I & II	LNP (say)	2.85 to 4.1	3.5	1.25	25	12 to 15.3	13.65	3.3
II & III	...	...to...	...	...	...	...to...	...	...
etc.	etc.	etc.	etc.	etc.	etc.	etc.	etc.	etc.

[E] Calculation of μ, r_p and g_m

Table 3.20

Curve employed	Triangles employed	$\mu = \dfrac{\Delta e_b}{\Delta e_c}$	$r_p = \dfrac{\Delta e_b}{\Delta i_b} \times 10^3$ in ohm	$g_m = \dfrac{\Delta i_b}{\Delta e_c} \times 10^{-1}$ in mho	$r_p \cdot g_m$ (= μ)	Remark
I & II	LNP	$\dfrac{25}{1.25} = 20$ (say)	7575	26.4×410^{-4}	...	Thus $\mu = r_p \cdot g_m$ is verified
II & III	...	...	...	...	...	
III & IV	...	...	...	...	...	
etc.	etc.	etc.	etc.	etc.	etc.	

Discussion

(i) Special care is to be taken to avoid parallax when readings of voltmeter and ammeter are taken.

(ii) Since valve parameters are constant only at the approximate linear portions, so for their calculations the linear regions are chosen.

N.B.: Other discussions are the same as in the previous experiment.

[**Special works :** *To find the nature of the variation of* μ, r_p *and* g_m *with variation of plate current*]

For this, the operating plate current is chosen from the non-linear region of the different curves (say, at 1 mA) and for each graph (for 190 V, 165 V, etc.) a dotted line is drawn. About this dotted line (with operating current 1 mA) different triangles are symmetrically drawn. For each triangle μ, r_p and g_m are calculated. For a second operating point (say, taking current at 3 mA) similarly, μ, r_p and g_m are determined. The observation may be repeated for other values of current. Then μ, r_p and g_m can be calculated.]

Some extra works with triode valve

(1) The students may find the values of μ, r_p and g_m in the non-linear region and may check that the relation, $\mu = r_p \cdot g_m$ is not valid.

(2) The students may find the values of μ, r_p and g_m for different operating currents, starting from the lowest non-linear region to the highest linear region of the characteristic curves. From these, the curves g_m versus i_p ; r_p versus i_p and μ versus i_p can be drawn.

(3) The students may calculate the valve parameters by drawing the e_b versus e_c curve for a fixed value of i_b, such curves are drawn for different fixed plate currents. These curves are straight lines.

3.6 To Draw Characteristic Curves of Tetrode and Pentode

Circuit Diagram

The circuit diagram for studying the characteristics is shown in Fig. 3.16. Valve EF 89 may be suitably used for the purpose whose base connection is separately given. In the figure, G_1 is the control grid, G_2 is the screen grid and G_3 is the suppressor grid. For studying the characterisitcs of a tetrode G_2 and G_3 are connected to each other while for studying the characteristics of a pentode G_3 is connected to cathode *K*. The necessary connections are made by a two-way key *K'*. Here V_{g_1}, V_{g_2} and G_{g_3} are the three voltmeters connected respectively to the control grid, screen grid and the suppressor grid. Two milliammeter mA_1 and mA_2 are connected to the plate and the screen grid respectively for noting the corresponding currents.

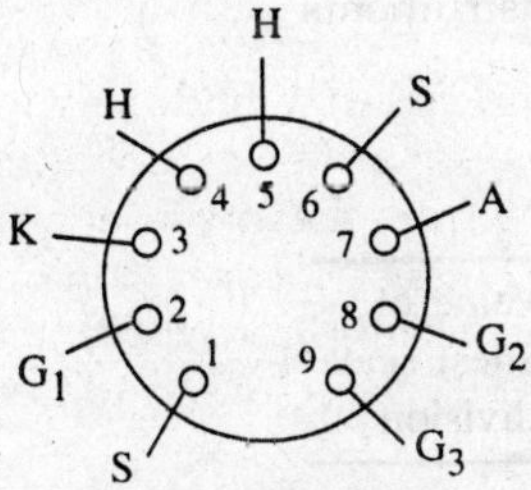

Valve EF 89 type
Base connection

Fig. 3.16(a)

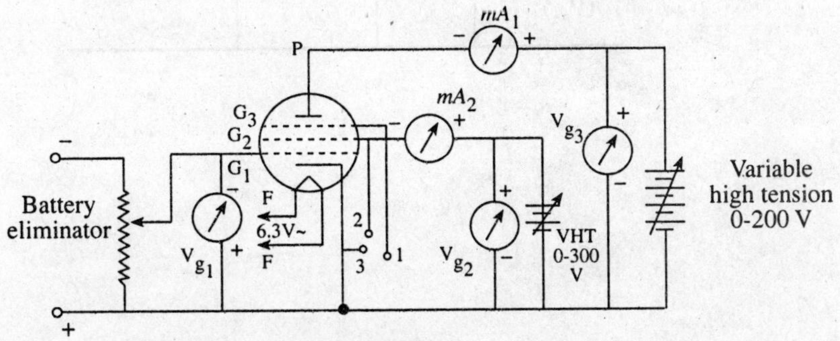

Fig. 3.16(b)

Working Formula

In order to get the characteristic curves of a tetrode, the screen grid voltage (e_{c_2}) is kept at a fixed high value. Then for a particular control grid voltage (e_{c_1}) the plate voltage (e_b) is increased in steps and the corresponding screen grid current i_{c_2} and the plate current i_b are noted. In a tetrode with fixed control grid and screen grid potentials the total space current which is the sum of i_b and i_{c_2} is practically constant.

Again, as in the case of a tetrode, the screen grid is kept fixed at a high potential for getting the characteristics of a pentode. Then for a particular control grid voltage, the plate voltage is altered in regular steps and the corresponding screen grid and the plate currents are noted. For a pentode also if the space current ($i_b+i_{c_2}$) is plotted against the plate voltage shows a constant value.

Experimental Procedure

(1) The connections are made as shown in the circuit diagram. The polarities of the different meters used are carefully checked. The two-way key *K'* helps to use the circuit for studying both tetrode and pentode characteristics.

(2) The screen grid potential is kept fixed at a certain value less than the value recommended in the tube manual. Then for a particular value of the control grid voltage, the plate voltage is changed and the corresponding value of plate current and the screen grid current are noted.

(3) For the same screen grid voltage, the control grid voltages are adjusted to different fixed values, say, –1, –1.5, –2, –2.5 volts, etc. and the operation (2) is repeated. For the observation, the plate voltages may be set to different ranges, say, up to 10 volts at the interval of 2 volts, from 10 to 100 volts at the interval of 5 volts and from 100 to 200 volts at the interval of 10 volts.

(4) The observations are continued up to a value of plate voltage which is higher than the

screen grid voltage. Keeping the screen grid potential constant to other fixed values one more set of observation may be taken.

(5) The curve are next drawn where anode current (i_b) and screen grid current (i_{c_2}) are plotted against the plate potential (e_b). The sum of the anode and screen grid currents ($i_b + i_{c_2}$) for a particular control grid voltage are also plotted on the same graph paper.

(6) For getting the pentode characteristics operations (2) to (5) are to be followed. In this case, however, the connection is made between G_3 and K, and not between G_2 and G_3 as in tetrode.

Experimental Results

[A] To find the smallest division of different measuring instruments

Table 3.21

Instrument employed	Range	Total no. of scale division	Value of smallest scale division
Milliammeter (mA_1)	...mA	...	... mA
Milliammeter (mA_2)	...mA	...	... mA
Voltmeter (Vg_1)	...volt	...	... volt
Voltmeter (Vg_2)	...volt	...	... volt
Voltmeter (V_{g3})	...volt	...	... volt

Table 3.22

Plate Voltage (eb) in volt	Grid voltage = −1.0 volt			Gric voltate = −1.5 volt		Grid voltage = −2.0 volt		Grid voltage = −2.5 volt	
	Plate current (i_b) in mA	Screen grid current (i_{c_2}) in mA	Total current ($= i_b + i_{c_2}$) in mA	Plate current (i_b) in mA	Screen current (i_{c_2}) in mA	Plate current (i_b) in mA	Screen current (i_{c_2}) in mA	Plate current (i_b) in mA	Screen current (i_{c_2}) in mA
0	0	11.8	11.8	C	9.0	0	7.7	0	6.0
2.	2.9	9.0	11.9	2.7	6.6	2.2	5.6	2.4	3.8
4	…	…	…	…	…	…	…	…	…
⋮	⋮	⋮	⋮	⋮	⋮	⋮	⋮	⋮	⋮
10	…	…	…	…	…	…	…	…	…
15	…	…	…	…	…	…	…	…	…
20	…	…	…	…	…	…	…	…	…
⋮	⋮	⋮	⋮	⋮	⋮	⋮	⋮	⋮	⋮
90	…	…	…	…	…	…	…	…	…
100	…	…	…	…	…	…	…	…	…
110	…	…	…	…	…	…	…	…	…
⋮	⋮	⋮	⋮	⋮	⋮	⋮	⋮	⋮	⋮
200	…	…	…	…	…	…	…	…	…

[Repeat the above observations for a second screen grid voltage]

[B] Data for tetrode characteristics

Filament voltage = 6.3 volts (ac)

Screen grid voltage = 90 volts (say)

[C] To draw the tetrode characteristics

Plate voltages e_b in volts is plotted along the X-axis while i_b, i_{c_2} and $(i_b + i_{c_2})$ in mA are plotted along the Y-axis. The nature of the curves obtained is as shown in Fig. 3.17.

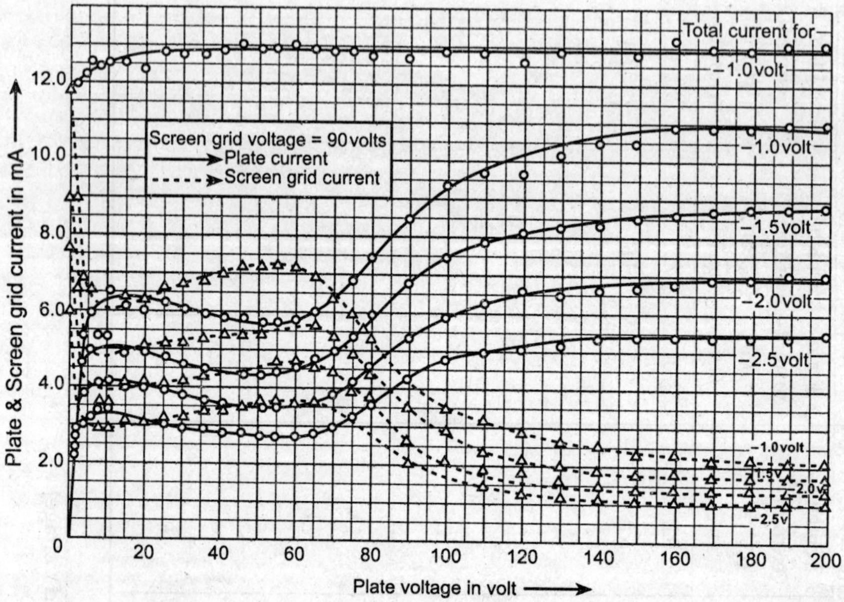

Fig. 3.17

Table 3.23

Plate voltage (e_b) in volt	Grid voltage = −0.5 volt: Plate current (i_b) in mA	Screen grid current (i_{c_2}) in mA	Total current ($= i_b + i_{c_2}$) in mA	Grid voltage = −1.0 volt: Plate current (i_b) in mA	Screen grid current (i_{c_2}) in mA	Grid voltage = −1.5 volt: Plate current (i_b) in mA	Screen grid current (i_{c_2}) in mA	Grid voltage = −2.0 volt: Plate current (i_b) in mA	Screen grid current (i_{c_2}) in mA
0	0	11.5	11.5	0	9.2	0	7.3	0	5.6
2	0.4	11.0	11.4	0.8	8.6	0.8	6.7	1.0	4.4
4	...	...	...	...	...	...	...	...	...
...	...	...	...	...	...	...	...	...	...
10	...	...	...	...	...	...	...	...	...
15	...	...	...	...	...	...	...	...	...
20	...	...	...	...	...	...	...	...	...
...	...	...	...	...	...	...	...	...	...
50	...	...	...	...	...	...	...	...	...
60	...	...	...	...	...	...	...	...	...
70	...	...	...	...	...	...	...	...	...
...	...	...	...	...	...	...	...	...	...
200	...	...	...	...	...	...	...	...	...

[Repeat the above observation for a second screen grid voltage.]

[D] Data for pentode characteristics

Filament voltage = 6.3 volts a.c.

Screen grid voltage = 100 volts

[E] To draw the pentode characteristics

Here also the various plate voltages e_b in volts are plotted along the X-axis and i_b, i_{c2},

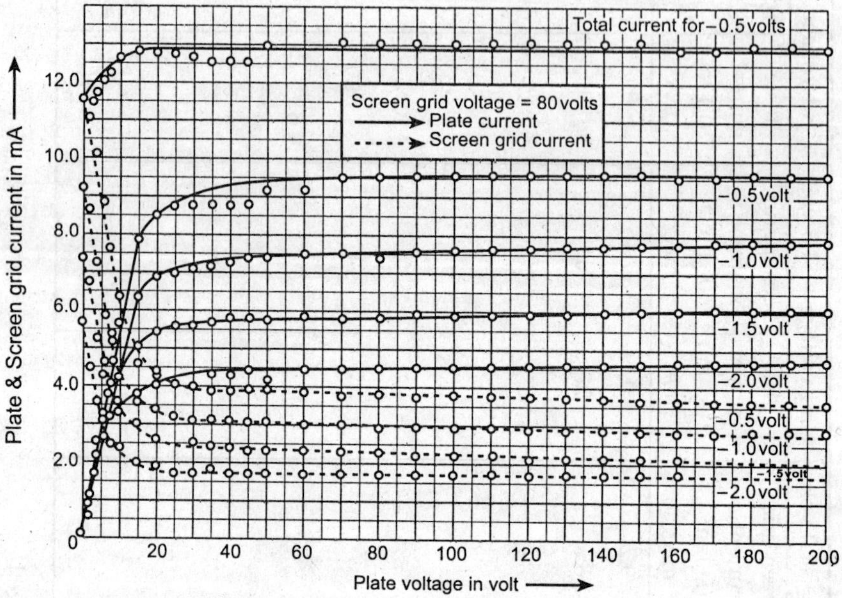

Fig. 3.18

$(i_b + i_{c2})$ in mA are plotted along the Y-axis. The nature of the curves obtained is as shown in Fig. 3.18.

Discussion

(i) Here the same tube is used for getting the tetrode and pentode characteristics. During tetrode characteristics the screen grid and the suppressor grid are connected together but during pentode characteristics the suppressor grid and the cathode are connected.

(ii) In the tetrode characteristics with the increase of plate voltage the plate current increases at first after which it goes to decrease. This means that the tube behaves as a 'negative resistance'. This phenomenon is known as 'dynatron effect'. When the plate current decreases, screen current begins to increase due to the emission of secondary electrons.

(iii) It is seen that the kink in the plate characteristics of a tetrode is eliminated in a pentode as the effect due to secondary emission has been suppressed.

Additional task : For studying the transfer characteristics of a tetrode, the screen grid potential is kept fixed and the plate potential is adjusted to a suitable value. The control grid voltage is now changed in small steps from a negative to a positive value, and the corresponding anode currents are recorded. The observations are repeated for different fixed grid voltages. The following table may be followed :

[F] Variation of anode current with grid voltage

Screen grid voltage = ... volt

Table 3.24

Control grid voltage (e_{c1}) *in* volt	*Anode current* (i_b) *in* mA *when*			
	e_b=...volt	e_b=...volt	e_b=...volt	e_b=... volt
−...volt	...	...	...	...
−...volt	...	...	...	...
−...volt	...	...	...	...
⋮	⋮	⋮	⋮	⋮
+...volt	...	...	...	...
+...volt	...	...	...	...

For a pentode, the screen grid voltage is kept fixed. A number of sets of observations are then noted for various anode currents for different plate potentials. The tabulation may be made in a manner as shown below.

[G] Variation of anode current with anode voltage

Table 3.25

Anode voltage (e_b) in volt	*Anode current* (i_b) *in* mA *when*			
	e_c=...volt	e_c=...volt	e_c=...volt	e_c=... volt

ORAL QUESTIONS WITH ANSWERS

I. **What do you mean by the following terms?**

(1) *A-battery*

Ans. Batterry for supplying power to the filament. This is also known as LT battery.

(2) *Air-core*

Ans. Without iron in its magnetic circuit.

(3) *Alternating voltage (or current)*

Ans. It is that voltage or current which alternates its direction regularly with an average value of zero.

(4) *Amplification*

Ans. Increase of signal voltage or power.

(5) *Anode*

Ans. Positive electrode or plate.

(6) *Axial leads*

Ans. Leads from the ends of capacitors, resistors, etc.

(7) *B-battery*

Ans. Battery for supplying power to the plate and screen circuits. This is also known as HT battery.

(8) *B-supply*

Ans. Plate and screen voltage source for an amplifier or a receiver.

(9) *Bias*

Ans. Application of a voltage to the grid to get the required operating point.

(10) *Bleeder resistor*

Ans. Resistor used across a voltage source as an additional load in order to provide a good regulation.

(11) *Blocking*

Ans. Cutting off the plate current by a high negative bias on the grid.

(12) *By-pass condenser*

Ans. Allows alternating current to by-pass part of a circuit.

(13) *C-battery*

Ans. Battery for supplying voltage to the grid. This is also known as bias battery.

(14) *Cascade*

Ans. In an amplifier circuit if the output of one valve is applied to control the grid circuit of another valve, then the valve are said to be in cascade.

(15) *Cathode*

Ans. Negative electrode.

(16) *Cathode current*

Ans. Total electronic current passing through the cathode.

(17) *Choke coil*

Ans. Inductance.

(18) *Co-axial*

Ans. Having a common axis.

(19) *Coplanar*

Ans. In the same plane.

(20) *Core*

Ans. Centre of a coil.

(21) *Dissipation*

Ans. Loss of electrical energy.

(22) *Drain*

Ans. To take current from a voltage source.

(23) *Efficiency*

Ans. Percentage of output to input.

(24) *Electromotive force*

Ans. It is that force which tends to cause an electric current to flow in a circuit.

(25) *Equivalent circuit*

Ans. A comparatively simple circuit which may be used in calculations, having the same effect as in actual circuit.

(26) *Excitation*

Ans. Signal voltage applied to the control electrode of the valve.

(27) *Frequency*

Ans. Reciprocal of the time period; the number of cycles per second.

(28) *Fundamental frequency*

Ans. Lowest frequency of a number of harmonically related frequencies.

(29) *Generator*

Ans. It develops either direct or alternating electrical voltage at any frequency.

(30) *Ground*

Ans. Electrical connection to the earth.

(31) *Harmonic frequency*

Ans. The frequency which is the multiple of fundamental frequency. Twice the fundamental frequency is called the second harmonics and so on.

(32) hertz

Ans. Cycles per second.

(33) *High level*

Ans. A relative term indicating the final level of an amplifier.

(34) *Honeycomb winding*

Ans. A coil winding in which spaces are left between turns so as to give the appearance of a honeycomb.

(35) *Intensity*

Ans. The strength of a quantity like pressure, voltage of current.

(36) *Layer winding*

Ans. A coil winding in which one layer is wound over the other.

(37) *Lead*

Ans. Connecting wire.

(38) *Line voltage*

Ans. Mains voltage, i.e., the voltage of the power lines.

(39) *Litz wire*

Ans. A conductor composed of a number of fine, separately insulated strands.

(40) *Load*

Ans. A device which absorbs electrical power.

(41) *Low level*

Ans. A relative term indicating the level of the early stages of an amplifier.

(42) *Low-pass filter*

Ans. A filter which passes all frequencies below a critical value.

(43) *Lower sideband*

Ans. The sideband containing all the frequencies below the carrier frequency which is amplitude-modulated.

(44) *Manual*

Ans. Adjusted by hand.

(45) *Metal-film resistor*

Ans. A resistor constructed by coating a high temperature insulator, such as quartz, ceramic or mica with a metal-film by evaporation or firing.

(46) *MeO*

Ans. Abbreviation for million electron volt.

(47) mho

Ans. Reciprocal of ohm. It is the unit of conductance or admittance.

(48) microfarad

Ans. One millionth of a farad.

(49) *Monitoring amplifier*

Ans. An amplifier designed to amplify signals with the minimum distortion. It is used in audio engineering to check programme performance or the level of recording signals.

(50) *Mutual inductor*

Ans. Inductor used to change the mutual inductance between two circuits.

(51) *Negative bias*

Ans. Voltage applied to the control grid of a valve to make it negative with respect to cathode.

(52) *Negative electrode*

Ans. The cathode of a valve.

(53) *Negative glow*

Ans. In a discharge tube the luminous glow between cathode dark space and Faraday dark space.

(54) *Negatron*

Ans. Negative electron.

(55) *Neon*

Ans. A chemical element with atomic number 10, an inert gaseous element of the noble gas group. It has a characteristic red or fluorescent glow when ionized. In electronics neon tubes and lamps are extensively used as indicator.

(56) *Neon oscillator*

Ans. Oscillator based on neon lamp and a capacitor sometimes with a resistor in the circuit.

(57) *Network*

Ans. An electrical circuit. A system of interconnected admittances.

(58) *Network analyser*

Ans. An assembly of circuit elements so as to form models of electrical networks. These are widely used to solve problems in power distribution systems.

(59) *Node*

Ans. A point on a stationary wave which has zero amplitude.

(60) *Noise*

Ans. Undesired electrical fluctuations accompanying and tending to obscure an electrical signal causing disturbance within the useful frequency band.

(61) *Noise level*

Ans. Value of noise integrated over a specified frequency range. It is expressed in decibels relative to a standard level.

(62) *Non-linear*

Ans. Distorted.

(63) *On-off control*

Ans. A control system of only two choices.

(64) *Open circuit.*

Ans. A circuit in which continuity is incomplete and therefore carry no current.

(65) *Open grid*

Ans. Floating grid.

(66) *Oscillation*

Ans. Any quantity is in a state of oscillation when the value of that quantity changes continually between maximum and minimum values.

(67) *Output stage*

Ans. The voltage, current or power delivered by a circuit at the final stage.

(68) *Parallel circuit*

Ans. A circuit in which components are in parallel so that the current divides between them.

(69) *Paschen's law*

Ans. The p.d. which initiates discharge in a gas is the product of the electrode separation and gas pressure.

(70) *Photocurrent*

Ans. An electric current which varies with illumination.

(71) *Potential*

Ans. The potential at a point is the p.d. between that point and the earth.

(72) *Pulse*

Ans. Variation of a quantity whose value is normally constant. The characteristics of a pulse are a rise, a finite duration and a decay.

(73) *Q-factor*

Ans. The ratio of reactance to resistance of a coil, condenser or resonant circuit.

(74) *Quiescent current*

Ans. The current flowing to or from an electrode under specified normal conditions when the signal is absent to the input electrode.

(75) *Radial lead*

Ans. A lead at 90° to the axis.

(76) *Root-mean-square (r.m.s.) value*

Ans. The square root of the mean (average) value of the squares of the instantaneous values taken over a complete cycle.

(77) *Secondary battery*

Ans. A battery made up of storage cells.

(78) *Secondary emission*

Ans. Ejection of electrons due to the impact of charged particles.

(79) *Self-bias*

Ans. Bias developed due to the flow of the valve current through a resistor in the cathode circuit or through a grid resistor in the grid circuit and not by battery.

(80) *Short-circuit*

Ans. An accidental low resistance connection across any part of a circuit causing an excessive current flow.

(81) *Signal*

Ans. An electrical impulse or radio wave which gives information by varying in some manner.

(82) *Sinusoidal*

Ans. The form of a sine wave.

(83) *Space current*

Ans. Cathode current.

(84) *Symmetrical*

Ans. Having some characteristics for conduction in both the directions.

(85) *Tap*

Ans. Connecting point other than the end.

(86) *Terminal*

Ans. Point to which connections are made.

(87) *Tolerance*

Ans. Maximum permissible variation from an assigned value.

(88) *Toroidal coil*

Ans. A coil wound in the form of a toroidal helix.

(89) *Transducer*

Ans. An electrical transducer is an electrical network by which energy may flow from one transmission system to the other.

(90) *Unidirectional current*

Ans. Current which may change its value but never reverses its direction.

(91) *Voice coil*

Ans. Moving coil of a dynamic loudspeaker.

(92) *Voltage regulator*

Ans. An electronic device which regulates the voltages across a load.

(93) *VTVM*

Ans. Abbreviation for vacuum-tube-valve voltmeter.

(94) *Work-function (thermionic)*

Ans. The thermionic energy required to eject an electron from a hot metal.

(95) *X-band*

Ans. Radar frequencies from 5,200 to 10,900 MHz.

(96) *Zero signal*

Ans. Having no signal input voltage.

II. (1) *How are the vacuum tubes classified?*

Ans. Vacuum tubes are classified in a number of ways as discussed below :

(i) *Depending on the number of electrodes :* (a) Diode has two electrodes : an anode and cathode. (b) Triode has three electrodes : anode, cathode and control grid. (c) Tetrode has four electrodes : anode, cathode, control grid and screen grid. (d) Pentode has five electrodes : anode, cathode, control grid, screen grid and suppressor grid. Similarly, hexode has six electrodes, heptode has seven electrodes and so on.

(ii) *Depending on the presence or absence of gas :* (a) High vacuum tubes : After mounting of the electrodes in the tube envelope, the gas is removed from the electron tube with a vacuum pump of the order of 10^{-8} atmosphere. (b) Gas tubes : After evacuation an inert gas is inserted within the tube envelope.

(iii) *Depending on the type of cathode used :* (a) Thermionic tubes : In these tubes cathode is heated for emission of electrons. (b) Electron multiplier tubes : Electron emission at a number of electrodes, called dynodes, takes place due to secondary emission. (c) Cold cathode tubes : Electron emission is obtained by high field. (d) Photo tubes : Cathode emits electrons when light falls on it.

(2) *What is valve?*

Ans. The term 'di-ode' means two electrodes. The valve consists essentially of a cathode which serves as an emitter of electrons surrounded by a plate or anode which acts as a collector of electrons. Both the electrodes of the valve are sealed inside an evacuated glass or metallic envelop. The emitter may be either directly or indirectly heated type.

(3) *What are the static characteristics of a diode?*

Ans. The curves obtained on plotting anode current along the axis against anode voltage along the X-axis is called the current voltage characteristics of a diode. Since current and voltage involved pertain to the tube alone, irrespective of the associated circuits, the characteristics thus obtained usually referred to as 'static characteristics'.

(4) *Give the names of three electrodes of a triode valve?*

Ans. The triode valve, as the name implies, contains three electrodes : a cathode of either filament or indirectly heated type, a control grid and an anode.

(5) *Who first introduced the control grid? How is it made?*

Ans. The control grid was first introduced in 1907 by Lee de Forest. It is usually made by a spiral or fine wire placed between the cathode and anode, usually close to the cathode.

(6) *What is the function of a control grid?*

Ans. Since the control grid is usually placed closed to the cathode, the potential applied to the grid has a greater effect on the electric field inside the tube. The grid, therefore, has controlling effect on the plate current. The control in operation is kept at a negative potential with respect to the cathode. As a result, it does not draw any current and consume power.

As the voltage applied to the grid is made gradually negative, the space current also reduces. Ultimately a certain value is reached when the space current and hence the plate current becomes zero. Such a grid voltage at which plate current vanishes is known as 'cut-off grid voltage'.

(7) *What are the variable quantities of a triode valve?*

Ans. In a triode valve we have three variable quantities : (i) plate current (i_b), (ii) plate voltage (e_b) and (iii) grid voltage (e_c).

(8) *Give the names of triode characteristics.*

Ans. Three types of characteristics are there :

(a) Plate current/plate voltage characteristics, or plate characteristics. (b) Plate current-grid voltage characteristics or 'mutual characteristics'. This is also known as 'transfer characteristics'. (c) Plate voltage/grid voltage characteristics or 'constant current characteristics'. This is further known as 'amplification characteristics'.

(9) *What are valve constants? Give their names.*

Ans. The design factors of a valve are summarized by a series of numbers, called the valve constants. Valve manufacturers publish a manual in which these constants are listed along with the characteristic curves and other ratings for each valve.

The three most important constants of a valve are : (i) the anode slope resistance or the ac plate resistance, (ii) the transconductance or the mutual conductance, and (iii) the amplification factor.

(10) *Define plate resistance. Why is it also called anode slope resistance?*

Ans. Plate resistance (r_p) is defined as the ratio of a small change in plate voltage to the corresponding change in plate current, when the grid voltage is kept constant. Mathematically,

a.c. plate resistance,

$$r_p = \frac{\delta e_b}{\delta i_b} = \left.\frac{\Delta e_b}{\Delta i_b}\right|_{e_c\ =\ \text{constant}}$$

where Δe_b = a small change in plate voltage,

Δi_b = a small change in plate current.

Since r_p is a ratio of two small quantities, it is correctly called 'incremental plate resistance' or the 'dynamic plate resistance'. Again as it represents the slope of anode characteristic, it is also known as 'anode slope resistance'.

(11) *Define transconductance and the amplification factor of a valve.*

Ans. Transconductance (g_m) is defined as the ratio of a small change in plate current to the corresponding change in control grid voltage, when the plate voltage is kept constant. Mathematically,

Transconductance,

$$g_m = \frac{\delta i_b}{\delta e_c} = \left.\frac{\Delta i_b}{\Delta e_c}\right|_{e_b\ =\ \text{constant}}$$

where Δi_b = a small change in plate current,

Δe_c = a small change in grid voltage.

Amplification factor (μ) is defined as the ratio of a small change in plate voltage to the corresponding change in grid voltage, when the plate current is kept constant. Mathematically,

Amplification factor,

$$\mu = \frac{\delta e_b}{\delta e_c} = \left.\frac{\Delta e_b}{\Delta e_c}\right|_{i_b\ =\ \text{constant}}$$

where Δe_b = a small change in plate voltage,

Δe_c = a small change in grid voltage.

(12) *Why the transconductance is also known as the mutual conductance?*

Ans. Since the transconductance relates a quantity on the plate side to a quantity on the grid side it is also called 'mutual conductance'.

(13) *What is the relation between r_p, g_m and μ of a valve?*

Ans. The mathematical form of the amplification factor is

$$\mu = \frac{\delta e_b}{\delta e_c} = \frac{\delta e_b}{\delta i_b} \times \frac{\delta i_b}{\delta e_c}$$

$$= r_p \times g_m.$$

That is, the amplification factor is equal to the product of plate resistance and mutual conductance. The three valve constants are thus interrelated in accordance with their definition.

(14) *What is a tetrode valve?*

Ans. A tetrode valve has two grids in addition to cathode and anode. The first one, as mentioned before, is called the control grid while the second grid is known as the screen grid. In the construction, the screen grid is placed between the control grid and the plate, and thus avoids the undesirable grid-to-plate capacitance.

(15) *Explain the functions of screen grid.*

Ans. The screen grid of a tetrode valve is usually maintained at a high potential value, even sometimes greater than that of the plate. The screen grid of a tetrode valve has two important functions. First, it being maintained at a positive potential, helps to overcome the space charge effect and hence increases the plate current and mutual conductance. Secondly, when the valve is used as an amplifier, the screen grid eliminates the evil effect of interelectrode capacitance between the control grid and the anode.

(16) *What is a pentode valve?*

Ans. A pentode valve has five electrodes. These are : (i) cathode, (ii) control gird, (iii) screen grid, (iv) suppressor grid, and (v) plate. In the construction of a pentode the suppressor grid is placed between the screen grid and the plate.

(17) *Explain the functions of suppressor grid.*

Ans. In a pentode, the suppressor grid is connected directly to the cathode and thus maintains a negative potential with respect to anode. The secondary electrons emitted by the anode are, therefore, repelled by the negative suppressor grid and are driven back to the plate again. In this fashion the grid eliminates the effect to secondary electrons, and hence the plate current rises smoothy with the increase of plate voltage.

The addition of suppressor grid also increases the shielding action between the plate and the control grid, and thus reduces the grid to plate capacitance. Because of high screening, the plate current is little affected by the plate voltage and so both the plate resistance and the amplification factor of a pentode become extremely high.

(18) *What is a beam-power tube?*

Ans. Beam-power tube is a very useful hybrid between a tetrode and a pentode. Like a tetrode it has two grids, yet it behaves like a pentode by arranging special electrodes so as to produce a negative space charge close to the plate. Secondary electrons are repelled by this negative space charge and back to the plate again. Therefore, it is similar to the behaviour of a suppressor grid in a pentode.

Beam-power tube is superior to that of a pentode as it gives higher plate efficiency and higher power output for a given harmonic distortion. Moreover, it gives a small ratio of screen grid current to the plate current than in a pentode.

(19) *Why is rectification required ?*

Ans. Power companies generally supply a.c. power because it can be easily generated and also economical to transmit. On the other hand, d.c. supply usually serves the power requirements for various electronic devices. This demands a rectification of a.c. into d.c. voltages and currents.

(20) *What is a rectifier?*

Ans. A rectifier must provide a substantially one-way path for the flow of electric current. It changes a.c. into pulsating d.c. by eliminating the negative half-cycle. This is called a unidirectional characteristic or unilateral conduction. The diode is admirably suited for rectification as it permits flow of current in one direction only.

(21) *Why selenium rectifiers or crystal diodes are widely used?*

Ans. Nowadays selenium rectifiers or crystal diodes are widely used in various electronic circuits as no filament supply is required there and so they have become increasingly popular in the field of radio and television receivers, in particular.

(22) *What are the two different types of rectifiers?*

Ans. Rectifiers may be broadly put into two categories depending on the period of conduction. These are (a) *half-wave rectifier* and (b) *full-wave rectifier.*

(23) *What is half-wave rectifier?*

Ans. A half-wave rectifier (one-diode) converts the applied alternating voltage to a pulsating voltage using half-cycle of the applied voltage. That is, the conduction takes place during one half-cycle only.

(24) *What is a full-wave rectifier?*

Ans. A full-wave rectifier (two-diode) converts the applied alternating voltage to a pulsating voltage using full-cycle of the applied voltage. That is, the conduction takes place by the one rectifier element during one half and by the other element during the other half-cycle.

(25) *Define ripple factor.*

Ans. The rectified output is unidirectional in nature. But because of the continuous changes in amplitudes or pulsations the current is not a direct (d.c.) one, rather these pulsations contain both d.c. and a.c. components.

The ripple factor, denoted by γ, may be defined as

$$\gamma = \frac{\text{effective value of the ac components of current (or voltage)}}{\text{dc value of current (or voltage)}}$$

$$= \frac{I_{\text{a.c.}}}{I_{\text{d.c.}}}.$$

(26) *What do you mean by the efficiency of a rectifier?*

Ans. The efficiency of the rectifier, denoted by η, may be defined as

$$\eta = \frac{\text{d.c. output power}}{\text{a.c. input power}}$$

$$= \frac{P_{\text{d.c.}}}{P_{\text{a.c.}}}.$$

(27) *Make a comparative study of a half-wave rectifier with a full-wave rectifier.*

Ans. We may compare a half-wave rectifier with a full-wave rectifier by considering the following important points :

(i) Ripple factor of a half-wave rectifier is greater than that of a full-wave rectifier.

(ii) Rectifier efficiency of a full-wave rectifier is double than that of a half-wave rectifier.

(iii) A half-wave rectifier has a low transformer efficiency and a low output voltage than a full-wave rectifier.

(iv) In half-wave rectification current through the secondary of a transformer flows in the same direction while in full-wave rectification equal currents flow in the opposite direction through the centre-tapped secondary.

(v) A half-wave rectifier has low cost in comparison to a full-wave rectifier.

(28) *What is achieved by filter network?*

Ans. The output voltage obtained by using a rectifier circuit includes ripple which must be minimized to use it as a d.c. source. The required smoothing action is achieved by filter networks.

(29) *What are the filter elements?*

Ans. Choke, capacitor and resistor are usually filter elements.

(30) *What are the functions of a bleeder resistor?*

Ans. The bleeder has the following important functions : (i) It helps to maintain a constant output voltage by drawing a minimum current at all times. (ii) When the receiver tubes or other loads in the circuit are heating up, the bleeder resistor maintains a minimum load across the rectifier. Without the resistor R_B, there would be an initial high voltage surge as soon as the rectifier is turned on which might damage the circuit. (iii) After the rectifier has been turned off, the bleeder discharges the capacitors and thus helps to prevent dangerous shocks.

(31) *What is the function of an amplifier?*

Ans. The function of an amplifier is to increase the level of the current, voltage or power at its output.

(32) *What is the gain of an amplifier? When is it called a voltage or a power amplifier?*

Ans. The amount of increase of voltage or power is called the amplification or gain of the amplifier. It is called a voltage amplifier if the output voltage becomes greater than the input voltage. A power amplifier, on the other hand, is used to supply power (i.e., voltage times current) to the output or load circuit.

(33) *What is a cascade amplifier?*

Ans. If a number of amplifiers are connected in series so that the output of one becomes input to the next, then the early stages build up the voltage to a high level while the last stage builds up the power to a level suitable for operating a speaker or other similar devices. This amplifier is known as a cascade amplifier.

(34) *How are the amplifiers classified according to the type of load used?*

Ans. According to the type of load used amplifiers are classified as : (i) *untuned amplifiers* and (ii) *tuned amplifiers.*

Untuned amplifiers are again divided into two groups. These are audio frequency (up to about 15 kHz) and video frequency amplifiers (up to a few MHz).

The tuned amplifiers, also referred to as radio frequency amplifiers, are used to amplify an electrical signal containing a radio frequency (a frequency of above 30 kHz).

(35) *How are amplifiers classified according to the number of stages and methods of coupling?*

Ans. According to the number of stages and methods of coupling, amplifiers are classified as : (i) *single-stage amplifiers* and (ii) *cascade amplifiers.*

A single-stage amplifier is suitable for cases where small amplification is necessary, while in a cascade amplifier considerable amount of gain is achieved. Depending on the method of coupling, cascade amplifiers may again be divided as : Resistance Capacitance (RC) coupled amplifier, Transformer coupled amplifier, Direct coupled or dc amplifier and LC coupled amplifier.

(36) *How are the amplifiers classified according to the choice of the operating point?*

Ans. According to the choice of the operating point the vacuum tube amplifiers are classified as : (i) *Class A*, (ii) *Class AB*, (iii) *Class B* and (iv) *Class C amplifiers.*

In a class A amplifier the grid bias and the input signal are so adjusted that the plate current flows during the entire cycle of the input signal. A class AB amplifier is designed such that the plate current flows for more than half but less than the complete cycle of the input voltage. A class B amplifier is designed to have its grid bias almost equal to the cut-off voltage so that the plate current flows for half the a.c. cycle while in a class C amplifier the plate current flows over less than the half period of the input signal. In all the above classes of amplifiers, a subscript 1 is added to letter or letters to indicate that the grid current does not flow during any part of the input cycle (e.g., class A_1), while a subscript 2 is given to indicate that the grid current flows during a part of the input cycle (e.g., class A_2).

(37) *What is meant by distortion in amplifiers?*

Ans. The amplifier output in an ideal amplifier is the enlarged version of the input signal. But in practice, a difference is noticed always between the output and input waveforms. The amplifier is then said to have a disortion.

(38) *What are the three main types of distortion? Explain.*

Ans. The three main types of distortions are the amplitude distortion, frequency distortion and phase distortion.

Amplitude distortion—The distortion is caused by non-linear operations, such as plate current cut- off and grid current flow. This distortion result when harmonic frequencies, not present in the input signal, appear at the output.

Frequency distortion—This distortion is caused by the amplifier coupling elements when signals of different frequencies are amplified separately. In general, both amplitude and frequency distortions may be present simultaneously in an amplifier.

Phase distortion—This distortion is due to the unequal delay of different audio frequencies by phase shifts in the amplifier coupling elements.

(39) *What is meant by feedback in an amplifier?*

Ans. Technically, the term feedback means transferring a portion of the output energy of any device to its input causing distortion.

(40) *Find the difference between positive and negative feedbacks.*

Ans. If the feedback voltage in an amplifier adds to the input signal voltage, it increases the amplification of the amplifier and it is called a positive or regenerative feedback. On the other hand, if the feedback voltage subtracts from the input signal voltage, the amplification decreases and the feedback is called a negative feedback or degenerative feedback.

Additional Oral Question

(1) In a vacuum diode, there are two electrodes—cathode and anode; but when connecting in a circuit we give connections to four pins of its base. Explain why. **(2)** What are the three most commonly used cathode materials? Give important properties of each. **(3)** Explain what do you mean by space charge-limited and temperature-limited regions with reference to a vacuum diode. **(4)** There are a large number of free electrons moving randomly in a metal. Why do they not come out of the surface of the metal at room temperature? **(5)** Explain why any change in the grid potential of a triode causes a change in its plate current. **(6)** Discuss why we do not permit the grid of a triode to be at a positive potential with respect to the cathode. **(7)** Explain why a triode valve is not able to work as an amplifier for high frequency signals. **(8)** How does the screen grid of a tetrode reduce the inter-electrode capacitance between plate and control grid? **(9)** Why do the kinks occur in the plate characteristics of a tetrode? In what why are these kinks harmful? **(10)** Explain the purpose of each of the three grids of a pentode. **(11)** Mention the advantages of a pentode over a tetrode while walking as an amplifier at (a) audio frequency, (b) radio frequencies. **(12)** Draw the symbols of diode, triode, tetrode and pentode. Label each electrode. **(13)** Discuss how the effect of secondary emission is suppressed by the suppressor grid in a pentode. **(14)** What is the difference between the mutual characteristics of a remote cut-off pentode and a sharp cut-off pentode? Which of the two will have variable mutual conductance with grid bias? **(15)** Explain how a pentode can act as a constant current source. **(16)** Name some important applications where solid state devices have not been able to replace vacuum tubes as yet.

CHAPTER 4

EXPERIMENTS ON SEMICONDUCTOR DEVICES

4.1 To Draw the Static Characteristic Curves of a Semiconductor Diode, and hence to Calculate the Static and Dynamic Impedances (or Resistances)

Circuit Diagram

The circuit diagram for studying the forward and reverse characteristic curves of a semiconductor diode are shown in Fig. 4.1 and Fig. 4.2 respectively.

The indices are given below :

mV→ a millivoltmeter of suitable range

mA→ a milliammeter of suitable range

μA→ a microammeter of suitable range

R→ a resistance box having resistance from 0 to 5000 ohms

Rh→ a rheostat of suitable power ratings

pn→ the semiconductor diode

E→ a regulated d.c. power supply or a battery

K→ a plug key or a switch.

Working Formula

(i) *For diode characteristics.* When a semiconductor diode is biased in the forward or reverse direction, then holes and electrons flowing in opposite direction cross the junction and constitute a flow of current in the semiconductor.

The current voltage relationship for a semiconductor diode is given by

$$I = I_s\left[\exp.\left(\frac{eV}{kT}\right)-1\right]. \tag{4.1}$$

In the above equation, I represents the value of the current flowing in the semiconductor when a voltage V is applied across the diode and I_s represents the reverse saturation current, k is the Boltzmann constant, T is the absolute temperature of the surroundings at the time of experiment and e is the electronic charge.

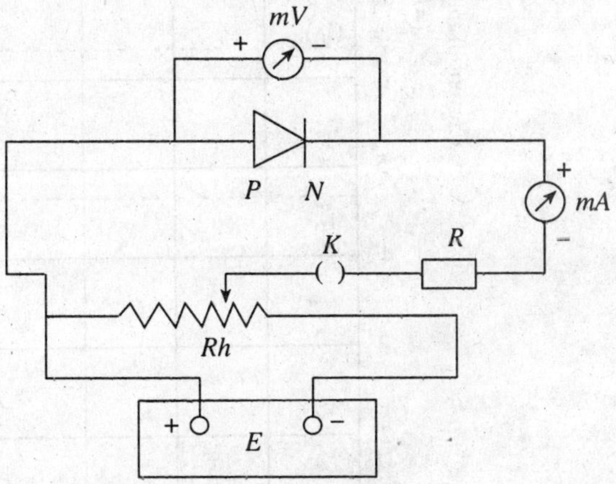

Fig. 4.1

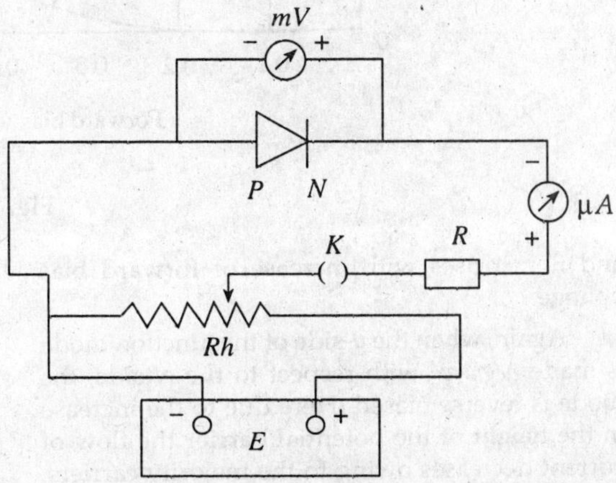

Fig. 4.2

When the p-side of the semiconductor diode is *made positive* in the closed circuit, then it is said to be *forward biased*. Here, the holes move from p to the N region while the electrons flow from n to p region (provided avalanche breakdown does not take place). This gives rise to a current flow

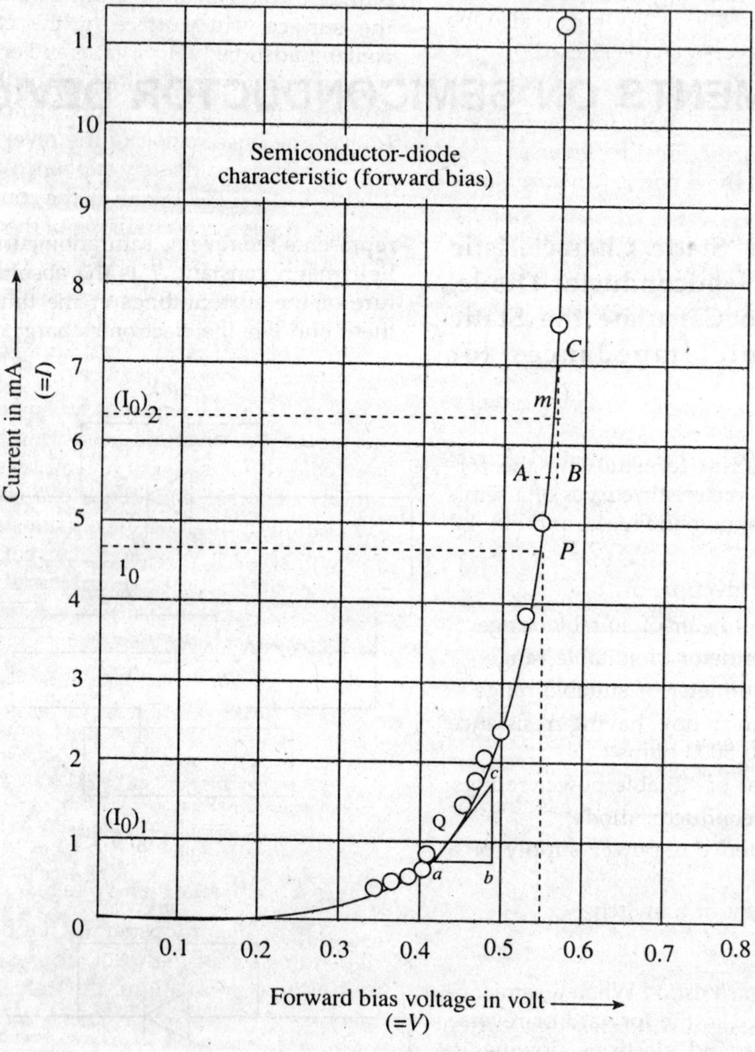

Fig. 4.3

and it increases with increase of forward bias voltage.

Again, when the p-side of the junction diode is made *negative* with respect to the N-side, the diode is reverse biased. Here due to the increase in the height of the potential barrier the flow of current decreases owing to the majority carriers. From Eq (4.1) it is evident that for a large negative value of the applied voltage, $I = -I_s$.

The static impedance Z of a semiconductor diode, when it is conducting, is the resistance offered under d.c. condition, and is given by,

$$Z = \frac{V}{I}. \tag{4.2}$$

The forward static resistance (or impedance) of a diode is different for different operating points.

The dynamic impedance Z or a.c. resistance of a diode is the resistance offered under dynamic condition. This dynamic impedance measures the resistance offered by that diode when a small a.c. signal is superimposed and is represented by Z_D. It is expressed as

$$Z_D = \frac{\Delta V}{\Delta I}. \tag{4.3}$$

This dynamic impedance also changes with change of operating point. For non-linear region it is obtained by drawing tangent at the operating point while for the linear region the value is directly obtained by drawing triangle about the operating point (see Fig. 4.3). It can also be obtained from the reverse characteristic.

Special note

Eq. (4.1) is applicable both for the forward bias voltage and partly for the reversed bias voltage. Assume that the diode is forward biased and hence V is positive. Also, the quantity $\frac{kT}{e}$ has the dimension of voltage (generally termed as the temperature-dependent voltage) and is represented by V_T. Hence, Eq. (4.1) is of the form $I = I_s\left[\exp\left(\frac{V}{V_T}\right)-1\right]$. The value of V_T at T = 300 K takes the value 26 mV. So, if the applied voltage V = 0.1 volt, then $I = I_s\left[\exp\left(\frac{0.1}{0.026}\right)-1\right]$

or, $I - I_s$ [exp.(4) – 1].

Hence, the quantity I in the above equation can be neglected in comparison to the first term. Thus, $I = I_s$ [exp.(4)]. Therefore, in the forward biased condition a very small voltage gives an appreciable current and the current increases exponentially with the voltage.

Near zero-bias, we can safely assume that $V \simeq V_T$, hence,

$$e^{(V/V_T)} = \frac{V}{V_T} + 1.$$

Therefore, $I = I_s \frac{V}{V_T}$.

The above equation indicates that current linearly increases with the applied voltage, when the biasing-voltage is small enough.

In the reverse bias condition V is negative, and so the current flow is very small. For a large value of negative applied voltage, $I = -I_s$ and so the current is independent of the applied voltage. But this is not *strictly* true, because at the surface of the junction a small leakage current flows. With high negative reverse voltage, avalanche breakdown occurs. The actual leakage current is generally controlled by the conditions existing at the surface boundary of the junction. Whatever may be the manner of forming the junction, it must end at some surface. The surface behaves as a discontinuity in the crystal structure and *it can collect foreign elements.* Hence the property of the bulk of the crystal is quite different from those at the surface. The surface of the crystal diode contains adsorbed solids, gases and ordinary water vapour and hence the life-time of the surface is governed by the above factors. Hence, the behaviour of the knee-region of the reverse characteristic cannot be competely explained by the above simple theory. The poorer is the condition of the surface, the lesser is the slope of the knee-region.

As the reverse biasing is increased more than that existing at the knee-region a different phenomenon takes place. With increase of the reverse voltage (i.e., the field applied), the velocities of the carriers crossing the depletion layer increase. Here the electrons are knocked off from the atoms of crystal and so electron-hole pairs are created. With increase of reverse biasing the pairs rapidly separate and hence causes further pair generation through other collisions. At the breakdown voltage a very large current is produced due to a slight additional increase of the reverse voltage. The current I in this region is governed by the multiplication factor m, and is given by $I = m \cdot I_s$.

The multiplication factor m is given by

$$m \simeq \frac{1}{1-\left(\frac{V}{V_b}\right)^n}. \tag{4.4}$$

where V_b = breakdown voltage.

The value of n generally lies between 3 to 6 depending on the nature of the crystal. For silicon and *N*-type germanium, the value of n may be taken as 3.

Experimental Procedure

(1) The *type number* of the semiconductor diode is noted. From the manufacturer's semiconductor manual the following information are recorded in the laboratory note-book :

(i) Semiconductor type (germanium or silicon)

(ii) Number of the semiconductor

(iii) Peak inverse voltage (PIV)

(iv) Maximum forward voltage drop and power rating

(v) Maximum reverse voltage and power rating

(2) *Recording for forward characteristics* :

(i) The circuit diagram as shown in Fig. 4.1 is drawn in the laboratory note-book and finally the connections are made (*if closed box instrument is not supplied*). *E* is a regulated d.c. supply source. A battery may also be used in place of the regulated power supply. For a regulated power supply source, the input terminals of the power supply is connected to a.c. mains. The circuit should be approved by the instructor before switching on the circuit.

(ii) Supply voltage is set to zero volt by regulating the knob (or by adjusting the resistance *R* in the circuit). The supply voltage is increased in small voltage steps, and in each step the corresponding current is noted. *The recordings should be such that the knee region of the forward characteristic can be properly shown on the graph paper.* The maximum voltage applied to the diode for drawing the forward characteristic should be such that the *maximum rating of forward current for the diode supplied is not exceeded.* For each set of recording, observations I and II may be made and from these the mean value may be taken.

Special note : For forward characteristic a small forward voltage *in millivolt range changes the forward current in milliampere range.* Hence the increment of voltage should be such that the current increases by one smallest division of the milli-ammeter supplied from the preceding value (for knee-region). Current change in the knee-region should be about 1 mA at each step. But for the *nearly linear region* of the forward characteristic, the current change in each step may be taken to about 5 mA from the previous one. Hence the voltage should be increased in such a way that the successive difference between the recorded currents is about 5 mA (for the nearly linear region).

(iii) A graph (Fig. 4.4) is drawn by plotting the forward voltage (in mV) along the *X*-axis, and the corresponding current in milliampere along the *Y*-axis.

(iv) From the graph (Fig. 4.4), the value of the static resistance and dynamic impedance are calculated at different operating points (using Eqs. 4.2 and 4.3).

(v) The circuit is switched off.

Extra task (*To be done if directed by the instructor)*

A 60-watt bulb (when incandescent) is placed in front of the semiconductor diode so that its distance is about 1.5 cm to 2 cm from the diode. Proceeding as before draw the forward characteristic curve and compare the curve with the former one. Interpret the result. Also calculate the d.c. resistance of the diode and note the effect of temperature on the forward d.c. resistance of the diode.

(3) *Recording for drawing reverse characteristic*

(i) The circuit diagram as shown in Fig. 4.2 is drawn in the laboratory note-book and finally the connections are made (if closed box instrument is not supplied). Before switching on the circuit, the connections should be approved by the instructor. At first, the reverse voltage is set to zero volt. The reverse voltage is increased in small steps and it will be found that the current recorded by the micro-ammeter is zero. The reverse voltage is increased in very small voltage steps and in each step the corresponding current is noted by the micro-ammeter (or a multimeter supplied). This increment of reverse voltage should be such that the maximum reverse voltage for the diode supplied is not exceeded.

(ii) A graph is drawn by plotting the reverse voltage (in mV) along the negative *X*-axis, and the corresponding current (in μA or in mA) along the negative *Y*-axis. The nature of the graph is shown in Fig. 4.4.

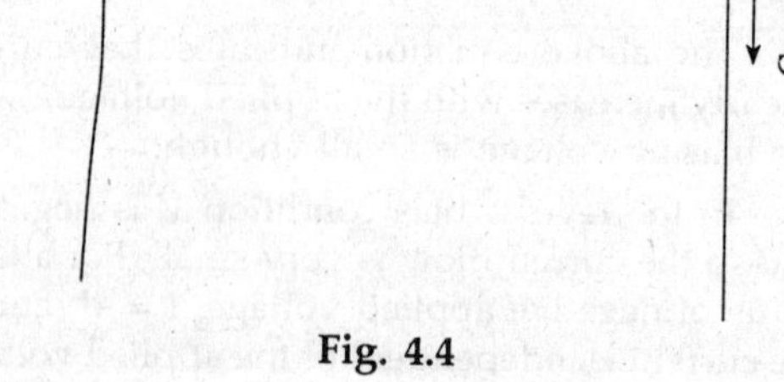

Fig. 4.4

(iii) From the reverse characteristic curve, the d.c. and a.c. resistance of the diode at a given reverse voltage (taken at the region about half the maximum reverse voltage value) are calculated.

Extra task (*To be performed if directed by the instructor*)

A 60-watt bulb (when incandescent) is placed in front of the semiconductor diode at a distance of about 1.5 cm to 2 cm. Care should be taken so that the hot bulb does not touch the diode. The distance should be such that the radiated heat can increase the temperature of the diode. Draw the reverse characteristic curve again as before for the heated diode. The effect of heat on the reverse characteristic, and also on the d.c. and a.c. resistance or impedance are to be noted.

(iv) Finally, from the knowledge of I_s, from the reverse characteristic, and the value of V from the forward characteristic, the theoretical formula given by Eq. (4.1) is compared with the experimental one.

N.B. : During recording of data for forward and reverse characteristic and the other constants involved therein, the temperature during the performance of the experiment must be noted. This is because, the nature of the characteristics are highly temperature-dependent.

Experimental result

N.B. The data shown in different tables should never be taken as the standard. The values differ from instrument to instrument. These are shown only for illustration.

[A] *Typical parameters of the diode*

Table I

Type	Construction	Peak inverse voltage in volt	Maximum rectifier current in ampere	Maximum forward voltage drop		Maximum reverse current	
				in volt	at ampere	in mA	at volt
	Silicon fabricated						

[B] *Range and graduation of the meters used*

Table II

Instrument used	Range	Value of one smallest division	Zero(or, intial) error (if any)
Voltmeter	...	...volt	...volt
Milli-ammeter	...	...mA	...mA
Micro-ammeter	...	...μA	...μA

[C] *Data for forward bias*

Table III

Room temperature ... °C

No. of obs.	Resistance R in ohm	Current in mA	Voltage in mV
1	5500	18	50
2	4000	24	65
3	3000	32	80
4	2800	34	84
⋮	⋮	⋮	⋮
19	950	98	142
20	900	104	144
21	850	110	148
22	800	117	150

[D] *Data for reverse bias*

Table IV

Room temperature ... °C

Current in mA	Voltage in volt
1	0.0069
2	0.0139
3	0.0308
⋮	⋮
5	0.3848
5	0.4248
5	0.5848

[E] *Calculation of static impedance (Z) and dynamic impedance (Z_D) (or dynamic resistance) from the forward characteristic.*

Note : For calculation of Z (the static resistance), a point *P* is considered on the forward characteristic (Fig. 4.3). Let the operating current be I_o mA. The corresponding voltage and current are noted at this operating current (I_o). Hence the value of *Z* is calculated with the aid of Eq. (4.2). The value of static resistance is determined at various operating currents.

For calculation of dynamic impedance (Z_D) in the non-linear region, say, at a point *Q* on the curve (Fig. 4.3), a tangent is drawn. From this, the triangle *abc* is constructed. Hence $Z_D = \frac{ab}{bc}$, can be calculated at this operating current $(I_o)_1$. For linear region, drawing of a tangent is not necessary. Here a triangle *ABC* is drawn about the operating point $(I_o)_2$ and the dynamic impedance (or resistance) is calculated simply by using the relation

$$Z_D = \frac{AB}{BC}.$$

In a similar way, the value of the static and dynamic impedances are calculated from the reverse characteristic curve (Fig. 4.4) at different operating points.

[F] *Calculation of static and dynamic impedances from forward characteristic.*

Table V

Room temperature = ... °C

Operating point	Value of *V* in volt	value of *I* in ampere	value of ΔV in volt	Value of ΔI in ampere	Value of $Z = \frac{V}{I}$ ohm	value of $Z_D = \frac{\Delta V}{\Delta I}$ in ohm
...	...	...	...	...	...	...
...	...	...	...	...	...	...
...	...	...	...	...	...	...
etc.	etc.	etc.	etc.	etc.	etc.	etc.

[G] *Calculation of static and dynamic impedances from reverse characteristic.*

Table VI

Make table similar to that of table V.

[H] *Comparison of the calculated value of I with the experimental value.*

Table VII

Set no.	Value of saturation current in ampere ($=I_S$)	Value of V for forward bias in volt	Temperature in degree kelvin	Value of k in erg/ deg. abs.	Value of I (from calculation) in ampere	Value of I from experiment in ampere
1.	5×10^{-6}	0.0109	305	1.381×10^{-16}	2.3115×10^{-6}	3×10^{-6}
2.	5×10^{-6}	0.0452	305	1.381×10^{-16}	22.36×10^{-6}	18×10^{-6}
etc.	etc.	etc.	etc.	etc.	etc.	etc.

Remark :

From the calculation of I at room temperature it is clear that the calculated and experimental values of I are nearly equal. A slight departure is due to the fact that the value of I cannot be very correctly obtained.

Discussion

(i) The specifications of the diodes should be properly noted. During recording the forward and reverse power ratings should never be exceeded.

(ii) The initial errors in different meters used should be eliminated, if possible. Otherwise, the corresponding corrections must be made.

(iii) During recordings, the parallax error should be eliminated. Each reading should be taken twice, and the mean value should be taken for correct procedure.

4.2 To Draw the Dynamic Characteristic Curves using *P-N* Junction Diode

Apparatus required : (1) Regulated power supply (*E*) from zero to 25 volts range, (2) a voltmeter (*V*) and a milliammeter (*mA*) of suitable range, (3) *p-n* junction diode, (4) a resistance box, (5) connecting wires, (6) key (*K*), etc.

Theory : When a resistance box (*R*) is inserted between the milli-ammeter and the source, then for a fixed value of load resistance (*R*), if we vary the plate voltage, the plate current will change. For each plate voltage, the corresponding plate current for that load resistance can be noted. Such curve is known as the dynamic characteristic curve. For different load resistances, dynamic characteristic curves are obtained.

Experimental procedure : (1) Circuit diagram is drawn in laboratory note-book and connections are made as shown in Fig. 4.5. The input terminals of the regulated power supply are connected to the a.c. main. Finally the circuit is approved by the instructor.

(2) Range of different meters used are noted.

(3) Different important characteristics (namely, maximum reverse power rating, maximum forward power rating, etc.) are noted from the semiconductor manual.

(4) At first the load resistance *R* is kept to a fixed value, say 100 Ω, for the diode supplied. By varying the regulated power supply (*E*), the voltage is gradually increased in steps of 0.2 volt (or to other values depending on the smallest division of the voltmeter *V*), and in each case, the corresponding plate current in milliampere is noted. This procedure is done up to the maximum forward voltage across the diode. In this case the supply voltage may be varied from zero volt to 22 volts.

(5) Now load resistance *R* is changed and kept to another fixed value, say 200 Ω. Similar to the procedure no. (4), the voltage is gradually increased by varying the power supply (*E*), and in each case, the corresponding plate current is noted. Again, the load resistance is kept fixed to 300 Ω (say) and similar procedure is done.

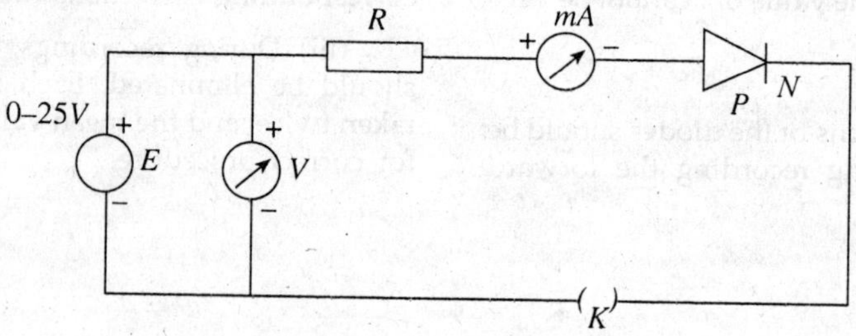

Fig. 4.5

(6) Now a graph is plotted with input voltage (V_0) along *X*-axis and corresponding current (I_0) along *Y*-axis. The graph will be nearly straight line passing through the origin. Actually the graph is slightly bent near the origin. The nature of the graph is shown in Fig. 4.6.

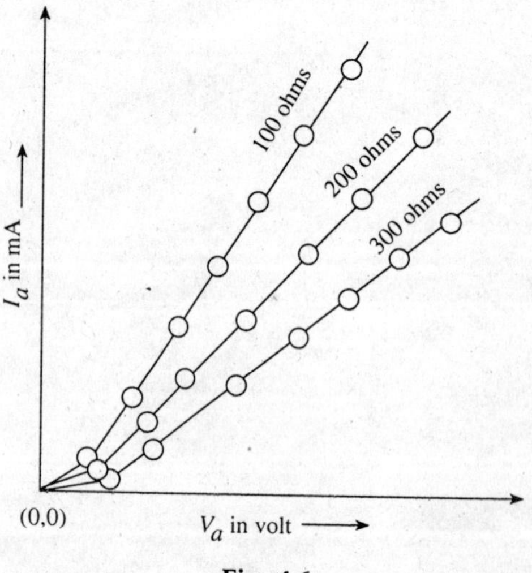

Fig. 4.6

Experimental results

N.B. : The data shown in different tables should never be taken as the standard. The values differ from instrument to instrument. These are shown only for illustration.

[A] *Value of graduation and the smallest divisions of different meters used.*

Table I

Instrument used	Range	Smallest division	Zero error (if any)
Milliammeter	0 to ... mA	... mA	... mA
Voltmeter	0 to ...volt	...volt	...volt

[B] *Data for dynamic characteristic curves.*

Table II

Resistance 100 Ω		Resistance 200 Ω	
Voltage (V_a) in volt	Current (I_a) in milliampere	Voltage (V_a) in volt	Current (I_a) in milli-ampere
0	0	0	0
0.2	0	0.2	0
0.4	0	0.4	0
0.6	0.5	0.6	0.5
0.8	3	0.8	1.25
10	5.5	1.0	2.25
12	7.25	1.2	3.5
—	—	—	—
—	—	—	—
—	—	—	—
—	—	—	—
19	130	19	90
20	190	20	195
21	200	21	100
22	210	22	105

N.B. : Record different values of I_a and V_a for other values of load resistances (say, 400 ohms, 500 ohm, etc.). Draw the I_a/V_a curve as shown in Fig. 4.6.

Discussion : *Same as that in the experiment no. 4.1.*

4.3 To Study the Semiconductor Diode as Rectifier in Power Supply

Circuit diagram

To use the diode as a half-wave rectifier the circuit as shown in Fig. 4.7(a) is to be made. In the figure, the input is an a.c. voltage applied across the series connected diode and the load R_L. To study the full-wave rectifier characteristics, circuit of Fig. 4.7(b) is used. In both the figures, C is a capacitor, mA is the milliammeter and R_L is the load resistance. The output voltage $V_{d.c.}$ is obtained across R_L.

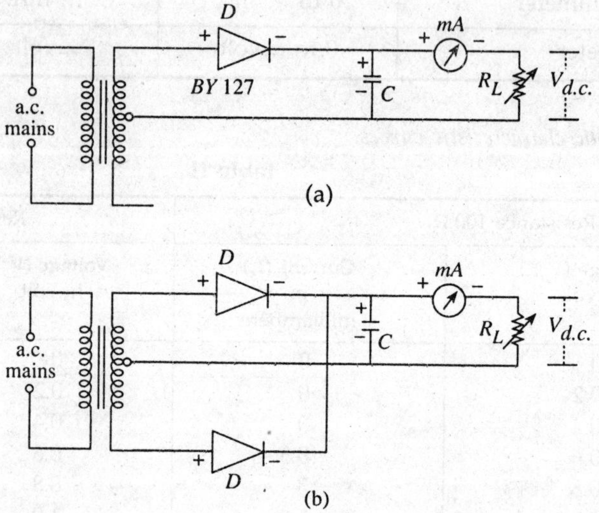

Fig. 4.7

Theory

A half-wave rectifier (consisting of one diode) converts the applied alternating voltage to a pulsating voltage using half-cycle of the applied voltage. In a half-wave rectifier during the positive half-cycle the *P*-side of the semiconductor is positive with respect to the *N*-side of the diode and the current flows in the circuit. But during the negative half cycle, the *P*-side of the diode is negative with respect to the *N*-side, and hence *no current flows* (provided the avalanche breakdown does not occur). On the other hand, a full-wave rectifier (consisting of two diodes) converts the applied alternating voltage to a pulsating voltage using full-cycle of the applied voltage. That is the conduction takes place by the one rectifier element during one half-cycle and by the other element during the other half-cycle of the alternating voltage.

Assuming a sinusoidal a.c. input of the form

$$V_i = V_m \sin \omega t,$$

the peak current in the load circuit for a half-wave rectifier is

$$I_m = \frac{V_m}{R_L + R_f}. \quad (4.5)$$

where R_f = forward resistance offered by the diode at the operating point, and R_L = value of the load resistance.

$$\text{Also, } I_{d.c.} = \frac{I_m}{\pi}. \quad (4.6)$$

Hence, the output voltage across the load resistance R_L for a half-wave rectifier is given by (from Eq. 4.5),

$$V_{d.c.} = \frac{I_m . R_L}{\pi} = \frac{V_m R_L}{\pi(R_L + R_f)}. \quad (4.7)$$

Similarly for a full-wave rectifier,

$$(I_{d.c.})_{\text{full-wave}} = \frac{2I_m}{\pi} \quad (4.8)$$

$$\text{and } (V_{d.c.})_{\text{full-wave}} = \frac{2I_m R_L}{\pi} = \frac{2V_m R_L}{\pi(R_L + R_f)}. \quad (4.9)$$

Voltage regulation : Voltage regulation for any rectifier is a measure of the ability of the rectifier to maintain the output voltage to a specified range though the load (or the supply current) varies. Percentage of voltage regulation ρ is defined as

$$\rho = \frac{V_{\text{no load}} - V_{\text{rated load}}}{V_{\text{rated load}}} \times 100\%, \quad (4.10)$$

where $V_{\text{no load}}$ = voltage across the diode when load is infinity (i.e., $I_L = 0$) and $V_{\text{rated load}}$ = load voltage for a particular current flowing through the circuit.

Experimental procedure

(1) The connections for the study of half-wave and full-wave rectifications are made as shown in Figs. 4.7(a) and (b) respectively. The a.c. is supplied by using a step-down transformer (30–0–30 volt, say).

(3) The same procedure as mentioned in procedure no. (2) is to be followed with a capacitance filter. If possible, instead of using only a capacitor filter or *L*-filter, (*L–C*) or a *p*-filter (*C–L–C*) may be suitably used, as shown in Fig. 4.8.

(4) The above operations are to be followed both for a half-wave and a full-wave rectifier circuits.

(5) Different values of d.c. output voltges are then calculated from the values of load resistance and d.c. load current.

(6) A graph is then drawn by plotting load current along X-axis and the corresponding load voltage along the *Y*-axis with and without the filter circuit for the half-wave rectifier. Another

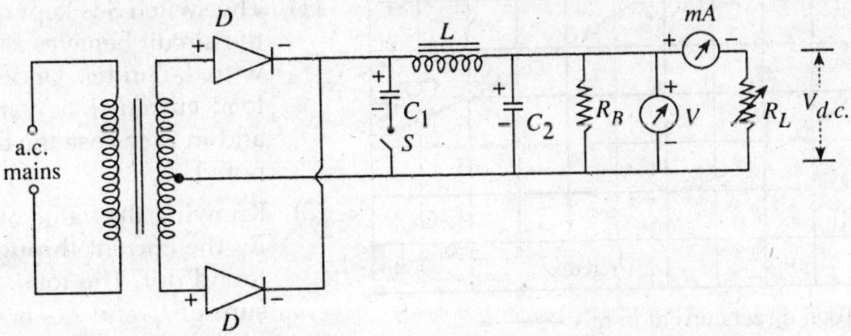

Fig. 4.8

(2) By varying R_L (in the resistance box or in a potentiometer) the load current $I_{d.c.}$ without the capacitance filter is noted. For each value of the load resistance R_L, two observations for load current are found out from which a mean $I_{d.c.}$ is determined.

similar graph is drawn on a separate graph paper for the full-wave rectifier.

Experimental results (*for a half-wave rectifier*)

Make similar tables as Tables I and II as in the experiment no. 4.1.

[C] Data for drawing half-wave rectification characteristics

Table III

Value of load resistance (R_L) in ohm	Without capacitance filter: Load current I_{dc} in mA				With capacitance filter: Load current $I_{d.c.}$ in mA			
	Obs. I	Obs. II	Mean	Output voltage $E_{d.c.}$ (= $I_{d.c.} \times R_L$) in volt	Obs. I	Obs. II	Mean	Output voltage $E_{d.c.}$ (= $I_{d.c.} \times R_L$) in volt
10000	1.45	1.35	1.4	14.0	4.0	4.0	4.0	40.0
9000	1.5	1.5	1.5	13.5	4.3	4.4	4.4	39.6
8000	1.7	1.7	1.7	13.6	4.9	4.9	4.9	39.2
⋮	⋮	⋮	⋮	⋮	⋮	⋮	⋮	⋮
1000	…	…	…	…	…	…	…	…
900	…	…	…	…	…	…	…	…
etc.	etc.	etc.	etc.	etc.	etc.	etc.	etc.	etc.

[D] Data for drawing full-wave rectificatin characteristics. [**Make a table similar to table III shown above. This is named as table IV,** (not shown).]

[E] To draw the rectification characteristics.

Current through the load ($I_{d.c.}$) in mA is plotted along the *X*-axis and the output voltage ($E_{d.c.}$) in volt is plotted along the *Y*-axis. The *upper curve* represents the variation with C-L-C filter. The *lower curve* represents the variation with L-C filter. The nature of the curves is shown in Fig. 4.9.

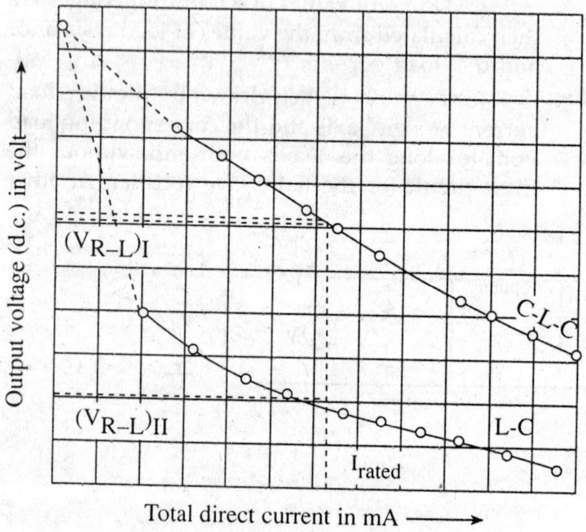

Fig. 4.9

[F] The percentage of voltage regulation as is calculated discussed in table VII for the full-wave rectifier.

Discussion

(i) While changing the resistance in the box, care should be taken so that the current greater than the amount prescribed for the diode may not pass through it. Because in that case the diode may be damaged.

(ii) Since the values of the output voltage in the rectification characteristic are found to be irregular in lower current region, so a mean curve is to be drawn. One must be careful enough to note the diode current in each case.

Extra work : To study the regulation characteristics of a full-wave rectifier using either an L-C filter or a π-filter (C–L–C). Fig. 4.8 may be employed. In the figure, R_B is the bleeder resistance and R_L is the load resistance. When the switch *S* is made off, *L* and C_2 come in the circuit and it behaves as an *L-C* filter; but when the switch *S* is made on, then C_1, *L* and C_2 all come in the circuit and so it behaves as a π-filter.

The voltage regulation is calculated by using the relation given below and is expressed in percentage. We have

percentage voltage regulation

$$= \frac{E_{\text{no load}} - E_{\text{rated load}}}{E_{\text{rated load}}} \times 100\%,$$

where $E_{\text{rated load}}$ is the value of the load voltage at a certain specified current.

Procedure

(1) The switch *S* is kept open (Fig. 4.8) so that the circuit behaves as a full-wave rectifier with *L-C* filter. Under this condition the load current I_L is varied in regular steps and in each case the output d.c. voltage is noted.

(2) Knowing the value of the bleeder resistor R_B the current through the bleeder I_{BB} is found out. The total current which is the sum of I_L and I_{RB} is thus determined.

(3) The switch *S* is now kept on (Fig. 4.8) so that the circuit behaves as a full-wave rectifier with π-filter. The above two operations are then repeated.

Experimental results

[A] Full-wave rectifier using *L* filter (when the switch *S* is off).

Table V

No. of obs.	Load current (I_L) in mA	Output d.c. voltage ($E_{d.c.}$) in volt	Current through the bleeder resistance R_B $\left[I_{R_B} = \frac{E_{d.c.}}{R_B}\right]$ in mA	Total plate current $I = I_L + I_{R_B}$ in mA

Table VI

No. of obs.	Load current (I_L) in mA	Output d.c. voltage ($E_{d.c.}$) in volt	Current through the bleeder R_B $\left[I_{R_B} = \frac{E_{d.c.}}{R_B}\right]$ in mA	Total current $I = I_L + I_{R_B}$ in mA

[B] Full-wave rectifier using π-filter (when the switch S is on).

[C] A graph is drawn by plotting load current in milli-ampere along the X-axis and output d.c. voltage in volt along the Y-axis for the L-C and π-filter supplied. The nature of the graphs are shown in Fig. 4.10.

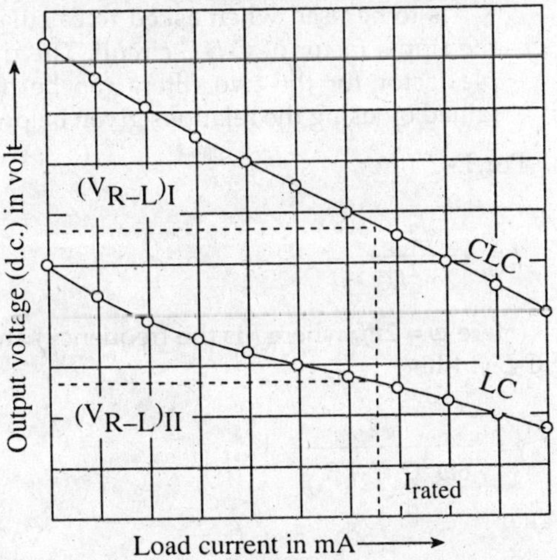

Fig. 4.10

[D] Calculation of voltage regulation.

(i) From load current and load voltage.

(Shunt regulation)

Table VII

Filter used	$E_{no\ load}$ in volt (from graph)	$E_{rated\ load}$ in volt (from graph) at current = ... ampere	Voltage regulation $= \frac{E_{N.L.} - E_{R.L.}}{E_{R.L}} \times 100\%$
L-C	...	...	...
C-L-C	...	...	...

N.B. : The percentage regulations are to be calculated at various rated currents for different operating points.

(i) From total plate current and load voltage.

(Line regulation)

Table VIII

Filter used	$E_{no\ load}$ in volt (from calculation)	$E_{rated\ load}$ in volt (from graph)	Voltage regulation $= \frac{E_{N.L} - E_{R.L}}{E_{R.L}} \times 100\%$
LC	...	...	...
CLC	...	...	...

Special note : Percentage of voltage regulation changes with change in the operating current as the value of $E_{rated\ load}$ varies, while $E_{no\ load}$ remains the same. It must be remembered that the no-load voltage is the voltage across the diode rectifier when the load is infinity, i.e., load is cut-off from the rectifier circuit. Evidently, the no-load voltage has a fixed value for a particular diode rectifier. *The designer has to adjust the no-load voltage to such a level that it does not exceed the power ratings of the diode as mentioned in the semiconductor manual.*

In order to study the variation of percentage regulation of voltage with operating current, the value of $E_{rated\ load}$ is noted for a particular operating current.

The value of percentage regulation of voltage is then calculated. The operating point is taken to another value, and again the value of percentage regulation is calculated.

Finally a graph is drawn with operating load current along X-axis and the calculated value of the percentage voltage regulation along Y-axis. *The nature of the graph is different for different types of filter used, and hence it is not reasonable to show the graph in the book. For a poor regulator, the nature of the graph is fully unsymmetrical,* while for a good regulator, the graph is more or less symmetrical. For a pure resistance as the load, the regulation is very poor.

[E] Data for percentage regulation characteristics.

Table IX

Operating current	$E_{no\ load}$	$E_{rated\ load}$	Percentage of voltage regulation	Remark about the nature of the graph
...	...	...	...	
...	...	...	...	
...	...	...	...	
etc.	etc.	etc.	etc.	

[F] Recordings for I_{dc} and V_{dc} for a half-wave rectifier. (To be done if directed by the instructor)

Note : The value of R_f is calculated from the forward characteristics of the diode supplied. In this case an a.c. signal of proper magnitude is applied to the input of the rectifier (instead of the d.c. source). The value of V_m is measured by an *a.c. voltmeter.* The value of R_L is known. Hence I_m, $I_{d.c.}$ and $V_{d.c.}$ can be calculated from Eqs. (4.5), (4.6) and (4.7). Finally $I_{d.c.}$ and $V_{d.c.}$ are measured by a d.c. milliammeter and a d.c. voltmeter. The theoretical and experimental values are compared.

Table X

Value of R_f in ohm	Value of V_m in volt	Value of R_L in ohm	Value of I_m in ampere	Value of $I_{d.c.}$ in ampere (theoretical)	Value of $V_{d.c.}$ in volt (theoretical)	Value of $I_{d.c.}$ in ampere (experimental)	Value of $V_{d.c.}$ in volt (experimental)	Remarks for the differences, if any
...	...	...	...	...	...	...	...	

[G] Recording of $I_{d.c.}$ and $V_{d.c.}$ for a full-wave rectifier (To be done if directed by the instructor).

Make similar table as above.

Discussion

(i) For no-load current in the circuit the output d.c. voltage is almost equal to the peak value of the input voltage. The latter is measured by an a.c. voltmeter by the voltage drop between the anode and the the centre tap of the transformer multiplied by $\sqrt{2}$.

(ii) A fraction of total plate current is passed through the bleeder resistor R_B connected parallel to the load R_L. This effective load thus consists of a parallel combination of R_L and R_B, and their equivalent resistance R_{eff} is to be used when asked to calculate the ripple factor of C-L-C circuit. The ripple factor for the two filters can be obtained by using the relations given below :

For L-C filter,

$$\gamma = \frac{\sqrt{2}}{12\omega^2 LC}.$$

(Here $\omega = 2\pi f$, where f is the frequency) and for C-L-C filter,

$$\gamma = \frac{\sqrt{2}}{8\omega^3 C_1 C_2 L R_{eff}}.$$

4.4 To Draw the Characteristics of a Zener Diode and Study of its use as a Voltage Regulator

Circuit Diagram

The basic principle of the circuit diagrams for studying the characteristics of a Zener diode under forward bias and reverse bias arrangements are shown in Figs. 4.11(a) and (b) respectively. In the figures, E represents a variable *d.c.* source whose positive and negative terminals are connected with the Zener diode Z in a manner as shown. R is a constant resistance, mA is the milliammeter and V the voltmeter connected across the Zener diode.

The circuit diagram for studying the Zener diode as a voltage regulator is shown in Fig 4.11. In the circuit a fixed load resistance R_L and a variable load resistance R'_L are there to use the Zener diode for the purpose.

Working Formula

If a curve is drawn by plotting voltage across the Zener diode as the abscissa and the corresponding current passing through it as the ordinate, then the curve so obtained is called the static characteristic curve of the Zener diode. In the reverse bias region when the voltage increases beyond a certain value, the Zener breakdown is said to occur. In this region the voltage across the Zener diode changes by a very small amount with the change of Zener current and is called the Zener breakdown voltage.

The current-voltage relationship for a semiconductor diode is given by

$$I = I_o(e^{V/\eta V_T} - 1),$$

where I is the current at a voltage V and I_o is the leakage current. The symbol V_T stands for the volt equivalent temperature and is equal to $T/11{,}600$ while η is a characteristic constant which is unity for germanium and *nearly* 2 for silicon at rated current.

For the rest part of the working formula, see after the special remarks.

Special Remarks

For a silicon Zener diode we have

$$I = I_o(e^{V/2V_T} - 1) \quad [\because \eta = 2]$$

For forward bias arrangement, let

$V = 0.1$ volt and $V_T = 26$ mV $= 0.026$ volt; then we have

$$I = I_o(e^{1.0/2\times 0.026} - 1)$$
$$= I_o(e^2 - 1).$$

So the current increases exponentially with a very small slope at low forward voltage. As V increases, the current is also found to increase following the above relation.

Again, when $V = 0$, we get $I = 0$.

For reverse bias arrangement since V is negative, we get $I - -I_o$; i.e., the current then becomes independent of the applied voltage.

Log I is plotted against V. If one extrapolates this graph to $V = 0$, the value of I_o can be obtained. Of course, the measured value of I_o is always greater than the theoretically predicted value.

Working formula (continued)

The dynamic or a.c. resistance (impedance) of a Zener diode is defined as,

$$Z_{a.c.} = \frac{\Delta V_z}{\Delta I_z}$$

where ΔV_Z and ΔI_Z are the change in voltage and current in the region of *specified test current.*

The static or d.c. resistance (or impedance) of a Zener diode is expressed as,

$$Z_{d.c.} = \frac{V_z}{I_z},$$

where V_z and I_z refer to voltage and current respectively at the specified test current.

N.B. : The breakdown voltage known as Zener voltage is specified by the manufacturers in a specification sheet. Generally, $Z_{a.c.}$ and $Z_{d.c.}$ are specified at the semiconductor manual at a test current I_{ZT}. This is on the linear portion of the reverse characteristic and corresponds to approximately one-quarter of the maximum power dissipation capability of the diode with power rating P_Z at the test current I_{ZT}. The relation is given below :

$$I_{ZT} = \frac{1P_z}{4V_z}.$$

For instance, if $P_z = 400$ mW and $V = 5$ V, then $I_{ZT} = 20$ mA.

Zener diode can be used (i) as a shunt regulator (or load regulator) and (ii) as a line regulator. To study the Zener diode as a load regulator, a variable load resistance is connected across the Zener diode. *The supply voltage is kept constant.* If a curve is drawn taking load current (I_L) as abscissa and corresponding voltage (V_L) across the Zener diode as ordinate, then the curve such obtained is termed as the load regulation curve.

The load regulation factor ρ can be defined as

$$\rho = \frac{V_{NL} - V_L}{V_L} \times 100\%,$$

where V_{NL} = voltage across the Zener diode with no load (i.e., $I_L = 0$)

and V_L = voltage across the Zener diode with a specified current, and hence it is termed as rated-load voltage.

For studying the line regulation, the load resistance is kept fixed while the supply (i.e., line) voltage or current is varied. Here also (V_L/I_L) curves are drawn for different load resistances.

N.B. : We have

$$\rho = \left(\frac{V_{NL}}{V_L} - 1\right) \times 100\%.$$

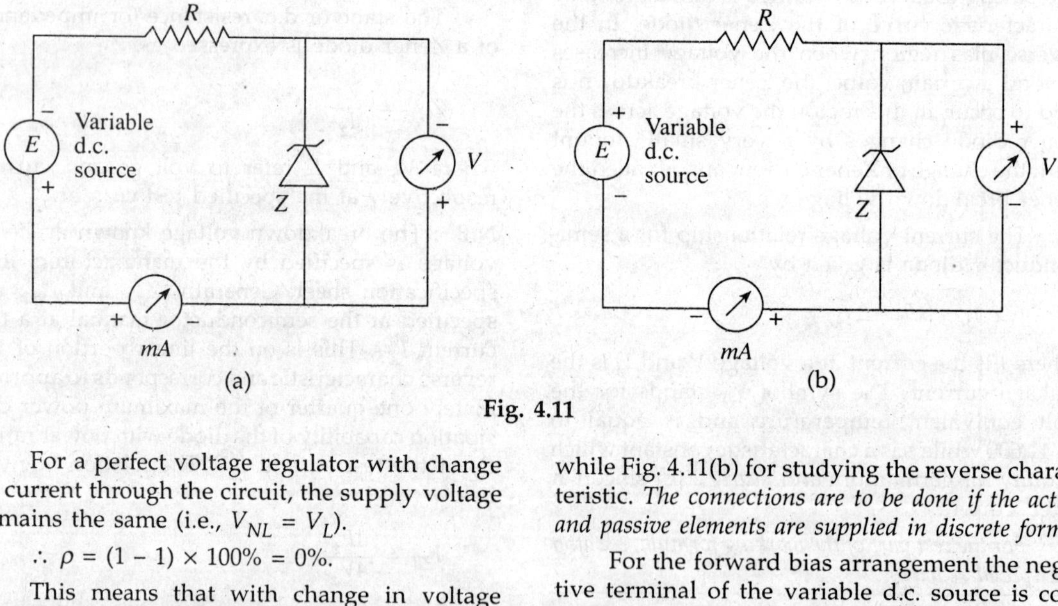

Fig. 4.11

For a perfect voltage regulator with change in current through the circuit, the supply voltage remains the same (i.e., $V_{NL} = V_L$).

$\therefore \rho = (1 - 1) \times 100\% = 0\%.$

This means that with change in voltage drop, i.e., ($V_{NL} - V_L$) with change in current through the Zener diode is negligible for a perfect voltage regulator. Hence the smaller the value of ρ, the greater is its stabilization capability. Of course, there is a certain regulation voltage spread up to which the Zener diode acts as a good regulator which is the characteristics of the Zener diode.

For a poor regulaton, the voltage drops with flow of current, i.e., ($V_{NL} - V_L$) has a certain positive value depending on the current range chosen. Hence, as the value of V_{NL}/V_L increases, the regulation becomes poorer.

Since in the working region of the Zener diode the current-voltage curve has a certain slope, the value of the regulation voltage depends on the choice of the operating point of the Zener diode.

Experimental procedure

(1) Specifications of the Zener diode from the semiconductor manual are to be noted, if possible. The main specifications for the purpose of the experiment are (i) power rating and (ii) maximum Zener current (d.c.), (iii) nominal test current, (iv) nominal Zener voltage. Otherwise, the values are to be collected from the instructor.

(2) Circuit connections (*showing the basic principles)* are to be drawn on the laboratory notebook and connections are made as shown in Fig. 4.11(a) for studying the forward characteristic, while Fig. 4.11(b) for studying the reverse characteristic. *The connections are to be done if the active and passive elements are supplied in discrete forms.*

For the forward bias arrangement the negative terminal of the variable d.c. source is connected through a resistance R to the negative terminal of the Zener diode. For the reverse bias arrangement, on the other hand, the polarity of the d.c. source and the meters are reversed, i.e., the positive terminal of the source is connected to the negative terminal of the Zener diode.

A milliammeter (mA) and a voltmeter (V) are connected in the circuit in the positions shown. Value of the smallest division of the different meters are noted.

(3) For drawing the forward characteristic, the applied voltage is kept *to a very low value so that the forward current can be recorded from the meter supplied.* The corresponding current is noted, *as discussed below* :

For each forward voltage the value of the current is to be noted avoiding parallax. This current is recorded as observation *I*. The forward voltage is changed slightly from the adjusted value and is again set to the previous forward voltage. Value of the corresponding current is cautiously noted under identical condition as before. This current is recorded as observation II. The mean of the two currents from observations I and II is the actual value of current for that forward voltage.

The forward voltage is increased in steps (depending on the calibration of the metre) and for each forward voltage the mean current as before is recorded.

Caution : Do not exceed the power rating.

(4) For studying the reverse characteristics the circuit connections are to be made (if closed box device is not supplied) as shown in Fig. 4.11(b). The reverse voltage is kept to the minimum value to obtain a small amount of current as permitted by the apparatus. The mean current for that reverse voltage is noted from observations I and II as discussed before.

The voltage is now increased in regular steps depending on the calibration of the apparatus and the maximum reverse Zener voltage and for each step the mean current is noted as before.

Caution : Do not exceed the power rating and maximum Zener current of the Zener diode supplied.

Study of Zener as a voltage regulator

(5) For studying voltage regulation the circuit connections as shown in Fig. 4.12 are made. The key K is to be closed.

For load regulation the supply voltage (i.e., line current) is kept fixed while the variable load R_L' is to be varied for obtaining the regulation. To use the Zener diode as a load regulator the combined load R_L and R_L' is kept to the maximum value by means of the rheostat. Supply voltage is now slowly increased so that the voltage across the Zener diode attains nearly the maximum value as obtained during the study of reverse characteristic. If now the load resistance is decreased, the change of Zener voltage becomes very small until the current through the Zener diode attains the knee-region of the reverse characteristic. This is the rough checking of the range up to which good regulation can be attained. This rough checking is to be made before recording the observations of regulation characteristics.

After this *rough checking,* the Zener voltage is set to the maximum reverse voltage as permitted by the apparatus (keeping the load resistance maximum). The load voltage and the corresponding load current are noted twice as observations I and II, and the mean load current for a particular load voltage is noted.

(6) The load resistance is now gradually decreased and in each case the mean load current is noted corresponding to a particular load voltage.

Caution : If the load resistance is made too low the current through the Zenner diode tends to the breakdown region and hence regulation becomes poor with decrease of load resistance.

(7) The line voltage or the line current is decreased by a small amount so that the initial maximum Zener voltage is slightly less than that adjusted in the previous operation. During this

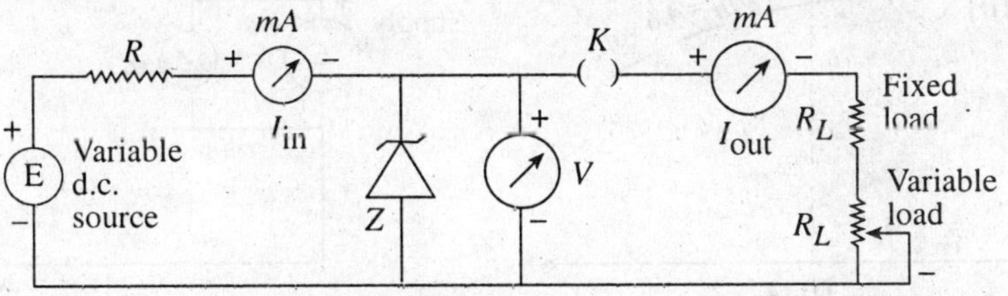

Fig. 4.12

adjustment the load voltage is kept at the maximum value. Now load is gradually decreased and for each load voltage the corresponding load current is noted.

(8) The above operation is repeated for other line voltage or line current.

(9) For each line voltage (or line current) the recorded values of voltages are plotted against the corresponding load currents. This gives the

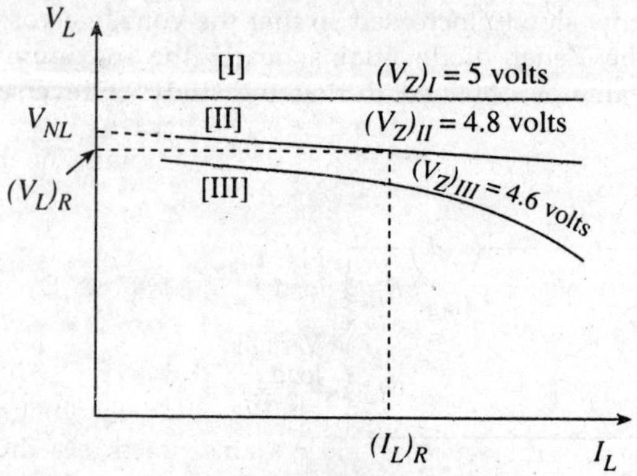

Fig. 4.13

load regulation characteristic *for that supply voltage*.

The load regulation characteristics for a Zener diode set at different initial *supply* voltage are shown in Fig. 4.13. Considering a particular regulation characteristic [say, $(V_z)_{II}$ = 4.8 volts], the value of the rated load voltage [i.e., $(V_L)_R$] for a particular rated current [say, $(I_L)_R$] is noted from the regulation characteristic (Fig. 4.13). Also the value of V_{NL} can be obtained either from the extrapolation or from the experimental data by opening the key *K* in the circuit (Fig. 4.12). Hence, the percentage regulation can be calculated using the theoretical relation. This is done by choosing *different operating rated currents on the same regulation characteristic.*

Also, the same procedure may be done using other *load-regulation characteristics*. In all cases, the percentage regulation of voltage is calculated for different rated currents.

(10) For studying the line regulation, the load is kept fixed but the supply voltage (i.e., line voltage) is changed. Here also for different fixed load resistances V_L–I_L data are noted for different initial line currents, and curves are drawn. The initial line current is to be chosen to such a value so that its change does not lead the Zener voltage to the knee-region.

Important points

(1) *Alternative circuit for studying the Zener characteristics.*

One of the methods for studying the reverse characteristic of Zener diode is shown in Fig. 4.11(b). Here, the voltages and currents can be directly read from the meters supplied. Sometimes the circuit shown in Fig. 4.14 may also be used.

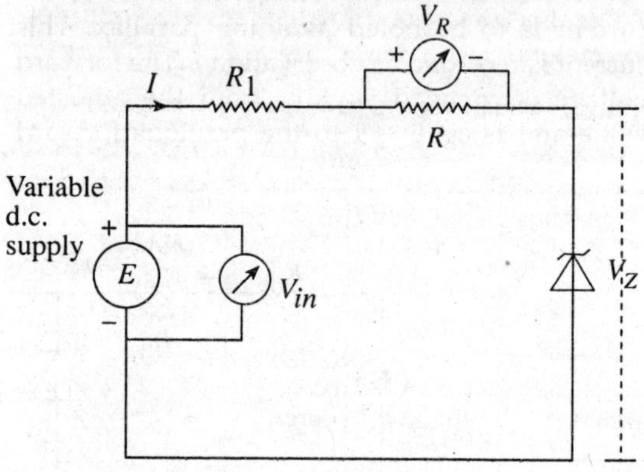

Fig. 4.14

Here, V_{in} can be measured, also, the voltage across *R* (i.e., V_R) can also be measured.

Hence, the current *I* flowing through the Zener diode can be calculated by using the relation, $I = \frac{V_R}{R}$. Also, the value of Zener voltage (i.e., V_Z) can be calculated by using the relation $V_Z = V_{in} - V_R$. Hence, the reverse characteristic can be drawn. *Similarly, the forward characteristic can also be drawn by reversing the Zener terminals.*

(2) *Works with a regulated power supply*

The power supply unit used for studying the semiconductor characteristics consists of the combinations of suitable step-down transformers and *proper* I.C.'s. Finally, the step-down a.c. voltages are converted to d.c. voltages with suitable rectifier unit. Hence, the power supply unit should be connected to a.c. mains. *It should never be connected to d.c. mains because it will damage the supply unit.*

After connecting the input terminals of the power supply unit to the a.c. mains, the connections are made as shown in Figs. 4.11(a) for forward characteristic, and 4.11(b) for reverse characteristic. Sometimes, some of the connections are made inside the black-box, while some connections are to be done by the experimentor. The basic principle is the same as shown in Figs. 4.11(a) and 4.11(b). *Before switching on the power supply unit, the circuit should be approved by the instructor.*

Experimental Results

[A] Evaluations of one smallest division of different meters.

N.B. : *Data are shown only for illustrations.*

Table I

Apparatus used	Total value for the full-scale division	Total no. of full-scale division	Value of one smallest scale division	Zerro error, if any
Milli-ammeter	50 milli-ampere	50	1 mA	
Voltmmeter	10 volt	50	0.2 volt	
Voltmeter (multimeter)	3 volt	60	0.35 volt	

[B] Zener voltage (V_Z) and the corresponding Zener current (I_Z) reading for forward biasing.

Supply voltage = ... volt, (15 volts, say)

Table II

Zener voltage (V_Z) in volt	Zener current (I_Z) in mA Obs. I	Obs. II	Mean
...	...	...	...
...	...	...	...
...	...	...	...
etc.	etc.	etc.	etc.

[C] Zener voltage and the corresponding Zener current reading for reverse biasing.

(Data are given only for illustrations for a 4.7 V Zener diode.)

N.B. : For a high quality Zener diode, the voltge across the Zener diode remains practically the same for a wide variation of Zener current in the Zener region.

Table III

Zener voltage (V_Z) in volt	Zener current (I_Z) in mA Obs. I	Obs. II	Mean	Region of the graph
3.5	...	...	0.5	Knee region
3.6	...	...	1.0	
3.7	...	...	1.5	
3.8	...	...	2.0	
3.9	...	...	2.5	Starting of Zener region
4.0	...	...	3.0	
4.1	...	...	4.0	
4.2	...	...	5.0	
4.3	...	...	7.0	
4.4	...	...	10.0	
4.5	...	...	14.0	
4.6	...	...	18.0	
4.6	...	...	21.0	Zener region (suitable for voltage regulation)
4.6	...	...	28.0	
4.7	...	...	32.0	
4.7	...	...	35.0	
4.7	...	...	38.0	
⋮	⋮	⋮	⋮	
etc.	etc.	etc.	etc.	

. *Alternative* table if Fig. 4.13 is used.

Table III(A)

Input voltage V_{in} in volt	Voltage across R in volt = V_R volt	Voltage across Zener = V_R volt = $(V_{in} - V_R)$ volt	$I = \frac{V_R}{R}$ in ampere	I in mA
etc.	etc.	etc.	etc.	etc.

Calculation of Zener dynamic impedance

For a non-linear region of reverse-characteristic curve, (say, at a point *P*) in Fig. 4.15, a tangent is drawn as shown by *mp*. Finally a triangle *mnp* is constructed. Hence, $Z_D = \frac{mn}{np} \times 10^3$ ohm. For the linear region, say, at an operating current i_o given by $i_o = \frac{i_1 + i_2}{2}$, a triangle is drawn (say, *abc*) from which Z_D is again calculated. In this way Z_D is calculated at different operating currents from the reverse characteristic.

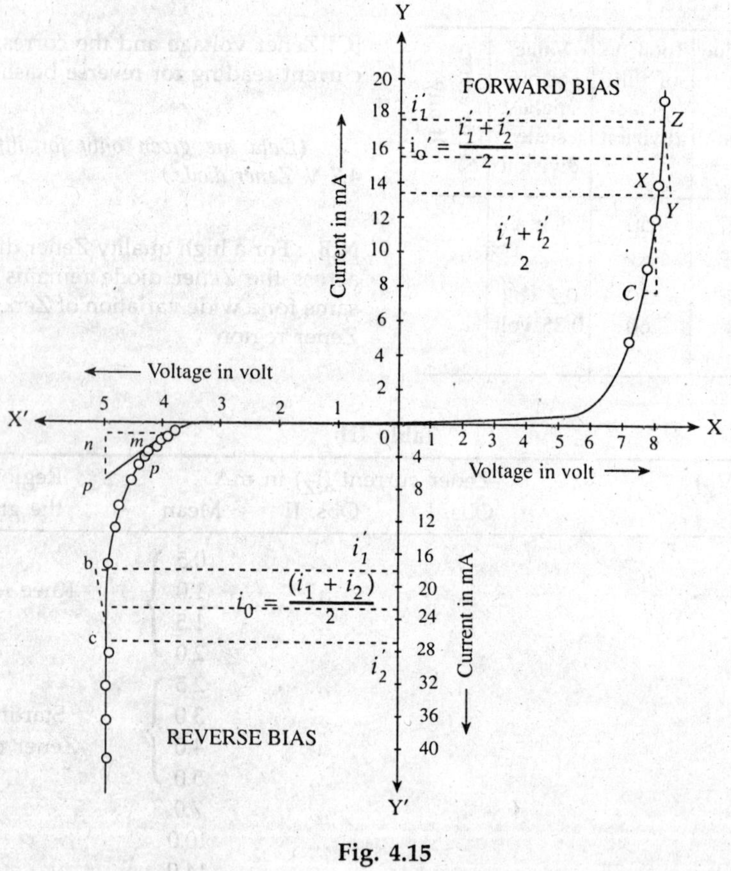

Fig. 4.15

[D] *To draw* (V_Z/I_Z) *curve*

Zener voltage (V_Z) is plotted along the *positive X*–axis and the Zener current (I_Z) along the *positive Y*-axis for drawing the forward characteristic. For drawing the reverse characteristic, reverse voltage is plotted along the *negative X*-axis, while the corresponding current along the *negative Y*-axis. The graph is shown in Fig. 4.15.

Similarly, from the forward characteristic (Fig. 4.15), different triangles *efg*, *xyz*, etc. are considered at different operating currents. Finally Z_D is calculated at different operating currents.

N.B. : The rate of change of the value of the dynamic impedance is a relative measure of the circuit as a voltage regulator. If with change in current, the value of ΔV is zero (or nearly zero), the device can be treated as a good voltage regulator. For example, consider that a freshly charged battery under normal service condition is taken. Let the current change through the resistance in the battery circuit but the voltage across the battery remains the same. Hence, though ΔI changes, yet ΔV does not change. Therefore, such battery is a good voltage regulator. On this analogy, for a larger value of current, with a Zener diode, in the reverse characteristic, the Zener diode acts as a good voltage regulator.

This is because, $Z_D \rightarrow 0$ as current increases. Sometimes, the forward characteristic curve is also used as a voltage regulator for smaller current with the same Zener diode. In this case also, the value of $Z_D \rightarrow 0$ is the criterion for its use as a voltage regulator. But, it is rarely done. A graph of Z_D versus operating current may be drawn using reverse characteristic. The variation of Z_D with operating current with change of temperature may also be studied.

Table V

(For load regulation)

[E] Load voltage (V_L) and the corresponding load current (I_L) reading for the study of the Zener diode as a voltage regulator (see Fig. 4.13).

Zener voltage initially set at 5.0 volts. Supply or line voltage set at = 12 volts (say)		Zener voltage set at 4.8 volts. Supply or line voltage set at ...... volt	
Current through the load (I_L) in mA	Voltage across the load (V_L) in volt	Current through the load (I_L) in mA	Voltage across the load (V_L) in volt
1.5	3.8	1.5	4.0
2.0	3.8	2.0	4.0
3.0	3.65	3.0	3.9
4.0	3.5	4.0	3.8
5.0	⋮	5.0	⋮
⋮	⋮	⋮	⋮
14.0	1.4	14	1.8

Table IV

Data are taken from the forward and reverse characteristics curves.

Curve employed	Region of the curve	Voltage change from ...volt to ...volt	Current change from... mA to ... mA	Voltage change (=ΔV) in volt	Operating Zener current in mA	Change in Zener current (= ΔI) in mA	Zener dynamic impedance $Z_D = \frac{\Delta V}{\Delta I} \times 10^3$ ohm	Remarks
Reverse characteristic	Non-linear	V_2 to V_1	I_2 to I_1 (say)	$V_2 - V_1$	$\frac{I_2+I_1}{2}$	$I_2 - I_1$		At non-linear region, the value of Z_D is large
		= ... etc.	= ... etc.	= ... etc.	= ... etc.	= ... etc.	= ... etc.	
	Linear	etc.	etc.	etc.	etc.	etc.		For linear region, with increase of Zener current Z_D tends to zero
Forward characteristic	Linear	... ... etc.	... ... etc.	... ... etc.	... ... etc.	... ... etc.	etc.	For linear region, with increase of current Z_D tends to zero

N.B. : The region up to which V_L remains nearly the same is the region for good voltage regulation.

[F] To draw (V_L/I_L) graph and hence to find the percentage regulation of voltage (for load regulation).

I_L is plotted along the positive X-axis and V_L along the positive Y-axis for each set. The nature of the curves is as shown in Fig. 4.13. From the graph for a specific load current $(I_L)_2$ value of $(V_L)_R$ is noted and the following table is made.

Note : *Make similar table for line regulation.*

Table VI

(Data are given only for illustrations)

Supply voltage in volt	$V_{\text{no-load}}$ (V_{NL}) in volt	Correspon-ding V_{LR} for $(I_L)_R$ in volt	$V_{NL}-V_L$ in volt	Percentage regulation $\rho = \frac{V_{NL}-V_L}{V_L} \times 100\%$
12	5.0	5.0	...	...
12	...	(for I_{LR}=... mA)	...	...
12	...		...	...
11	4.8	4.7	...	...
11	...	...	...	...
11	...	...	...	...
etc.	etc.	etc.	etc.	etc.

Discussion

(i) Before starting the experiment, the forward and reverse biasing condition for the Zener diode supplied should be checked. If the polarities cannot be identified properly, then after connecting the Zener diode in the circuit, the voltage is gradually increased and the maximum voltage across the Zener diode is noted (say, V_1 volt). Next the terminals of the Zener diode connections are changed and again the supply voltage is slowly increased. Let now the maximum voltage across the Zener diode be V_2 volt. If $V_2 > V_1$, then the second arrangement of the Zener diode gives the reversed biasing condition.

(ii) Special diodes with sharp (vertical) change in current-votage relationship when operated in reverse-bias condition are called Zener diode. Such diodes which are designed with adequate power dissipation capacities to operate in the breakdown region can be used as a constant voltage device.

(iii) The value of $\frac{\Delta V_z}{\Delta I_z}$ is zero in ideal cases for a perfect voltage regulator and it is exhibited by a Zener diode at high reverse current region.

(iv) Zener diode is specially manufactured in the breakdown region. Practically all Zener diodes are made of silicon because of the superior thermal qualities, i.e., they exhibit high voltage stabilization even if function temperature is high enough.

(v) The manufacturers specify the value of the breakdown voltage known as Zener voltage (V_z). This is on the linear portion of the reverse characteristic and corresponds to approximately one quarter of the maximum power dissipation capability of the diode. As for example, for a 5-V Zener diode if the maximum power rating is 400 mW, then the maximum safe current is 80 mA. For such diode V_Z is taken at 20 mA.

(vi) Both reliability and performance of a Zener diode can be improved considerably with the oxide-passivation process. Zener diodes may be used even up to 200 °C.

(vii) Some Zener diode are available with Zener voltage below 2 V. Usually diodes are then used in the forward direction.

(viii) For a high voltage reference, it is preferable to use two or more Zener diodes in series rather than using only one. Because in that case the combination allows higher dissipation, lower dynamic resistance and also lower temperature coefficient.

(ix) The Zener voltage regulators are used in general for applications of low power without going into much complexity.

(x) In order to measure the Zener voltage, the voltmeter to be used must be highly sensitive. If a voltmeter has a high input impedance, then practically a very small current is lost in the voltmeter circuit and it can be ignored. Hence the resistance of the voltmeter must be high enough.

Extra work for Zener diode

(1) Place a hot filament lamp at a distance of about 3 cm from the Zener diode and draw the

forward and reverse characteristics. Remark for the difference in readings.

(2) Some of the transistors like BO 147, BO 148 and BO 107 possess the breakdown voltage near about 8 volts between their base-emitter junction. Hence they can be used as V-Zener. For such transistors, the base and the collector terminals are shorted by a connecting wire. They behave as a Zener diode when the emitter and base terminals are considered. Perform the same experiment with such converted Zener diode.

4.5 To Draw the Static Characteristics of a P-N-P Transistor under *C-B* Configuration

Circuit diagram

The circuit arrangement for studying the characteristics under common-base configuration is shown in Fig. 4.16. Note that for the common-base configuration two milliammeters are connected to the emitter and the collector terminals as shown in the figure.

$$\alpha = \left.\frac{\Delta I_C}{\Delta I_E}\right|_{V_{CB}} . \tag{4.11}$$

Note 1 : In a common-emitter circuit, the input signal is applied to the base. This gives another important control characteristic designated by β, where β is defined as the ratio of the change in collector current (δI_c) to the change in base current (δI_B), when the collector voltage (V_{CE}) is kept constant.

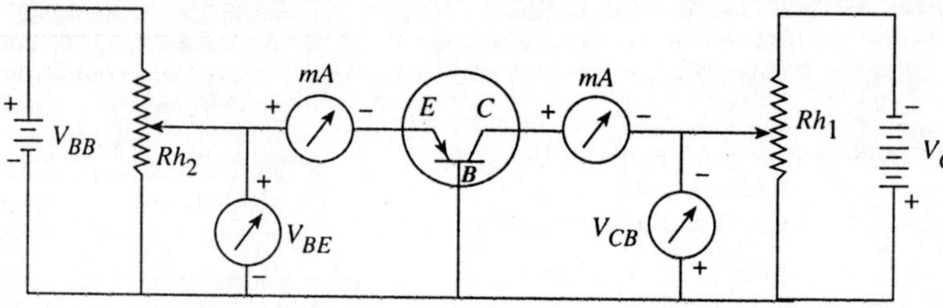

Fig. 4.16

Working Formula

A *P-N-P* transistor is a combination of two *P-N* junctions. It is formed by sandwiching a thin *N*-layer between two *P* layers. Of the two *P-N* junctions one is forward biased while the other is reverse biased. The *N*-layer in the middle is the base, the forward-biased *P*-layer on the left is the emitter while the reverse-biased *P*-layer on the right is the collector.

In a common-base circuit the collector current is controlled by the variation of emitter current. This gives one important control characteristic of a transistor designated by *α*, where *α* is defined as the ratio of the change in collector current (δI_C) to the change in emitter current (δI_E), when the collector voltage (V_{CB}) is maintained at a constant value.

The *a.c. current gain* (α) is defined as

$$\alpha = \left(\frac{\delta I_C}{\delta I_E}\right)_{V_{CB}} \quad \text{or, } \alpha = \left.\frac{\delta I_C}{\delta I_E}\right|_{V_{CB}} .$$

Since the collector and the emitter current changes are taken to *be finite and measurable* for the determination of α, hence the above expression is written as

Mathematically, the *a.c. current gain* for common-emitter configuration is expressed as

$$\beta = \left.\frac{\delta I_C}{\delta I_B}\right|_{V_{CE}} . \tag{4.12}$$

Since the experimental results are concerned with finite and measurable quantities, it is customary to write it as

$$\beta = \left.\frac{\Delta I_C}{\Delta I_B}\right|_{V_{CE}} . \tag{4.13}$$

For constant V_{CB} and constant V_{CE} we may write,

$$\beta = \frac{\alpha}{1 \pm \alpha} \text{ or, } \alpha = \frac{\beta}{1+\beta} \tag{4.14}$$

Experimental Procedure

(1) The range and the value of one smallest division of the meters supplied are noted. The transistor number and other necessary specifications are also noted from the semiconductor mannual.

(2) Before performing the experiment, the given two milliammeters are compared to see whether they are calibrated with the same scale.

(3) Circuit is now made as shown in the common-base configuration to get the data of collector current for different collector-to-base voltage (V_{CB}), when the emitter current is kept fixed. For each of value V_{CB} the collector current (I_C) is observed twice and the mean is taken.

(4) For different fixed values of emitter current (I_B), the operations as indicated in procedure no. (3) are repeated.

(5) A graph is now drawn showing the variation of collector current with different collector-to-base voltage. Another graph is to be drawn to show the variation of collector current with emitter current for a fixed V_{CB}. From the graph the current gain α is calculated which is the ratio of increase in collector current to a small change in emitter current for a given collector-to-base voltage.

Experimental Results

The data given are only for illustrations.

[A] *To find the ranges and smallest division of meters used.*

Table I

Types of the meters	Range	Total no. of divisions	Value of smallest division
Voltmeter	0–10 volts	...	...
Milliammeter (emitter ckt.)	0–10 mA	...	...
Milliammeter (collector ckt.)	0–100 mA	...	...

[B] *Comparison of the two milliammeters.*

Table II

No. of obs.	Readings in one milliammeter in mA	Corresponding readings in the other milliammeter in mA	Remarks
1	0.5	0.5	Scales of the milliammeters are same.
2	1.0	1.0	
3	1.5	1.5	
⋮	⋮	⋮	
etc.	etc.	etc.	

[C] *Data for drawing transistor characteristics with common-base configuration.*

Table III

Emitter current (I_E) in mA	Collector to base voltage in volt (V_{CB})	Corresponding collector current (I_C) in mA		
		Obs. I	Obs. II	Mean
0.5	1	0.48	0.48	0.48
	2	0.48	0.48	0.48
	3	0.48	0.48	0.48
	4	0.48	0.48	0.48
	5	0.48	0.48	0.48
	6	0.48	0.48	0.48
	7	0.48	0.48	0.48
1	etc.	etc.	etc.	etc.
⋮	⋮	⋮	⋮	⋮
5	etc.	etc.	etc.	etc.

[C_1] **To draw (V_{CB}/I_C) graph.**

A graph is drawn with V_{CB} (in volt) along *X*-axis and I_C (in mA) along *Y*-axis for fixed I_E.

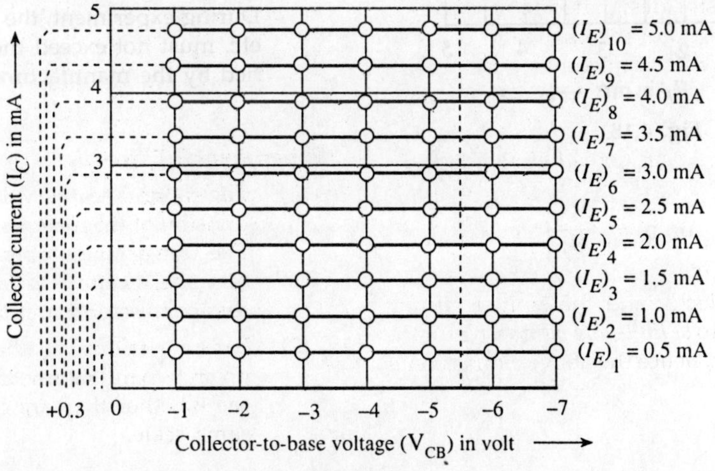

Fig. 4.17

The value of I_E is changed to different values, depending on the nature of the transistor specifications, and in each case V_{CB} versus I_C graph is drawn. The nature of the graph is shown in Fig. 4.17.

[D] *Data for drawing* I_C/I_E *graph.*

Collector-to-base voltage = 6 volts (say)

Table IV

Emitter current (I_E) in mA	0.5	1	1.5	2	2.5	3	3.5	4	4.5	5
Corresponding collector current (I_C) in mA	0.48	0.95	1.48	1.95	2.48	2.98	3.45	3.9	4.45	4.9

To draw I_C/I_E *curve.*

A graph is drawn by plotting I_E (in mA) along X-axis and I_C (in mA) along Y-axis for a fixed value of V_{CB}. The nature of the graph is shown in Fig. 4.18.

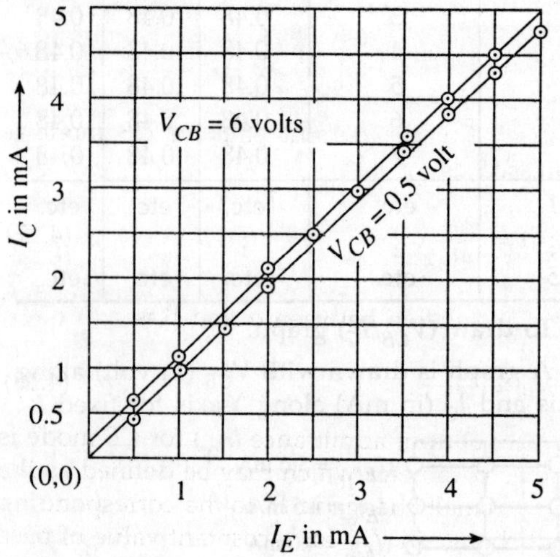

Fig. 4.18

The value of V_{CB} is adjusted to another value and the above procedure is performed to get another set of values of I_C and I_E. A separate graph is again drawn for this value of V_{CB}.

[E] *Determination of* α

Table V

Change in I_C, i.e., δI_C in mA	Change in I_E, i.e., δI_E in mA	Mean operating current in mA	$\alpha = \left.\frac{\delta I_C}{\delta I_E}\right\|_{V_{CB}}$
3.9 (say)	4.0	...	$\left.\frac{3.9}{4.0}\right\|_{6\text{ V}} = 0.93$
etc.	etc.	etc.	etc.

Discussion

(1) For a common base configuration, the collector current is found to be almost independent of the collector voltage for a fixed emitted current.

(2) From the relation $\alpha = \beta/(1+\beta)$ it is clear that as α approaches towards unity, β becomes increasingly larger. So during experiment α must be measured very accurately to avoid error in β.

(3) From the semiconductor manual, the maximum values of different variables (i.e., voltage, current, etc.) are to be noted. During experiment, the values of I_B, V_{CE}, etc. must not exceed the values as specified by the manufacturer.

4.6 To Draw the Static Characteristics of a Transistor (under CE mode) and to Calculate h_{fe}, h_{oe}, etc.

Circuit Diagram

The basic principle of the circuit diagram for this purpose is shown in Fig. 4.19 using *N-P-N* transistor. The indices are as follows :

Q→Transistor (*E* : emitter, *B* : base; *C* : collector); *μA*→Microammeter; *mA*→Milliammeter; Rh_1, Rh_2→Rheostats; V_{BE}, V_{CE}→Volmeters.

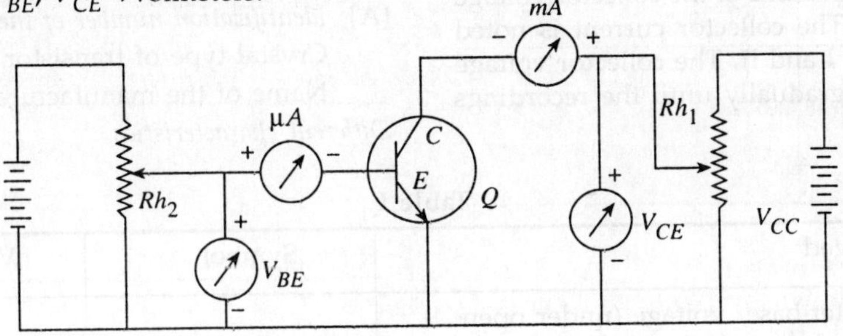

Fig. 4.19

The emitter of the transistor is forward biased and the collector is reverse biased with the help of external sources. The sources may be batteries or a regulated power supply unit.

N.B. : In some circuits, the voltmeter marked by V_{BE} is not supplied. This voltmeter is connected in the curcuit only when the drawing of input characteristic curves is necessary.

Working Formula

The collector current versus collector voltage curves are generally known as the *output* characteristic curves of a transistor. To draw the characteristic curves of a transistor in the *CE*-mode, the base current (I_B) is kept to a constant value and the collector currents are noted by varying the collector voltages. The operation is repeated for different base currents and then the curves are drawn by plotting collector current versus collector voltage.

(i) The a.c. current gain for a *CE* configuration for small signal mode is denoted by β, which is the ratio of small change in collector current to the small change in base current when collector-emitter voltage is maintained at a constant value.

Mathematically, the a.c. current gain for a common-emitter configuration is expressed as

$$h_{fe} = \beta = \left(\frac{\partial I_C}{\partial I_B}\right)_{V_{CE}}, \text{ or, } \beta = \left.\frac{\delta I_C}{\delta I_B}\right|_{V_{CE}}$$

Since the experimental results are concerned with finite and measurable quantities, hence it is customary to write it as

$$h_{fe} = \beta = \left.\frac{\Delta I_C}{\Delta I_B}\right|_{V_{CE}} \quad (4.15)$$

It is a dimensionless quantity.

(ii) The d.c. current gain for *CE*-mode is given by

$$\beta_{d.c.} = h_{FE} = \left.\frac{I_C}{I_B}\right|_{V_{CE}} \quad (4.16)$$

(iii) The a.c. current gain for *CB* mode is expressed as

$$\alpha = \left.\frac{\Delta I_C}{\Delta I_E}\right|_{V_{CB}} \quad (4.17)$$

The relation between α and β is

$$\alpha = \frac{\beta}{\beta + 1}. \quad (4.18)$$

(iv) Output admittance (h_{oe}) for *CE* mode is another parameter which may be defined as the ratio of small change in I_C to the corresponding small change in V_{CE} for a cosntant value of base current (I_B), i.e.

$$h_{oe} = \left.\frac{\Delta I_C}{\Delta V_{CE}}\right|_{I_B}. \quad (4.19)$$

Experimental Procedure

(1) The circuit connections are made as shown in Fig. 4.19. Before switching on the circuit, it should be approved by the instructor.

(2) The type number and different characteristics of the transistor supplied are noted from the manual (if available).

(3) The value of the smallest division of different meters are evaluated.

N.B. : Zero (or initial) error, if any, for any one of the meters supplied are minimised by adjusting the base-screw of the pointer, (if possible). Otherwise, proper correction should be done.

(4) For a particular value of the base current (say, at 10 μA) the collector voltage is increased in regular steps (e.g., V_{CE} = 1, 2, 3 volts, etc.) and the corresponding collector current at each value of the collector voltage is recorded. The collector current is noted twice as obs. I and II. The collector voltage is increased gradually until the recordings for the active region of the curve may be taken. If possible, the recording for the *saturation region* should also be taken for lower values of V_{CE}.

(5) The base current is fixed to other values and similar procedure as described in operation (4) is repeated.

(6) V_{CE} is kept to a fixed value and the variations of I_B and V_{BE} are noted. This is done for different values of V_{CE}.

(7) V_{CE} is kept to a fixed value and the variations of I_B and I_C are noted.

Experimental Results

N.B. : Do not exceed the maximum power rating during recordings. Data shown in different tables should never be taken as the typical. These are shown only for illustrations.

[A] *Identification number of the transistor* = ...
Crystal type of transistor = ...
Name of the manufacturer = ...

Different characteristics

Table I

Quantities involved	Symbol	Value
Maximum collecter-base voltage (under open emitter condition)	$-V_{CBO}$	...V
Maximum collector-emitter voltage (under open base condition)	$-V_{ICEO}$	...V
Maximum emitter-base voltage (under open collector condition)	$-V_{EBO}$	...V
Maximum collector current	$(-I_C)_{max}$	...A
Total power dissipation at ambient temp. 25 °C	P_{tot}	...W
d.c. current gain at ambient temp. 25 °C	h_{FE}	...

N.B. : This should be filled up if the semiconductor manual is available. This may also be filled up as directed by the instructor.

N.B. : This table is important in the sense that during recording the values V_B, I_B, V_C, I_C, V_E, I_E, etc. should be kept to such a value that the transistor is not damaged.

[B] *Evaluation of the smallest division of different meters.*

Table II

Instrument employed	Total value for full-scale deflection	Total number of divisions for full-scale deflection	Value of one smallest scale division	Zero (or initial) error, if any
Voltmeter	10 volts	05	0.2 volt	...
Microammeter	500 μA	50	10 μA	...
Milliammeter	30 mA	60	0.5 mA	...

[C] *Collector current and collector voltage reading for different constant values of base current.*

Table III

Base current (I_B) in μA	Collector voltage (V_{CE}) in volt	Collector current (I_C) in mA		
		Obs. I	Obs. II	mean
	1	...	...	...
	2	...	...	...
	3	...	...	...
20	4	...	...	...
(say)	5	...	...	...
	6	...	...	...
	7	...	...	...
	8	...	...	...
	9	...	...	...
	etc.	etc.	etc.	etc.

N.B. : The maximum collector voltage will be set at a value as directed by the instructor.

[The above operations are to be repeated for other fixed base currents and similar tables as shown in table III are to be made.]

[D] *To draw (I_C-V_{CE}) curve*

For each constant value of I_B (in μA) different V_{CR} (in volt) are plotted along the X-axis and the corresponding values of I_C (in mA) along the Y-axis using Table III. The nature of the curves obtained is as shown in Fig. 4.20.

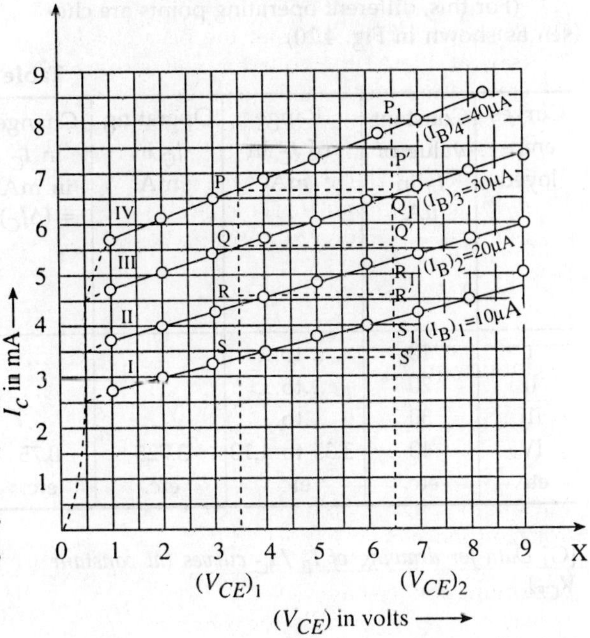

Fig. 4.20

[E] *To find the a.c. current gain, or a.c. β (i.e., h_{fe}) from graph.*

Table IV

Constant V_{CE} in volt at which β is to be determined	Curves employed	Range of I_C in mA	Operating I_C in mA	Change in I_C, i.e., $-(\Delta I_C)$ in mA	Range of base current I_B in μA	Change of base current in μA $-(\Delta I_B)$	Value of current gain $\beta = \left.\frac{\Delta I_C}{\Delta I_B}\right\|_{V_{CE}}$	Value of $\alpha = \frac{\beta}{\beta+1}$
	I & II	...to...	...	...	...to...	...	...	...
3.5	II & III	...to...	...	...	...to...	...	...	...
$(V_{CE})_1$	III & IV	5.55 to 6.70	6.125	1.15	60 to 70	10	115	0.9913
in	I & III	4.45 to 5.55	5.0	1.10	50 to 60	10	110	0.9910
Fig. 4.20	1 & IV	3.35 to 4.45	3.90	1.10	40 to 50	10	110	0.9910
	...			...	...	...	...	...
6.5	etc.	etc.	etc.	etc.	etc.	etc.	etc.	etc.

N.B. : ΔI_C and ΔI_B are to be converted to the same unit before calculation of h_{fe}. Different operating points are considered to calculate the hybrid parameters as shown in Fig. 4.20.

[F] *To find the output admittance (h_{oe}) from graph.*

(For this, different operating points are chosen as shown in Fig. 4.20)

Taking different operating points the hybrid parameter h_{oe} are determined. The mean value of it is considered. The value of h_{oe} can also be determined from the graph as shown in Fig. 4.20. It should be noted that these parameters are quiescent point dependent.

Table V

Curves employed	Constant value of I_B in μA	Range of I_C in mA	Operating I_C in mA	Change in I_C in mA $= (\Delta I_C)$	Range V_{CE} in volt	Operating V_{CE} in volt	Change in V_{CE} in volt (ΔV_{CE})	Output admittance $h_{oe} = \left(\frac{\Delta I_C}{\Delta V_{CE}}\right)_{I_B} \times 10^{-3}$ in mho
I	10	...to...	...	...	...to...	...	...	...
II	20	...to...	...	...	...to...	...	...	...
III	30	...to...	...	...	...to...	...	...	...
IV	40	3.35 to 4.10	3.725	0.75	3.5 to 6.5	5	3	...
etc.	etc.	etc.	etc.	etc.	etc.	etc.	etc.	etc.

[G] *Data for drawing of I_B / I_C curves (at constant V_{CE}).*

N.B. : This table is to be made if directed by the instructor.

Table VI

Constant value of V_{CE} in volt	Base current (I_B) in μA	Collector current (I_C) in mA		
		Obs. I	Obs. II	Mean
	10	...	...	...
	20	...	...	...
0.5	30	...	...	...
	40	...	...	...
	etc.	etc.	etc.	etc.

[Repeat the above type of observations for other higher constant values of V_{CE} (say, 4 V, 6 V, etc.)]

[H] *To draw (I_B / I_C) curve*

A graph is drawn by plotting I_B in μA along X-axis and the corresponding value of I_C in mA along Y-axis for fixed V_{CE} (in volt). The nature of the curves obtained using Table VI is shown in Fig. 4.21.

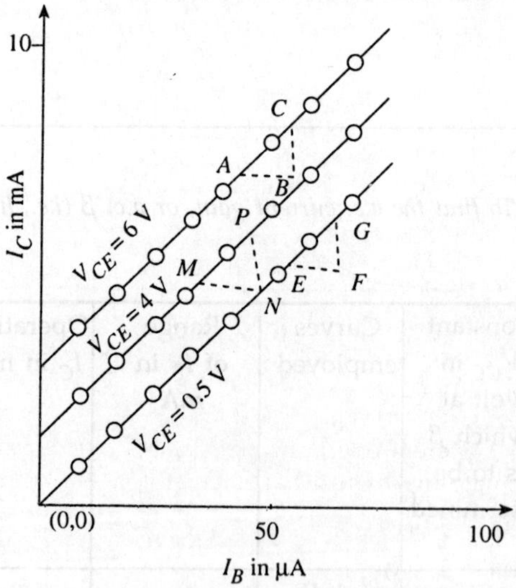

Fig. 4.21

[I] *Data for drawing of V_{BE} / I_B curve at constant V_{CE}.*

[These recordings are to be done if a voltmeter is supplied to measure V_{BE}. This part of the experiment should be done if directed by the instructor.]

Table VII

Constant value of V_{CE} in volt	I_B in mA	V_{BE} in volt		
		Obs. I	Obs. II	Mean
	10	...	...	...
	20	...	...	...
2	30	...	...	...
	40	...	...	...
	etc.	etc.	etc.	etc.

N.B. : This procedure is followed for different values of V_{CE}. Do not exceed the maximum power rating.

[J] *To draw* (V_{BE}/I_B) *curve.*

A graph is drawn by plotting I_B in μA along Y-axis and corresponding V_{BE} along X-axis for fixed V_{CE}. The nature of the curves obtained using Table VII is shown in Fig. 4.22.

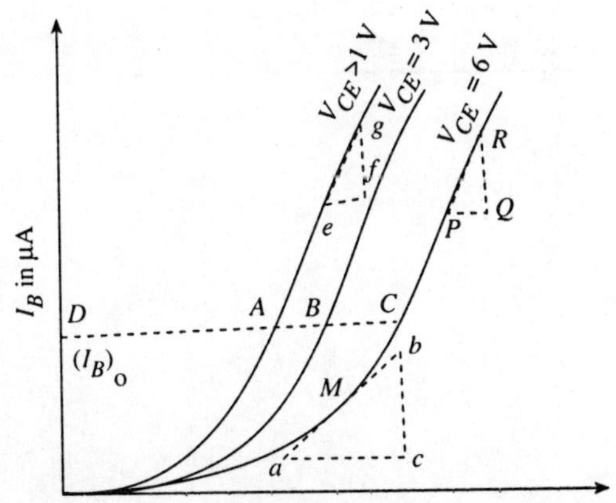

Fig. 4.22

Different operating points are considered to calculate the hybrid parameter (h_{ie}). We know

$$h_{ie} = \left.\frac{\Delta V_{BE}}{\Delta I_B}\right|_{V_{CE}}.$$

For non-linear region, tangents are drawn, while for the linear region, different triangles are drawn to calculate the value of h_{ie} at different operating points.

Discussion

(i) The current gain for the common-emitter configuration is very large in comparison to that for the common-base configuration.

(ii) The operating points must be recorded since the values of hybrid parameters change with change of operating points in the active region for the same transistor.

(iii) The output characteristics of a transistor are not parallel to each other due to slight variation of α with V_{CB}. But because of early effect, a slight change in α (say, 5%), β may vary 34% or so. Output characteristics for larger values of collector current are crowded together for constant base current increments.

(iv) For calculating the mean value of the current, observation I and observation II should be taken under identical equilibrium conditions.

Extra work on *P-N-P* transistor

To be performed if directed by the instructor.

(1) From the output characteristics (i.e., I_C-V_{CE} curves) calculate the variation of (i) a.c. current gain (h_{fe}) and (ii) d.c. current gain h_{FE}, when the operating collector current varies from 10 mA to 50 mA for the transistor supplied. Comment on the result.

Also, show graphically the variation of h_{FE} and h_{fe} with I_C and V_{CE}.

(2) Place a hot electric bulb (60 W, 220 V) at a distance of about 3 to 4 cm from the transistor so as to heat the transistor. Draw the output and the transfer characteristics, and investigate the effect of temperature rise on collector current. Comment on the result. Also, plot h_{FE} versus I_C at the two temperatures on the same graph paper.

(3) Investigate the variation of h_{re}, h_{oe}, h_{ie} and h_{fe} with change of the operating collector current. Plot the curves showing the variations of h_{re}, h_{fe}, h_{ie} and h_{oe} with *operating* collector current *on the same graph paper.* Use the *output characteristic and transfer characteristics* for this purpose.

Note : The value of h_{FE} at first rises to a maximum value with increase of I_C. The extent of the maximum region and its value depend on the value of h_{FE} of the transistor supplied. For a transistor with a high d.c. β value, the position of maximum is very sharp, while for the low d.c. β value, the maximum is not prominent.

(4) Calculate the value of I_{CEO} (i.e., cut-off current) at different collector current using the relation

$$I_C = \beta I_B + I_{CEO}.$$

(5) For the transistor supplied, keep the base current at 50 μA. Plot the collector voltage along

the X-axis and the base voltage along Y-axis (keeping the base current constant at 50 μA). In this way draw the V_{BE}/V_{CE} curve at a constant I_B. Change the value of I_B to other values and obtain the characteristic curves. Calculate h_{re} and h_{ie} from these characteristic curves. Compare these values with those obtained before.

4.7 To Draw the Drain Characteristics of a FET and hence to Calculate the FET Parameters

Circuit Diagram

The circuit diagram required for the experiment is shown in Fig. 4.23. Two power supplies, one milliammeter and two voltmeters are required for the purpose. In the figure, V_{GG} is used as the gate bias supply and V_{DD} as the drain voltage source.

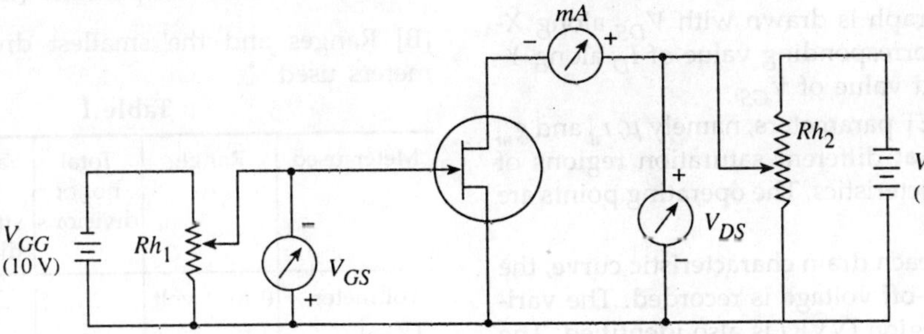

Fig. 4.23

Working Formula

The curve obtained by plotting drain current (I_D) along Y-axis against drain-to-source voltage (V_{DS}) along X-axis for fixed gate bias voltage (V_{GS}) is called the drain characteristic of a FET. A group of such drain characteristics may be drawn by setting V at different fixed negative values.

The FET parameters can be defined as follows :

(i) *Dynamic drain resistance*, considering finite change, is given by

$$r_d = \left.\frac{\Delta V_{DS}}{\Delta I_D}\right|_{V_{GS}} \tag{4.20}$$

(ii) *Mutual conductance or transconductance*, for finite change, is given by

$$g_m = \left.\frac{\Delta I_D}{\Delta V_{GS}}\right|_{V_{DS}} \tag{4.21}$$

(iii) *Amplification factor* (μ)

Amplification factor can be defined as the ratio of change of drain voltage to the ratio of change of gate voltage (all measured with respect to source) to keep the drain current constant. Hence,

$$\mu = \left.\frac{\partial V_{DS}}{\partial V_{GS}}\right|_{V_{DS}} \tag{4.22}$$

Negative sign indicates that the change of V_{DS} and V_{GS} are in opposite sense.

Since the voltage changes are finite, so *considering the magnitude of* μ, we can write,

$$\mu = \left.\frac{\Delta V_{DS}}{\Delta V_{GS}}\right|_{I_D} \tag{4.23}$$

The above three parameters are related by the equation,

$$\mu = r_d g_m \tag{4.24}$$

Experimental Procedure

(1) Symbolic diagram, *showing the basic principle*, is drawn in the laboratory note-book. The source, drain and gate are to be properly identified.

(2) The type no. of the *FET* is noted, and the maximum value of drain-to-source voltage, drain current, maximum gate-to-source voltage, maximum power rating are recorded from the semiconductor manual.

(3) Connections are made as shown in Fig. 4.23. *Before switching on the power supply, the circuit should be approved by the instructor. Do not connect the input terminals of the power supply unit to* d.c. *mains supply. Connect the input terminals of the power supply unit with* a.c. *mains.*

(4) The value of gate-to-source voltage is adjusted to zero value. The corresponding values of V_{DS} and I_D are noted (avoiding parallax error). The drain-to-source voltage is increased in regular steps of 0.5 V (or the other value, depending on the smallest division of the voltmeter used in the drain-to-source circuit), and the corresponding drain current is noted keeping V_{GS} at the fixed adjusted value (say, V_{GS} = 0 V). The value of V_{DS} is again increased to another value and the corresponding drain current is noted. This increment of V_{DS} is done up to the maximum value as specified by the manufacturer.

For BFW-11 JEFT, the maximum value of V_{DS} = 15 volts. The maximum power rating should never be exceeded. The data should be taken from the semiconductor manual.

(5) The reverse bias of the gate-to-source voltage is increased. For example, if BFW-11 type of FET is used, then the V_{GS} should be made more negative than before. This is because the above type is an *n*-channel FET.

Hence, V_{GS} is adjusted to −0.4 V (say). V_{DS} is increased in steps of 0.5 V and in each case the corresponding drain current is noted as before.

(6) The graph is drawn with V_{DS} along X-axis and the corresponding value of I_D along Y-axis for a fixed value of V_{GS}.

(7) The FET parameters, namely μ, r_d and g_m are calculated at different saturation regions of the drain characteristics. The operating points are also noted.

(8) From each drain characteristic curve, the value of pinch-off voltage is recorded. The variable voltage region (VVR) is also identified. The saturation region for each value of V_{GS} are also recorded. The value of I_{DSS} is also recorded.

Experimental Results

The data shown in different tables are given only for illustrations. They should not be taken as typical. The results will vary from apparatus to apparatus using different FETs.

[A] Specification

(1) Type number of the FET = ...
(2) Type of channel = ...
(3) Maximum drain current (I_{DSS}) = ... mA
(4) Maximum drain-to-source voltage $(V_{DS})_{max}$ = ... volt
(5) $(V_{GS})_{max}$ = ... volt
(6) ± V_{DSS} = ... volt
(7) $(I_{DSS})_{max}$ = less than ... mA
(8) P_{total} = ... mW
(9) Ambient temperature (max) = ... °C

[B] Ranges and the smallest divisions of the meters used.

Table I

Meter used	Range	Total no. of divisions	Value of one smallest division	Zero (or, initial) error, if any
Voltmeter (V_{DS})	0 to... volt	...	...volt	
Voltmeter (V_{GS})	0 to... volt	...	...volt	
Milliammeter (*mA*)	0 to...mA	...	...mA	

[C] Data for drawing (V_{DS}/I_D) characteristic, with V_{GS} as a parameter.

Table II

V_{DS} in volt	V_{GS} = − 0 V corresponding I_D in mA	V_{GS} = −0.4 V; corresponding I_D in mA	V_{GS} = − 0.8 V; corresponding I_D in mA	V_{GS} = −1.2 V; corresponding I_D in mA	V_{GS} = −1.6 V; corresponding I_D in mA	V_{GS} = −2.0 V; corresponding I_D in mA
0						
0.5						
1.0						
1.5						
2.0						
2.5						
3.0						
⋮	⋮	⋮	⋮	⋮	⋮	⋮
11.0						
12.0						
13.0						
14.0						
15.0						

Note : The maximum value of V_{GS} for the FET is to be adjusted as directed by the instructor (or from the semiconductor manual). Also, the maximum value of V_{DS} should not exceed the value as specified by the manufacturer. For *BFW* 11, the maximum value of V_{DS} is 15 volts.

[D] *To draw drain characteristics.*

For drawing the drain characteristics of the FET, data obtained in Table II are used. For different fixed values of V_{GS}, a graph is drawn by plotting V_{DS} along *X*-axis and I_D along *Y*-axis. The nature of the curves such obtained is as shown in Fig. 4.24.

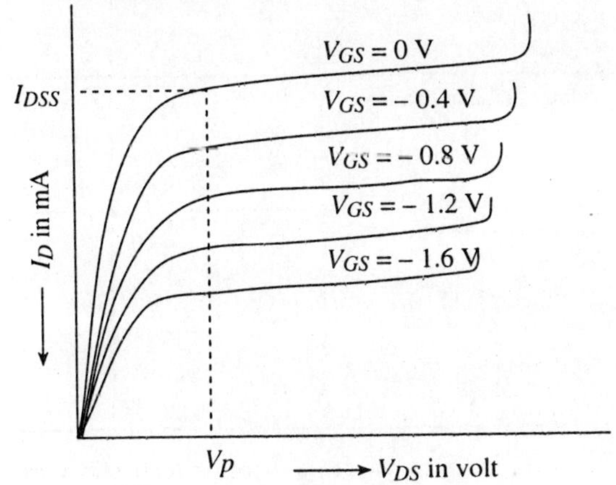

Fig. 4.24

From, the graph the following values are determined.

(i) Pinch-off voltage = ... volt.

(ii) The value of I_{DSS} = ... mA.

(iii) Maximum drain-to-source voltage = ... volt.

N.B. : Note that the drain characteristics are more crowded as the value of V_{GS} is increased. That is, the spacing between the successive curves decreases with increase of V_{GS}.

[E] *To find the JEET parameters.*

Discussion

(i) FET is a unipolar device, since its function depends only on one type of charge carier.

(ii) In an *n*-channel JEFT, the gate is given a negative bias with respect to the source and the drain is given positive potential with respect to the source (as we have shown in the circuit diagram). In the case of a *p*-channel JFET, the terminals of all the batteries are reversed.

(iii) V_{GS} may be made increasingly negative until I_D cut-off is reached. This value of V_{GS} is known as V_{GS} off.

(iv) If V_{DS} is increased beyond V_{DS} (max), there is an avalanche increase in I_D which may rapidly destroys the JFET.

Extra works

(1) The transfer characteristic of a JFET can be obtained by using the same circuit. To get the characteristic, V_{DS} is kept at a fixed value while

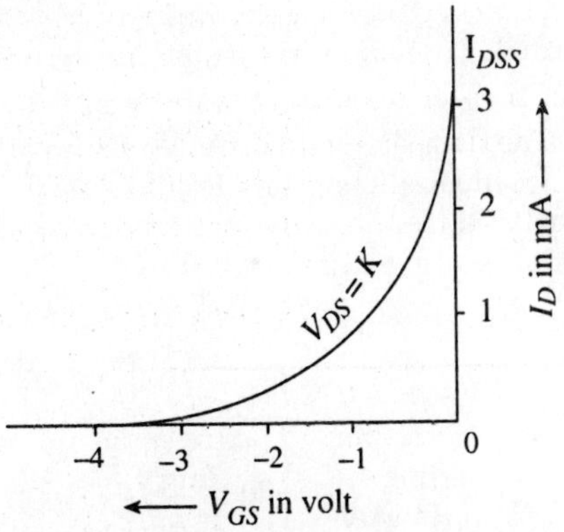

Fig. 4.25

V_{GS} is altered and the corresponding I_D is noted. The transfer characteristics may also be obtained

Table III

ΔV_{DS} in volt	ΔI_D in mA	ΔV_{GS} in volt	$r_d \left(= \frac{\Delta V_{DS}}{\Delta I_D}\right)$ in kΩ	$g_m \left(= \frac{\Delta I_D}{\Delta V_{GS}}\right)$ in mho	$\lvert\mu\rvert = \left.\frac{\Delta V_{DS}}{\Delta V_{GS}}\right\rvert_{I_D}$	$\mu = r_d g_m$	Remarks about the validity of $\mu = r_d g_m$

from the family of drain characteristics. For this a required value of V_{DS} is selected first. Then the corresponding values of V_{GS} and I_D are noted to get the co-ordinates of points on the I_D-V_S-V_{GS} curve. The nature of the transfer characteristics of a JFET is shown in Fig. 4.25.

(2) Calculate the FET parameters at different regions, e.g., at the left side of the pinch-off voltage, and also at the right side of the pinch-off voltage. From the data, show that the parameters depending on the choice of operating point.

(3) From the curves, shown in Fig. 4.24, calculate the value of μ for linear region and also for the V.V.R. Show that, $\mu \neq r_d g_m$ for V.V.R. Plot a graph showing (i) the variation of μ with I_D and also (ii) the variation of μ with V_{DS}.

4.8 To Measure the Input Off-set Voltage, Off-set Current and Input Bias Currents of the Supplied Op-Amp

Circuit Diagram

To measure the input off-set voltage [Fig. 4.26] to meausre the input off-set current [Fig. 4.27] and to measure the input bias current [Figs. 4.28(a) and (b)] are used. Here any one of the IC 741 group of OP AMP is selected for the purpose which is a monolithic silicon-integrated circuit with 11 resistors and 19 transistors.

[IC 741 OP-AMP consists of a differential input amplifier which effectively drives a gain and level shifting stage with an emitter-follower output. The two transistors of the first stage of an OP-AMP draw some base current for a suitable operating condition. OP-AMPs are so designed that the input bias current becomes very low and hence the input signal cannot be loaded.]

Working Formula

Due to the symmetry in the differential amplifier stages of an OP-AMP, the two bias currents should be theoretically same. But in practice, they are always unequal causing an off-set current defined as

$$I_{i_o} = I_{B1} - I_{B2}, \tag{4.25}$$

where I_{B_1} = bias current at inverting input,
and I_{B_2} = bias current at non-inverting input.
This off-set current would aways flow resulting in a non-zero output voltage without any input. This can be nulled (or balanced) by applying a low voltage at any one of the internal off-set balance terminals of the OP-AMP. Such low value of the d.c. input voltage is known as the input off-set voltage (V_{i_o}) which is obtained from the relation,

$$V_{o1} = \frac{R_1+R_2}{R_1} V_{i_o} \tag{4.26}$$

where V_{o1} is the output voltage in Fig. 4.26.

To find the input off-set current (I_{io}), the following relation may be employed :

$$I_{io} = I_{B_1} - I_{B_2} = \frac{V_{o2}-V_{o1}}{R_s\left(1+\frac{R_2}{R_1}\right)}, \tag{4.27}$$

where V_{o2} is the output voltage in Fig. 4.27.

Since the input bias current is half the sum of the separate bias currents of the two input terminals of the balanced amplifier, we can write it mathematically as

$$I_B = \frac{I_{B_1}+I_{B_2}}{2}, \tag{4.28}$$

$$\text{where } I_{B1} = \frac{V_{o3}-V_{o1}}{R_s\left(1+\frac{R_2}{R_1}\right)}$$

[where V_{o3} is the input voltage of Fig. 4.28(a)]

$$\text{and } I_{B2} = \frac{V_{o1}-V_{o4}}{R_s\left(1+\frac{R_2}{R_1}\right)}$$

[where V_{o4} is the output voltage of Fig. 4.28(b)]

Experimental Procedure

(1) Suitable circuit connections as shown are carefully made. Following the application of the power supply voltage there is normally an initial warm-up time drift of V_{io} and I_B. This warm-up time differs for different OP-AMPs. It may vary in the range 2–5 minutes in general but for some special cases it may be even greater than 20 minutes. So measurements are to be made after its initial warm-up has taken place.

(2) To measure the input off-set voltage the output voltage V_{o1} is noted (Fig. 4.26) for a set of

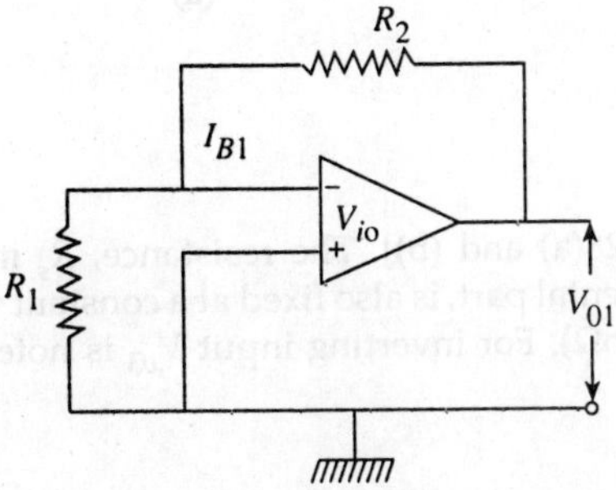

Fig. 4.26

resistances R_1 and R_2. The observation is repeated for different sets of R_1 and R_s. In each case, V_{io} can be calculated and hence a mean V_{io} is found out.

(3) To measure the input off-set current a constant value of the source resistance R_s is inserted (say, 1 mΩ) to both the inverting and non-inverting inputs. For a set of resistances R_1 and R_2 the output voltage V_{o2} (Fig. 4.27) is noted. Knowing V_{o2} one may calculate the input off-set current (I_{io}) in each case. A mean value of I_{io} is then determined.

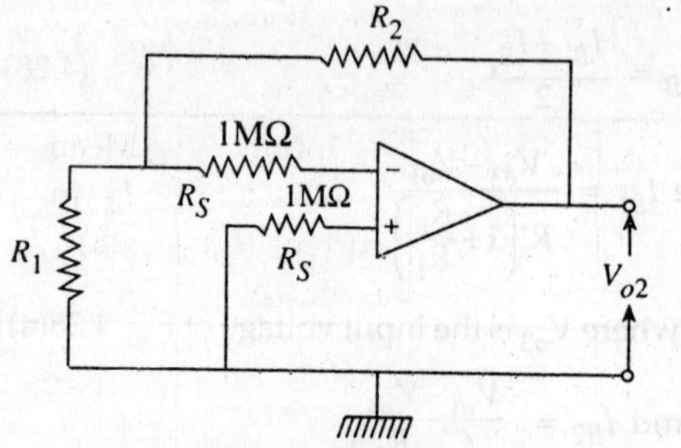

Fig. 4.27

(4) To measure the bias currents I_{B1} and I_{B2} at inverting and non-inverting terminals respectively two separate circuits are to be followed

different sets of R_1 and R_2. For non-inverting input V_{o4} is measured. A mean value of V_{o3} and also of V_{o4} are then calculated.

(5) The resistances R_1 and R_2 for the measurements of (2), (3) and (4) are so chosen that the output voltage does not attain close to ±12 volts which is the supply voltage.

Experimental Results

[A] Specifications of the OP-AMP used from the manual.

(Make a table showing the type of the OP-AMP and its details.)

[B] To find the range and the smallest division of the meters used for measurement.

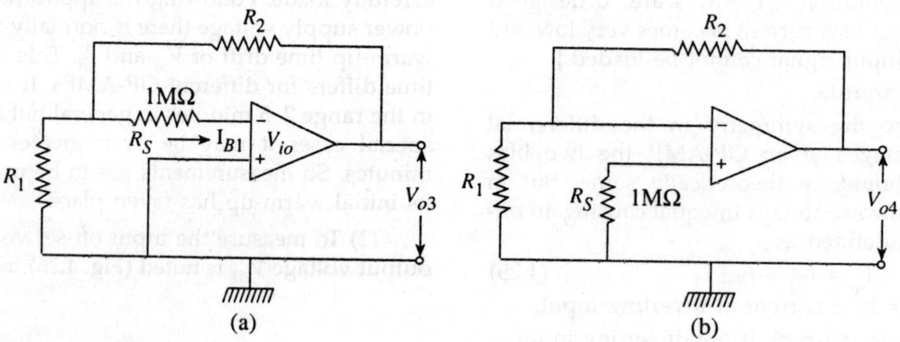

Fig. 4.28

[Figs. 4.28(a) and (b)]. The resistance, R_s in this experimental part, is also fixed at a constant value (e.g., 1 mΩ). For inverting input V_{o3} is noted for

[Make a table as tabulated in other experiments.]

[C] To measure the input off-set voltage.

Supply voltage = 12 volts (say)

Room temperature = ...°C

Table III

No. of obs.	Resistance R_1 in ohm	Resistance R_2 in ohm	Value of $\frac{R_1}{R_1+R_2}$	Output voltage (V_{o1}) in volt	Input off-set voltage (V_{io}) in millivolt	Mean V_{io} in millivolt

[D] To measure the input off-set current.

Table IV

No. of obs.	Resistance R_1 in ohm	Resistance R_2 in ohm	R_s in MΩ	V_{o1} in volt (from Table III)	V_{o2} in volt	Input off-set current (I_{io}) in μA	Mean I_{io} in μA

[E] To find the input bias current.

V_{o1} = ... volt (from Table III)

R_s = 1 MΩ (say)

Table V

No. of obs.	For inverting input				For non-inverting input				Mean (I_{B1}) in μA	Mean (I_{B2}) in μA	Input bias current ($I_2 = \frac{(I_{B1}+I_{B2})}{2}$ in μA
	R_1 in Ω	R_2 in Ω	Output voltage (V_{o2}) in volt	Bias current (I_{B1}) in μA	R_1 in Ω	R_2 in Ω	Output voltage (V_{o4}) in volt	Bias current (I_{B2}) in μA			

Discussion :

(i) IC 741 OP-AMP used in this experiment provides output short-circuit protection and internal phase compensation. It also provides latch-free operation, i.e., the output voltage can be latched or struck at some value regardless of the value of the input voltage.

(ii) Basically, the input bias currents are measured by forcing them to flow in large resistors of the order of 1 MΩ which may be bipassed to reject noise by making use of a capacitor (~0.01 μF).

(iii) For every change of the resistive network, off the circuit at first.

(iv) Input off-set voltage drifts over a period of time and also with temperature is observed. It is also found to depend on the supply voltage applied.

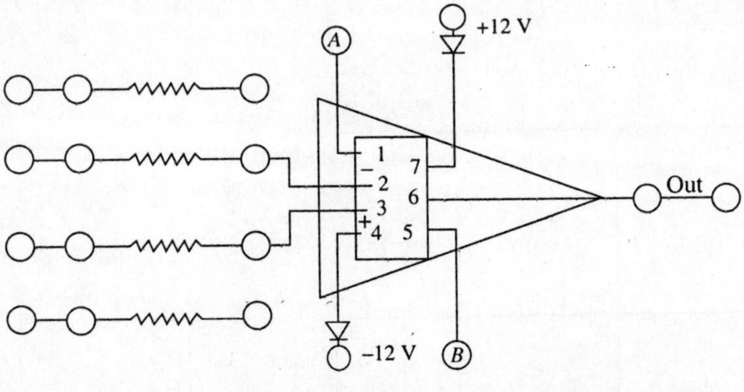

Fig. 4.29

Special Remarks

Adjustment of off-set null—The effect of the input off-set voltages are very low and hence can be ignored. But in certain cases it cannot be neglected. This can then be nulled or balanced by applying the followng procedures :

(a) Connections are made as shown in Fig. 4.26. Terminal 1 and 5 of the OP-AMP are connected to *A* and *B* (Fig. 4.29) of the 5 kΩ potentiometer whose centre is already connected internally to –12 V terminal. Next, the following steps are followed :

(b) At first R_1 is made equal to R_2 and their values are set at 10 kΩ and the potentiometer knob is adjusted to get zero output voltage.

Next R_1 is set at 1 kΩ and R_2 is set at 10 kΩ and the potentiometer knob is adjusted to get the zero output voltage.

Finally, the value of R_1 is set at 100 Ω, while R_2 is made equal to 10 kΩ. Next, the knob of the potentiometer is adjusted to obtain zero output voltage.

Caution i Do not disturb the knob such adjusted for the second experiment with OP-AMP.

4.9. To Find the Closed Loop Gain of an Op-Amp and also to use of an Op-Amp as an Adder and Differential Amplifier

Circuit Diagram

The circuit diagram for studying the gain of an inverting amplifier and non-inverting amplifier is shown in Figs. 4.30(a) and (b) respectively. In both the figures the output is fed back to the inverting input terminal through a resistor R_1. In the inverting amplifier the input voltage V_1 is applied to the minus(–) input while in the non-inverting amplifier the input voltage is applied to the plus(+) input.

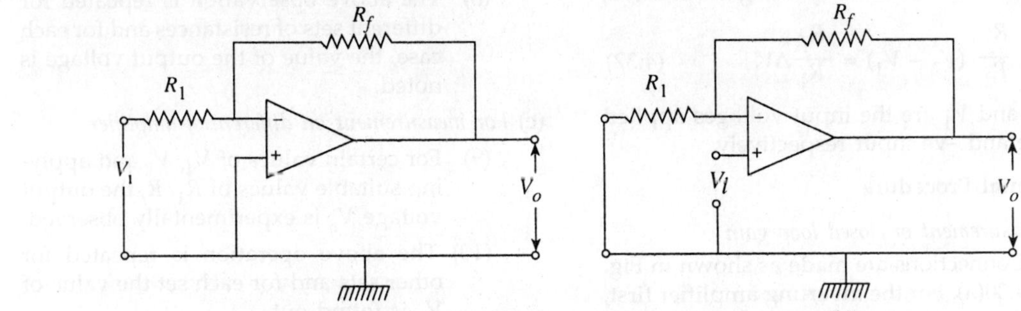

(a) (b)

Fig. 4.30

The circuit diagram of an adder (or summer) is shown in Fig. 4.31. In the figure V_1, V_2 and V_3

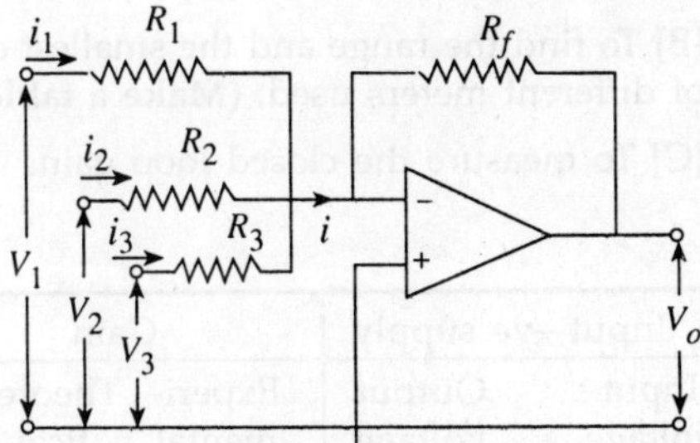

Fig. 4.31

are the input voltages connected to the minus input through the resistors R_1, R_2 and R_3 respectively.

The connection of the differential amplifier is shown in Fig. 4.32. Here V_1 and V_2 are the two input voltages connected through the same resistance R_1 to the minus and plus input. the inputs V_1 and V_2 are respectively connected to the points A' and B' whose other ends are earthed. A resistance R_f is connected between the output V_o and the point A while another resistance of equal value (i.e., R_f) is connected between the point B and the earth.

Working Formula

(i) The closed loop gain (i.e., with feedback) of an ideal inverting amplifier can be expressed as

$$A_{vf} = \frac{V_o}{V_i} = -\frac{R_f}{R_1}, \tag{4.29}$$

where V_o and V_i are the output and input voltages respectively. For a non-inverting amplifier the closed loop gain A'_{vf} can be written as

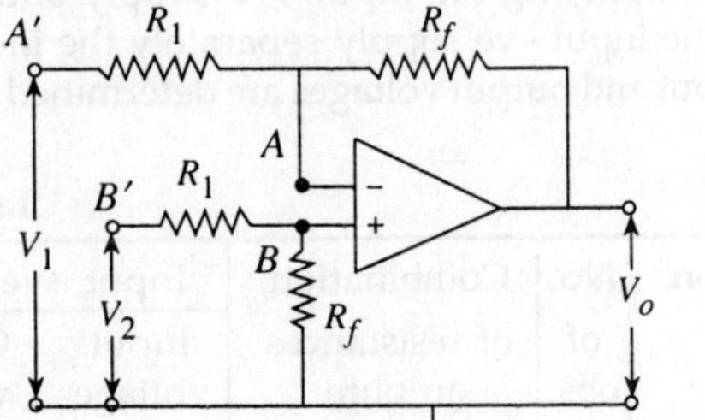

Fig. 4.32

$$A'_{vf} = 1 + \frac{R_f}{R_1}. \tag{4.30}$$

Adjusting R_1 and R_f any desired gain can be obtained using the above two equations.

(ii) Since in an adder circuit a virtual ground exists at the OP-AMP input, the sum of the currents in R_1, R_2 and R_3 becomes equal to the current in R_f,

i.e., $i_1 + i_2 + i_3 = i$

$$\text{or,} \quad \frac{V_1}{R_1} + \frac{V_2}{R_2} + \frac{V_3}{R_3} = -\frac{V_o}{R_f}$$

$$\text{or,} \quad V_o = -R_f\left(\frac{V_1}{R_1} + \frac{V_2}{R_2} + \frac{V_3}{R_3}\right). \qquad (4.31)$$

(iii) For a differential amplifier, the output voltage V_o can be obtained by using the relation,

$$V_o = \frac{R_f}{R_1}(V_2 - V_1) = \frac{R_f}{R_1}\Delta V. \qquad (4.32)$$

where V_2 and V_1 are the input voltages applied at the +ve and –ve input respectively.

Experimental Procedure

(a) *For measurement of closed loop gain*

(1) Connections are made as shown in Fig. 4.30(a). For the inverting amplifier first, the combination of resistances R_1 and R_f are properly chosen and for the +ve input supply, the input and output voltages are found out.

(2) For the same set of R_1 and R_f and making the supply at the input '–ve', the input and output voltages are noted.

(3) Operations (1) and (2) are repeated for different sets of R_1 and R_f.

(4) Connections are next made as shown in Fig. 4.30(b). For different sets of R_1 and R_f applying the input +ve supply and the input –ve supply separately the input and output voltages are determined.

(5) The gain of the inverting and non-inverting amplifiers are then calculated from the data obtained and compared with the theoretical data.

(b) *For measurements on adder circuit*

(6) The resistances R_1, R_2 and R_3 are made equal (say, 10 kΩ) and the value of R_f is set at 100 kΩ after the suitable circuit connections (Fig. 4.31).

(7) The actual values of the input voltages V_1, V_2 and V_3 are recorded, and the output voltage is calculated.

(8) The above observation is repeated for different sets of resistances and for each case, the value of the output voltage is noted.

(c) *For measurement on differential amplifier*

(9) For certain values of V_1, V_2 and applying suitable values of R_1, R_f the output voltage V_o is experimentally observed.

(10) The above operation is repeated for other sets and for each set the value of V_o is found out.

(11) Theoretical value of V_o is calculated in each case and this value is compared with the experimental value of V_o.

N.B. : For working with closed box instrument see discussion (iii).

Experimental Results

[A] Specification of the OP-AMP.

(Make a Table I)

[B] To find the range and the smallest division of different meters used. **(Make a table II)**

[C] To measure the closed loop gain.

Table III

Condition of the input terminal	No. of obs.	Combination of resistances in ohm	Input +ve supply		Input –ve supply		Gain	
			Input voltage (V_i) in volt	Output voltage (V_o) in volt	Input voltage (V_i) in volt	Output voltage (V_o) in volt	Experimental (= V_o/V_i)	Theoretical
Inverting amplifier								
Non-inverting amplifier								

[D] To measure the output voltage of an adder circuit.

Table IV

No. of obs.	Resistances in ohm	Voltage V_1 (actual) in volt	Voltage V_2 (actual) in volt	Voltage V_3 (actual) in volt	Output voltage (V_o) in volt	
					Experimntal	Theoretical
1	$R_1 = R_2 = R_3 = 10\ k\Omega$ $R_f = 100\ k\Omega$					

[E] To measure the output of the differential amplifier.

Table V

No. of obs.	R_1 in ohm	R_f in ohm	V_2 in volt	V_1 in volt	$\Delta V\ (=V_2 - V_1)$ in volt	V_o in volt	
						Experiment	Theoretical

Discussion

(i) The excessive gain of an amplifier can always be reduced to our choice by external circuitry.

(ii) In a differential amplifier the voltage difference ΔV should be so chosen that the input off-set voltage may be neglected in comparison to ΔV. Otherwise off-set null adjustment must be made before taking readings.

(iii) In some laboratories the circuit connections of the OP-AMP are kept in a closed box where only the binding tags are used by the students for performing the experiment. The internal connections are generally shown by lines on a panel board which, however, varies from one manufacturer to other. But in some other cases a part of the whole instrument is enclosed in a box while the remaining is kept outside which is to be properly connected following the basic principle of the circuits shown in different figures. The working principle, of course, remains same.

(iv) For OP-AMP using as an adder the output voltage of the operation amplifier is measured and hence the error from the theoretically predicated value and also the percentage error are calculated. For subtraction also from the output voltage measurement, the errors can be calculated.

(v) OP-AMP can also be used as an integrator and differentiator with proper change of passive elements. For using OP-AMP as an integrator the feedback resistance R_f in Fig. 4.30(a) is to be replaced by a capacitor C, the position of R_1 remaining same. For differentiator, the position of C and R_1 in the case of integrator are to be interchanged.

ORAL QUESTIONS WITH ANSWERS

Semiconductor

(1) *How can a semiconductor be distinguished from a conductor and an insulator?*

Ans. : Semiconductor is a solid whose resistivity lies between conductor and insulator. A conductor and an insulator are first distinguished by their extreme values of electrical conductivity. A conductor has a conductivity of 10^4 to 10^6

ohm^{-1} cm^{-1} while the conductivity of an insulator is less then 10^{-6} ohm^{-1} cm^{-1}. Moreover, the conductivity of a conductor decreases with the increase of its temperature whereas that for an insulator increases slightly with increasing temperature. In addition to conductors and insullators, there is a class of solids with intermediate values of electrical conductivity in the range 10^{-6} to 10^{3} ohm^{-1} cm^{-1} and whose conductivity increases more strongly with increasing temperature. Such solids are called semiconductors.

(2) *Give examples of some semiconductors.*

Ans. Transistors, metallic rectifiers, integrated circuits, junction diodes, Zener diodes, tunnel diodes, etc. are examples of semiconductors.

(3) *What are the uses of semiconductors?*

Ans. : They are used widely in radio and TV receivers, in computers and in various electronic instruments.

(4) *What are the advantages of a semiconductor over a vacuum tube?*

Ans. : The controlling functin of a vacuum tube is similar to that of a semiconductor but it is rapidly replacing vacuum tubes because of added advantages, some of which are :

(i) Semiconductor is solid. Therefore, there is a little chance of vibration as in vacuum tube.

(ii) Little power is required for a semiconductor and also it radiates less heat energy than a tube.

(iii) They do not need time for warm up and so operate immediately with the application of power.

(iv) A semiconductor does not undergo any chemical deterioration which occurs in a tube cathode.

(v) A semiconductor is small in size and light in weight.

(5) *What is meant by current carriers?*

Ans. : In vacuum tubes, negatively charged electrons are considered as the current 'carriers'. In order to explain current flow in semiconductor diodes and transistors, this concept must be modified by the addition of positive charge carriers, known as 'hole', which have mass, mobility and velocity. Current flow in semiconductors is thus carried on by the flow of both negative charges (free electrons) and positive charges (holes). Holes are usually attracted by free electrons and when they combine, the free electron 'fills' the hole neutralizing its charge. When this happens the hole and free electron both are lost as current carriers and form new current carriers at other points in the semiconductor.

(6) *Explain how p- and n-type semiconductors are formed?*

Ans. : Impurity like arsenic or antimony (pentavalent atoms) increases the conductivity of silicon, which has four valence electrons in its outer shell, by increasing the number of negative charge carriers. Silicon which has been doped here with arsenic or antimony is designated as '*n-type*'. Similarly, impurity like indium or gallium (trivalent atoms) increases the conductivity of silicon by increasing the number of positive charge carriers. For this reason, silicon which has been doped with indium or gallium is called '*p-type*'. Current flow in *n-type* and *p-type* silicon is carried on by 'free electrons' and 'holes' respectively, and these are known as majority carriers. Some holes, however, exist in *n-type* silicon and some free electrons in *p-type*. But these are in the minority and are, therefore, called minority carriers.

(7) *What is a junction diode ?*

Ans. : When *p-type* and *n-type* silicons are joined together a junction diode is formed. The diode thus created has a characteristic to pass current in one direction readily but not in the other. This unique characteristic may be explained by considering the theory of negative and positive carriers.

(8) *What do you mean by forward-bias arrangement?*

Ans. : A diode connected with positive battery-terminal to the *p-type* and negative battery-terminal to the *n-type* is called 'forward bias' and the diode is said to have a low forward resistance.

(9) *What is a Zener diode?*

Ans. : A junction diode is operated suitably by using forward-bias arrangement. In Zener diode, its unique reverse-bias characteristic provides some important different applications from those of the crystal diode.

(10) *What is the main use of a Zener diode?*

Ans. : Zener diode is used mainly as a regulator in power supply.

(11) *What is a tunnel diode?*

Ans. : A tunnel diode is a *p-n* junction device with a high concentration of impurities in its semiconductor elements. Due to a large number of impurities, the space charge region at the junction is very thin and as a result electrical charges move easily through it. This process is called 'tunnelling'.

(12) *What are the applications of a tunnel diode?*

Ans. : A tunnel diode has applications as an amplifier; oscillator, or switching element. These are, in fact, the advantages over a conventional diode which cannot act for such purposes.

(13) *What are the advantages of a transistor over a vacuum tube?*

Ans. : One of the great advantages of a transistor over a vacuum tube is that it is small in size and it does not require any filament power. The supply voltage of a transistor is also very low and it operates as soon as power is applied. In addition, the circuit connection of a transistor is much simpler than that of a tube.

(14) *How are the transistors classified?*

Ans. : Transistors are of many types which may be broadly classified as follows :

(i) *Depending on the basic materials from which they are formed.* In this category we have silicon and germanium transistors. Now-a-days most transistors are of silicon type.

(ii) *Depending on the number of elements used.* Here we have the three-element transistor and the four-element transistor.

(iii) *Depending on the ability to dissipate power.* Here we have a wide range from low power (below 50 mW) to high power (2 W and above) type.

(iv) *Depending on the process of construction.* In this category we have various junction transistors such as grown junction, alloy junction, mesa, epitaxial mesa, planar transistors, drift-field and point-contact transistors.

(15) *What is a junction transistor?*

Ans. : The junction transistor, a radically new type of transistor, was first developed by Shockley in 1949, just a year later of the discovery of the point contact transistor. A junction transistor, now widely used, is nothing but a combination of two *p-n* junctions.

(16) *What are p-n-p and n-p-n transistors?*

Ans. : A typical junction transistor is formed by sandwiching a thin *n-type* layer between two *p-type* layers. Such a transistor is called a *p-n-p* junction transistor or in short a *p-n-p* transistor. If, alternatively, there are two *n-type* layers on either side of *p-type* layer it is called *n-p-n* transistor.

(17) *Define emitter, base and the collector of a transistor.*

Ans. : There are three leads from individual metallic plates which make contact with the respective crystals. The *n*-layer in the middle called the 'base', the forward biased *p*-layer on the left is the 'emitter' while the reverse-biased *p*-layer on the right is the 'collector'.

(18) *Discuss how the leads of a transistor can be identified.*

Ans. : The emitter, base and collector leads of a transistor can be conveniently identified by the following two ways :

(i) The central lead corresponds to the base. The spacing between the emitter lead and the base lead is less than the spacing between the base lead and the collector lead.

(ii) If the leads are along the circumference of a circle, then the centre one is called base. The emitter and the collector leads are then identified by reading clockwise from the base. In such reading first comes the collector and then the emitter. The collector lead can also be identified by a dot which is pointed on the body of the transistor close to the collector.

(19) *How are the transistors manufactured by using alloy or fusion method?*

Ans. : In this method for producing *p-n-p* transistor two small pellets of indium of slightly different diameters are placed on either side of a wafer of *n*-type germanium, say, with the help of a suitable jig. This assembly is heated for a few minutes at a temperature of about 5000 °C as a result of which indium melts and combines with the germanium. After taking out from the furnace three wires are soldered for the emitter, base and the collector leads. The large *P*-layer forms the collector while the smaller one is the emitter of a transistor.

An *n-p-n* transistor can also be made by a similar process using *p*-type germanium wafer and taking arsenic or antimony in place of indium.

Since here the *p-n* junction is produced by alloying or fusion the transistors so formed are known as alloyed or fused junction transistors.

(20) *Discuss briefly the grown junction method for manufacturing transistor.*

Ans. : The grown *p-n-p* junction is made by adding appropriate acceptor and donor impurities alternately to a molten bath of a semiconductor material. The growth process is repeated at a regular interval of time and pulling the crystal simultaneously. The crystal is then cut into pieces parallel to the junction and this is subsequently diced. Transistors so produced are called grown junction transistors.

(21) *Define the common-emitter, common-base and the common-collector circuits of a transistor.*

Ans. : When a transistor is used in common-emitter configuration, the input signal is applied between base and emitter while the output is obtained from the collector to emitter. The emitter in such a case is thus common to both the input and the output circuits and hence the name common-emitter. Similarly, in a common-base or in common-collector circuit, the base or the collector serves as a common reference for both the input and the output signal.

(22) *Why is the common-emitter configuration used frequently in electronic circuits?*

Ans. : The common-emitter is most frequently used as it provides suitable voltage, current and power gain in different electronic circuits.

(23) *Define a.c. α and a.c. β factors of a transistor.*

Ans. : In a common-base circuit the collect or current is controlled by the variations of emitter current. This gives one important control characteristic of a transistor designated by a.c. α, where α is defined as the ratio of the change in collector current (ΔI_C) to the change in emitter current (ΔI_E), when the collector voltage (V_{CB}) is maintained at a constant value. Mathematically, this can be expressed as

$$\alpha = \left(\frac{\Delta I_C}{\Delta I_E}\right)_{V_{CB}}.$$

α is also sometimes designated by the letter h_{fb} and is called the forward-current-transfer ratio. In a common-base arrangement, since the collector current is slightly less than the emitter current, α will be slightly less than unity.

In a common-emitter circuit, the input signal is applied between the base and emitter. This gives another important control characteristic designated by a.c. β, where β is defined as the ratio of the change in collector current (ΔI_C) to the change in base current (ΔI_B) when the collector voltage (V_{CE}) is kept constant. Mathematically, this can be expressed as

$$\beta = \left(\frac{\Delta I_o}{\Delta I_B}\right)_{V_{CB}}.$$

The *a.c.* β is also sometimes designated by the letters h_{fe} and it is the current amplification factor in a common-emitter amplifier.

(24) *Why is the value of β always greater than unity?*

Ans. : Since small changes in base current produce large changes in collector current, the value of β will be always greater than unity.

(25) *What is the relation between α and β?*

Ans. : We may write,

$$\beta = \frac{\alpha}{1 \pm \alpha} \quad \text{or,} \quad \alpha = \frac{\beta}{1 \pm \beta}.$$

It is clear that as α approaches towards unity, β becomes increasingly larger. For a transistor whose α is 0.98, β becomes 49 while an α of 0.99 gives β of 99. So during experiment α must be measured very accurately to avoid formula error in β.

(26) *What is meant by biasing of a transistor?*

Ans. : By the term biasing of a transistor we mean application of an appropriate current to its base terminal. This is similar to the application of a voltage to the grid terminal of a valve.

(27) *What are the main features of a good biasing system?*

Ans. : The following are the main featues of a good biasing system :

(i) The bias related to the emitter base and the collector base junctions should be provided by one battery.

(ii) The gain of the transistor should be influenced by negative feedback.

(iii) The desired operating point must be stable irrespective of change in temperature.

(28) *Why is the equivalent circuit using h-parameters considered most useful?*

Ans. : The equivalent circuit using h-parameters is considered most useful because of the following advantages :

(i) The h-parameters are the easiest measure.

(ii) The h-parameters are more independent of each other and other variables like operating point, frequency, etc.

(iii) The values of h-parameters nearly correspond to actual values of input and output impedances and current gain for most of the applications.

(29) *Discuss the relative performance characteristics of the three modes of operation.*

Ans. : We may draw the following conclusions regarding the relative performance characteristics of the three modes of operation.

Out of the three amplifiers, viz., *CB*, *CE* and *CC* types, the common-emitter amplifiers are most efficient and flexible in character and so they are frequently used while the other two types have applications only for specialized cases, such as for impedance matching or as a buffer stage.

OP-AMP

(30) *What is an operational amplifier?*

Ans. : An operational amplifier, often referred to as OP-AMP, is a high-gain, direct-coupled differential amplifier whose response characteristics are controlled externally by negative feedback arrangement. Now-a-days, OP-AMPs are widely used in various branches of electronics particularly in linear and digital circuits. OP-AMPs are suitably used in computer technology to perform mathematical operations like summing, differentiation and integration. These are also used as video and audio amplifiers, oscillators, etc.

(31) *What are the characteristics of an OP-AMP?*

Ans. : An ideal operational amplifier has the following important characteristics which do not drift with temperature :

(i) Input resistance = ∞

(ii) Output resistance = 0

(iii) Bandwidth = ∞

(iv) Voltage gain = $-\infty$

(v) $V_{out} = 0$, when $V_1 = V_2$, independent of the magnitude of V_1.

(32) *What are the plus and minus inputs of an OP-AMP?*

Ans. : The amplifier has two inputs of which one is marked plus (+) and the other is marked minus (–). The plus input is the non-inverting input while the minus input is the inverting input. A signal applied to the plus input will be appeared in the same phase at the output. On the other hand, a signal applied to the minus input will be shifted in phase 180° at the output.

(33) *What is the gain of an operational amplifier?*

Ans. : The expression for the gain of this amplifier is :

$$\text{Gain} = R_f/R.$$

ADDITIONAL ORAL QUESTIONS

Semiconductor

1. State names of two conductors and hence give their order of conductivities.
2. Give names of two insulators and hence give their order of conductivities.
3. Give names of two commonly used semiconductors and give the order of their conductivities.
4. Using energy band diagram explain the difference in conductors, insulators and semiconductors.
5. When atoms share electrons, what type of bonding is it called?
6. If the temperature is increased what will happen to the number of electrons in the conduction band of a semiconductor?
7. Why the conductivity of germanium is greater than that of silicon at room temperature?
8. What is a hole? How do the holes move in an intrinsic semiconductor?
9. Explain why the temperature coefficient of resistance of a semiconductor is negative.
10. Sketch the two-dimensional crystal structure of intrinsic silicon at the absolute zero temperature and also sketch its energy band diagram.
11. What process is the opposite of thermal generation of electron-hole pairs?
12. Why the concentration of free electrons and holes is equal in an intrinsic semiconductor?
13. Why a pentavalent impurity atom is known as donor type impurity?
14. Which of the following atoms could be used as *p*-type impurities and which *n*-type impurities? (a) Antimony; (b) Arsenic; (c) Phosphorus; (d) Boron; (e) Aluminium; (f) Indium.
15. Discuss briefly how holes are created in a *p*-type semiconductor.

Transistor

16. Why is an ordinary junction transistor called bipolar?
17. What is done to the base region of transistor to improve its operation?
18. What causes collector current to flow when the emitter current is zero? What is this collector current caused?
19. Explain why the base current in a transistor is much smaller than I_E in active operation.

20. For a *P-N-P* transistor in the active region, what is the sign (+ve or –ve) of I_B, I_C, I_E, V_{CB} and V_{EB}?
21. What is the input terminal and what is the output terminal in the *CE* configuration?
22. Though the collector-base junction of a transistor operating in active region is reverse biased, the collector current is still quite large. Explain.
23. An *N-P-N* transistor is to be used in *CE* configuration. Show how will you connect the external batteries so that the transistor works in the active region. What is the order of the voltage required?
24. Explain how will you determine the voltage gain of the *CE* amplifier by plotting d.c. load line on the output characteristics of the transistor.
25. Explain what is meant by thermal run-away in a transistor.

FET :

26. Draw the circuit symbols of (a) an *n*-channel *JFET*, (b) a *p*-channel *JFET*.
27. Sketch the basic structure of *n*-channel *JFET*.
28. Show the biasing arrangement of an *n*-channel *JFET*.
29. What is meant by the term channel in a *JFET*.
30. Define the parameters of a *JFET*.
31. Explain the shape of the characteristics of the *FET* qualitatively.
32. Write the equation giving the variation of g_m with V_{GS}.
33. Draw the small signal model of the *FET* (a) at low frequencies and (b) at high frequencies.
34. Give the order of inter-electrode capacitances of a *JFET*.
35. Give the order of magnitude of g_m, r_d and μ for (i) *JFET* and (ii) *MOSFET*.

OP-AMP

36. What is the relationship between the polarity of the output and input voltages in your experimental OP-AMP? Refer to your data.
37. Does the maximum undistorted sine wave output of the OP-AMP change with the amplifier gain?
38. What is the relation between experimental inverting amplifier gain and R_f and R_1 ?
39. In which procedural step is the summer operated as a subtractor?

SECTION B

EXPERIMENTS WITH BRIEF PROCEDURE

EXPERIMENT 1

TO OBSERVE THE FORWRD AND REVERSE *V-I* CHARACTERISTICS AND BREAKOVER VOLTAGE OF A DIAC

Theory

A diac is a two-terminal, three-layer bidirectional device (Fig. 1.1), which can be switched from its OFF state to ON state for either polarity of applied voltage.

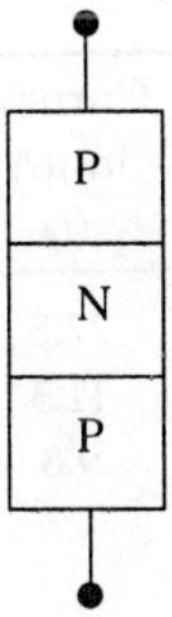

(a) Diac

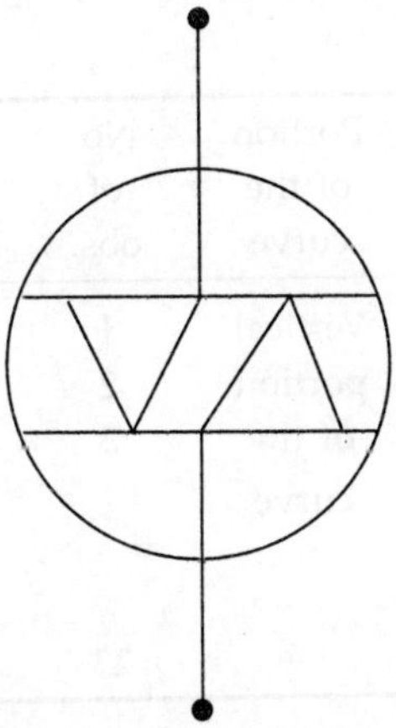

(b) Schematic symbol of a diac

Fig. 1.1

The diac is constructed in much the same way as dipolar transistor. The device has three alternatively doped semiconductor layers as shown in the figure. However, it differs from the bipolar transistor because the doping concentrations around both the junctions are equal, and leads are attached only to outer layers. No electrical connections are made to its middle region and thus the diac has only two leads. Therefore, the device appears as an ordinary PN junction diode though it sometimes packaged like a conventional bipolar transistor but only with two leads.

Since, both its junctions are equally doped the diac has the same effect on current flowing in either direction through its leads. In either direction one junction will always be forward-biased while the other is reversed. In each case the reverse-biased junction primarily controls the current flowing through the diac and the device operates as if it contains two PN junction diodes that are in series but are connected back-to-back.

The diac remains in an off state (conducts only a small leakage current) in either direction until the applied voltage in either direction is

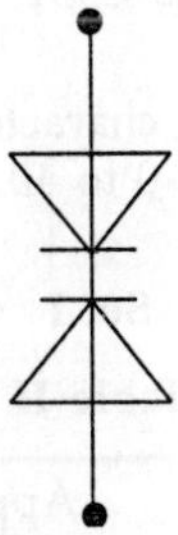

Equivalent circuit of diac

Fig. 1.2

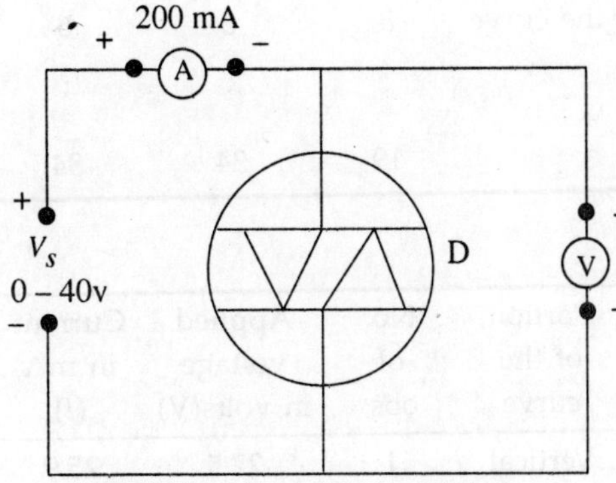

Circuit for diac characteristics

Fig. 1.3

high enough to cause its respective reversed-biased junction to break down. When this happens the device turns on and current suddenly rises to a value which is limited by the resistance in series with the device. The diac is therefore functions as a bidirectional switch, which will turn on whenever its breakdown voltage (in other direction) is exceeded. An equivalent circuit that is commonly used to represent the diac is shown in Figure 1.2, while the circuit diagram for studying diac characteristics is shown in Fig. 1.3.

Experimental Data :

A. Specification of the meter used :

Table I

Supply Voltage	Digital ammeter	Digital voltmeter
0-40 V	0-200 mA	0-50 V

B. Diac type : JT D83 C531

C. Data for the *V-I* characteristics varying the supply voltage from 0 to 40 V :

Set I

Table II

Portion of the curve	No of obs.	Applied voltage in volts(*V*)	Current in μA (*I*)
Horizontal portion of the curve	1	2	2
	2	4	4
	3	6	6
	⋮	⋮	⋮
	19	34	34

Portion of the curve	No. of obs.	Applied voltage in volts(V)	Current in mA (*I*)
Vertical portion of the curve	1	22.5	25.9
	2	22.6	25.5
	3	22.7	13.9
	⋮	⋮	⋮
	23	25.9	3.1

D. Data for the *V-I* characteristics varying the supply voltage from 0 to 40 V :

Set I

Table III

Portion of the curve	No. of obs.	Applied voltage in volts(*V*)	Current in μA (*I*)
Horizontal portion of the curve	1	2	2
	2	5	5
	3	7	7
	⋮	⋮	⋮
	19	35	35

Portion of the curve	No. of obs.	Applied voltage in volts(*V*)	Current in μA (*I*)
Vertical portion of the curve	1	23.0	27.8
	2	23.2	11.3
	3	23.4	9.8
	⋮	⋮	⋮
	23	25.9	3.3

E. Data for the *V-I* characteristics varying the supply voltage from 0 to 40 V.

Vary the applied voltages to different values and note the corresponding currents. It will give data for set II. Repeat it for set III.

F. V-I characteristics :

The *V-I* characteristics for a typical diac is shown in Figure 1.4. This gives the relation between the current flowing through the diac in either direction to the corresponding voltage across it.

The current through the diac remains at a low value until the voltage across it increases to a point where the device breaks down in either direction. The voltage required to break down the diac is designed as $+V_{BO}$ and $-V_{BO}$ and is called the breakover voltage. When breakover occurs, the resistance of the diac decreases rapidly as current increases and the net result is a decrease in voltage across it. Therefore, once breakover point is reached the diac exhibits negative resistance characteristics.

The diac is considered to be in the off state until the breakover voltage ($+V_{BO}$ & $-V_{BO}$) is reached. When $+V_{BO}$ or $-V_{BO}$ is reached, the device switches to the on state and then it is capable of conducting relatively high currents.

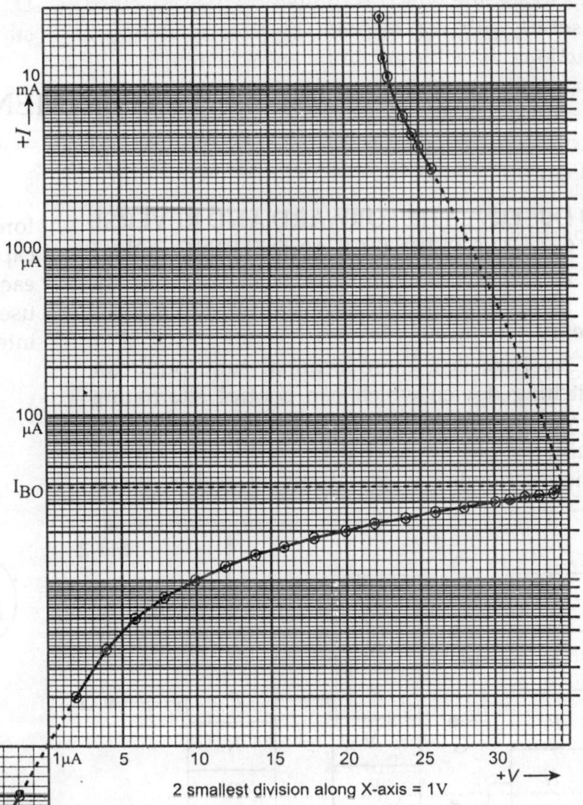

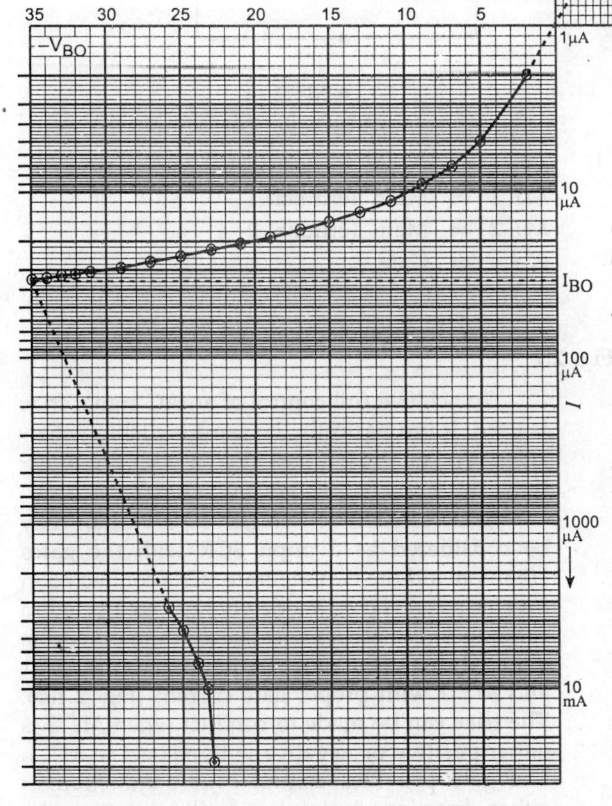

G. Value of diac breakover voltage from *V-I* characteristics :

Table IV

Set No.	$+V_{BO}$	$+I_{BO}$	$-V_{BO}$	$-I_{BO}$
I	34 V	38 μA	35 V	36 μA
II	34 V	39 μA	35 V	36 μA
III	31 V	35 μA	31 V	34 μA

Discussion

(i) Applied voltage across the diac should be increased step by step and when it attains a relatively high value the current through the diac is increased from a very low to a high value which is the diac breakover voltage.

(ii) The breakover voltage should be noted carefully for either direction (+ & –).

Fig. 1.4

EXPERIMENT 2

TO OBSERVE THE FORWARD AND REVERSE *V-I* CHARACTERISTICS AND FIRING POINT OF A TRIAC

Theory

A Triac is a three-terminal semiconductor switching device which can control alternating current in a load (Fig. 2.1). A silicon controlled rectifier (SCR) is capable of controlling current which is flowing in one direction and is, therefore, a unidirectional device. A single SCR is capable of operating on just one alternation of each AC input cycle. Thus it becomes essential to use two SCRs in parallel to convert each AC cycle into a DC signal.

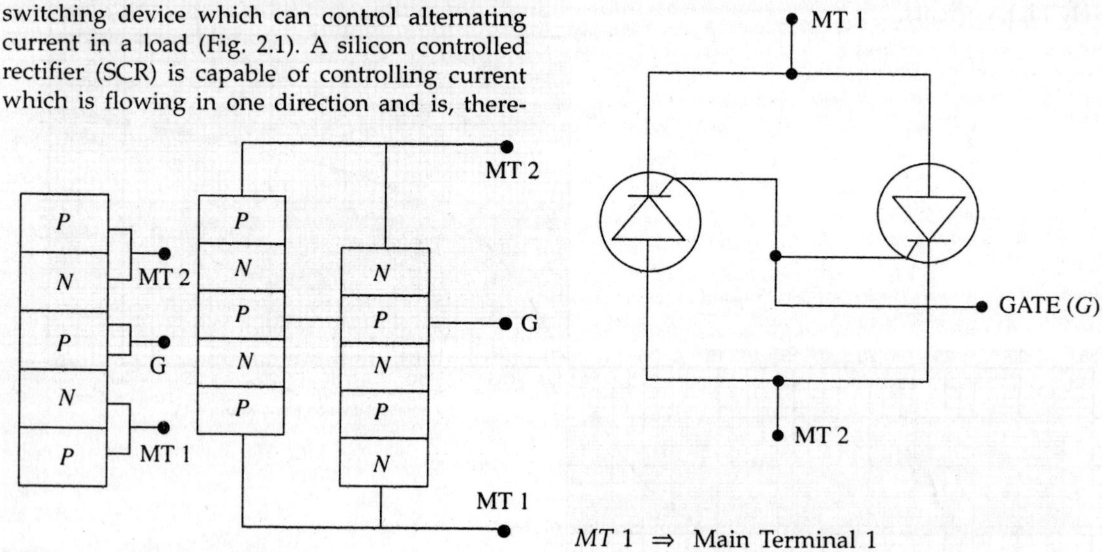

MT 1 ⇒ Main Terminal 1
MT 2 ⇒ Main Terminal 2
G ⇒ Gate

(a) Triac

MT 1 ⇒ Main Terminal 1
MT 2 ⇒ Main Terminal 2
G ⇒ Gate

(b) Equivalent circuit of a triac

Fig. 2.1

To achieve full control of an AC signal triac is used. It has basically the same switching characteristics as an SCR and also in both the directions.

Triac consists of two halves; each half may be considered as a pnpn SCR with the gates common and the two seperate SCRs connected in inverse parallel (i.e., anode of one is connected to cathode of the other) with gates common and is denoted by *G*. The other two terminals are *MT* 1 and *MT* 2 with proper gate current (Fig. 2.1(b). The triac can be made to conduct when *MT* 2 is either positive or negative with respect to *MT* 1. The gate provides control over conduction in either direction. A simple circuit for getting the characteristic of a triac is shown in Fig. 2.2.

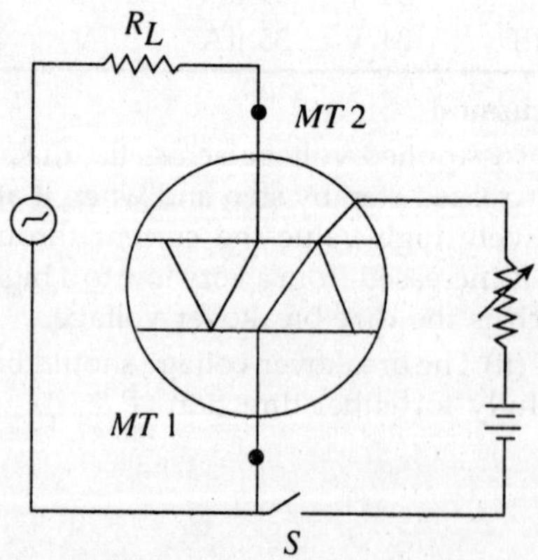

Fig. 2.2 : A simple circuit of a triac

Experimental Data

A. Specification of meter used :

Triac BT 136-600D PHm 0135 AI

Table I

Milliammeter I	Milliammeter II	Voltmeter
0-20 mA	0-200 mA	0-50 V

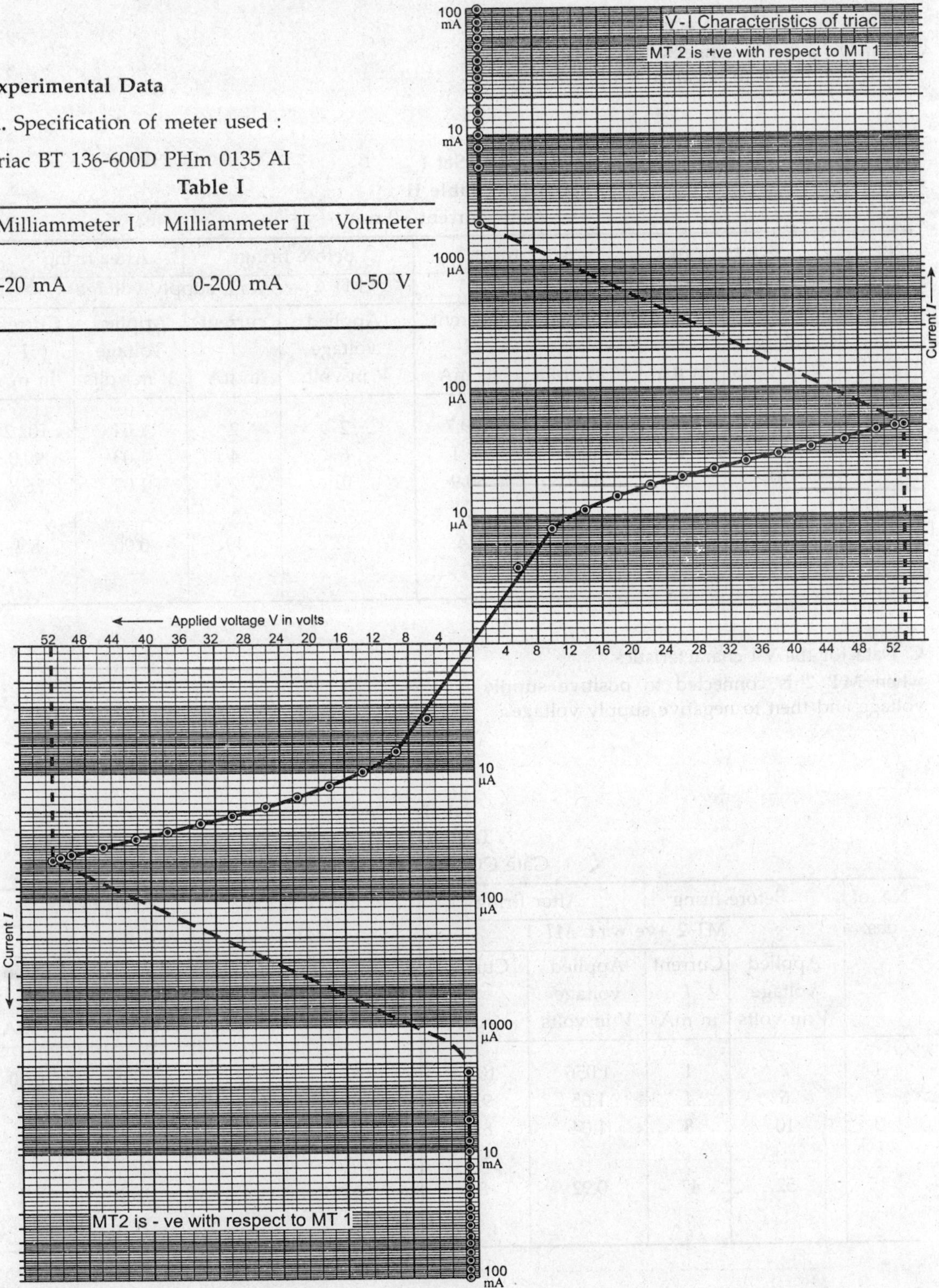

Fig. 2.3 (Set 1)

B. Data for the V-I characteristics when *MT* 2 is first connected to positive supply voltage and then to negative supply voltage :

Set I

Table II

Gate Current : 0 mA

No. of obs.	Before firing		After firing		Before firing		After firing	
	MT 2 +ve w.r.t. supply voltage				MT 2 –ve w.r.t. supply voltage			
	Applied voltage *V* in volts	current *I* in μA	Applied Voltage *V* in volts	Current *I* in mA	Applied voltage *V* in volts	current *I* in μA	Applied Voltage *V* in volts	Current *I* in mA
1	2	1	1.03	109.7	2	2	1.04	102.2
2	6	4	1.02	93.1	6	4	1.03	90.0
3	10	8	1.01	80.9	10	7	1.02	76.8
⋮	⋮	⋮	⋮	⋮	⋮	⋮	⋮	⋮
15	52	52	0.89	8.4	52	49	0.90	8.4
⋮	⋮	⋮	⋮	⋮	⋮	⋮	⋮	⋮

C. Data for the V-I characteristics when **MT** 2 is connected to positive supply voltage and then to negative supply voltage :

Set II

Table III

Gate Current : 1 mA

No. of obs.	Before firing		After firing		Before firing		After firing	
	MT 2 +ve w.r.t. *MT* 1				*MT* 2 –ve w.r.t. *MT* 1			
	Applied voltage *V* in volts	Current *I* in mA	Applied voltage *V* in volts	Current *I* in μA	Applied voltage *V* in volts	Current *I* in mA	Applied Voltage *V* in volts	Current I in mA
1	2	1	1.056	102.4	2	2	1.058	103.0
2	6	4	1.05	93.8	6	4	1.05	90.8
3	10	8	1.04	82.3	10	6	1.04	81.0
⋮	⋮	⋮	⋮	⋮	⋮	⋮	⋮	⋮
15	52	47	0.92	8.1	52	47	0.92	8.0
⋮	⋮	⋮	⋮	⋮	⋮	⋮	⋮	⋮

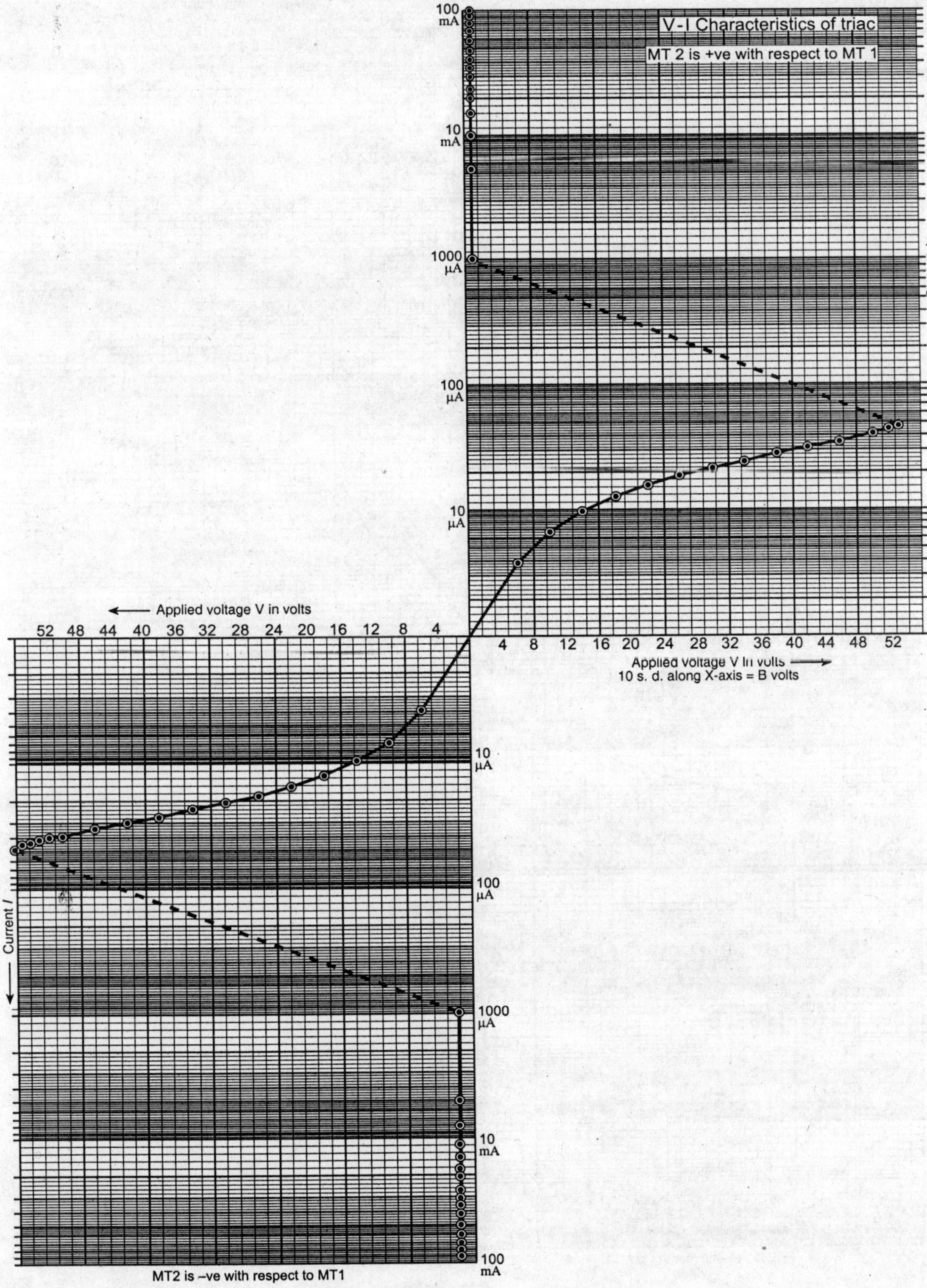
V-I Characteristics of triac
MT 2 is +ve with respect to MT 1
100 mA
10 mA
1000 μA
100 μA
10 μA
Current I
Applied voltage V in volts
52 48 44 40 36 32 28 24 20 16 12 8 4
4 8 12 16 20 24 28 32 36 40 44 48 52
Applied voltage V in volts
10 s. d. along X-axis = B volts
10 μA
100 μA
1000 μA
10 mA
100 mA
Current I
MT2 is –ve with respect to MT1

Fig. 2.4 (Set II)

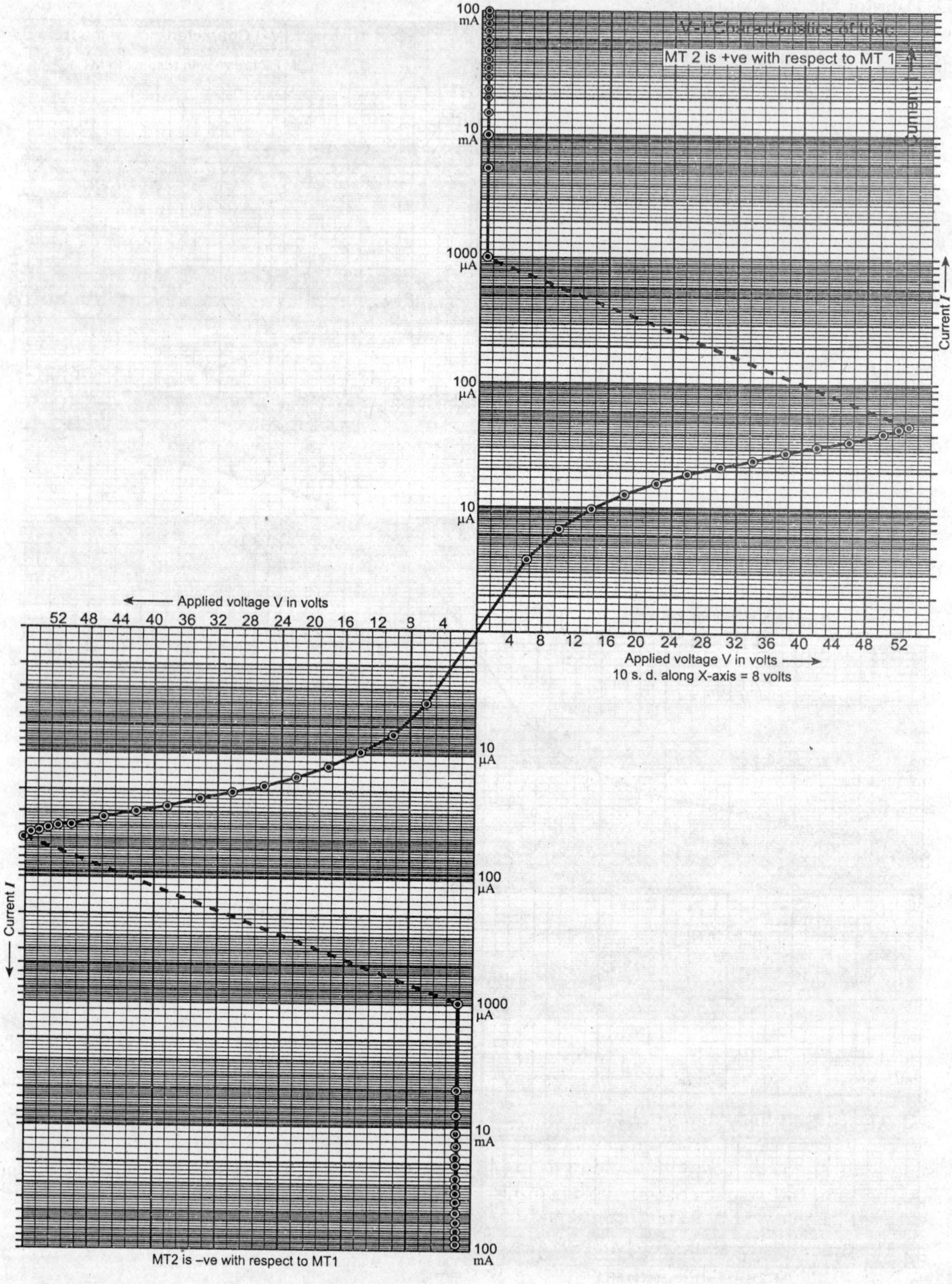

Fig. 2.5 (Set III)

D. Data for the V-I characteristics when **MT 2** is connected to positive supply voltage and then to negative supply voltage :

Set III

Table IV

Gate Current : 2 mA

No. of obs	Before firing		After firing		Before firing		After firing	
	MT 2 +ve w.r.t. *MT* 1				*MT* 2 –ve w.r.t. *MT* 1			
	Applied voltage *V* in volts	Current *I* in mA	Applied Voltage *V* in volts	Current *I* in mA	Applied voltage *V* in volts	Current *I* in mA	Applied Voltage *V* in volts	Current *I* in mA
1	2	1	1.064	103.1	2	2	1.067	103.3
2	6	4	1.06	97.0	6	4	1.06	91.4
3	10	7	1.05	84.0	10	7	1.05	81.0
⋮	⋮	⋮	⋮	⋮	⋮	⋮	⋮	
15	52	46	0.93	10.3	52	46	0.93	
⋮	⋮	⋮	⋮	⋮	⋮	⋮	⋮	⋮

E. V-I characteristics :

This gives the relationship between the current flowing through its main terminal in each direction to the voltage applied in each directon. Furthermore, it is plotted with no gate current flowing through the triac; *MT* I was used as the reference point for all voltages and currents.

When *MT* 2 is +ve w.r.t. *MT* 1 the current through the device remains at the low value until $+V$ rises to $+V_{BO}$ of the device. At this time the triac switches from the OFF state to the ON state and $+I_T$ is essentially limited by the external resistance of the circuit. Triac remains in the ON state until $+I_T$ drops below the holding current I_H. The same *V-I* characteristic is obtained also when *MT* 2 is –ve w.r.t. *MT* 1 (Fig. 2.3, 2.4 and 2.5).

The triac thus exhibits the characteristics of an SCR in either direction and it responds to either a +ve or a –ve gate voltage.

F. Calculation of breakover voltage and holding current : This is shown in Table V

Table V

Set No.	$+V_{BO}$ in V	$-V_{BO}$ in V	$+I_{BO}$ in µA	$-I_{BO}$ in µA	$+I_H$ in mA	$-I_H$ in mA	$+I_T$ in mA	$-I_T$ in mA
I	52.4	52	53	49	2.1	2.4	109.7	102.2
II	53	53	51	51	3.1	2.9	102.4	103.0
III	53	54	48	52	3.3	3.1	103.1	1003.3

Discussions

(I) Care should be taken in measuring the breakover voltage for either direction and near the firing point taking as many readings as possible.

(ii) After achieving the firing point the current increases very rapidly with a small increment in the applied voltage and so after achieving the firing point the voltage should be incrased slowly and by small amount.

(iii) The advantage of triac over an SCR is that it operates for both halves of the input cycle.

(iv) Triac can act as a switch.

EXPERIMENT 3

TO STUDY BASIC AND GATE CHARACTERISTICS OF SCR

Theory

If two diodes are connected in series the resulting PNPN 'sandwich' device is called an SCR. There are three terminals—Anode (A), Cathode (K) and Gate (G) as shown in Fig. 3.1. With no connection to the gate terminal, if a voltage is applied to the SCR anode and cathode, J_2 is reverse biased it prohibits current flow. If the polarity is reversed w.r.t. anode, and junctions J_1 and J_3 are reversed biased current flow is likewise prohibited. In

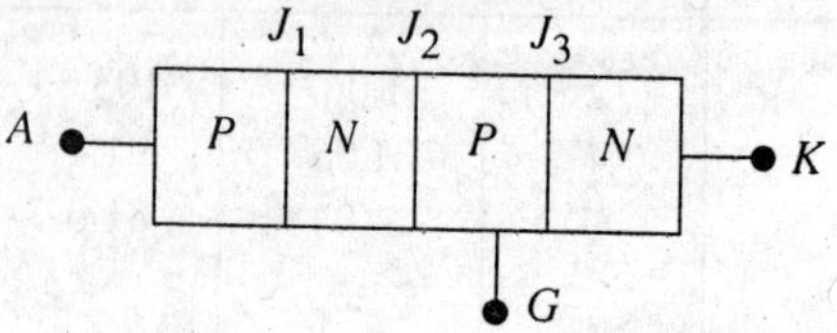

$A \Rightarrow$ Anode terminal
$K \Rightarrow$ Cathode terminal
$G \Rightarrow$ Gate terminal

J_1 = Junction 1
J_2 = Junction 2
J_3 = Junction 3

Fig. 3.1

$Q_1 \Rightarrow$ Transistor 1
$Q_2 \Rightarrow$ Transistor 2
$A \Rightarrow$ Anode terminal
$K \Rightarrow$ Cathode terminal
$G \Rightarrow$ Gate terminal

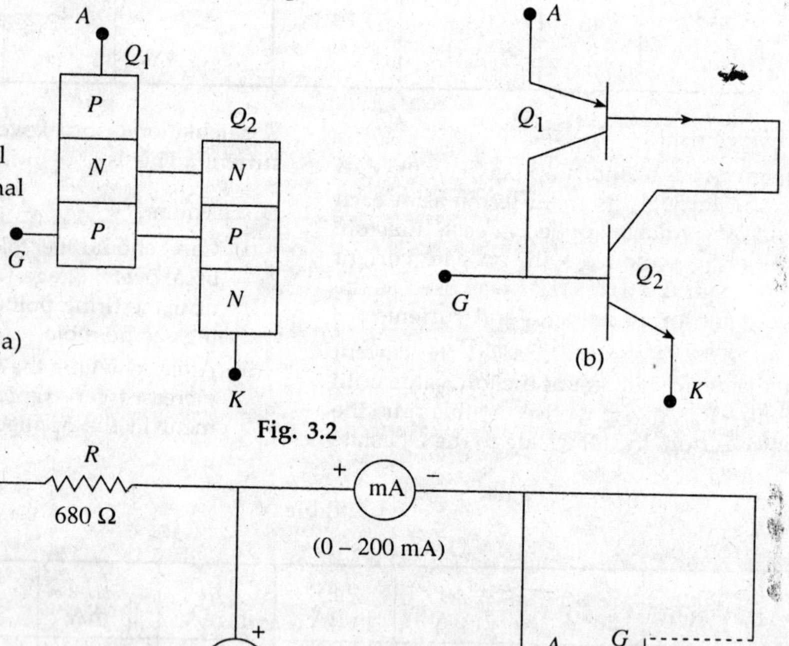

Fig. 3.2

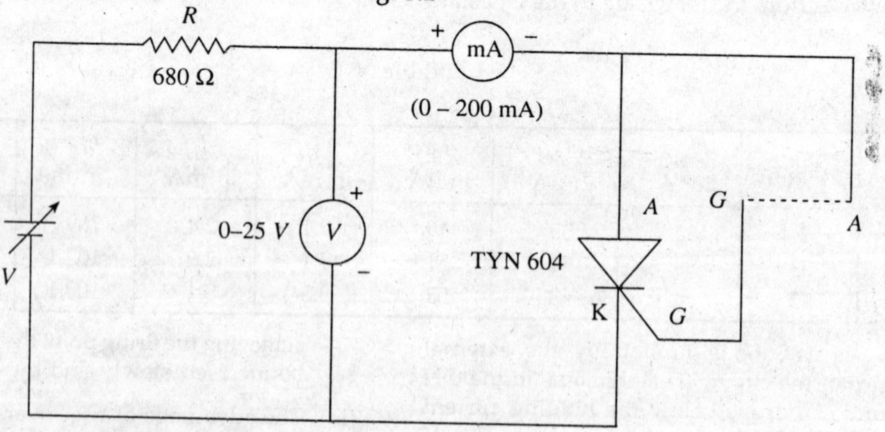

$V \Rightarrow$ Voltmeter (0–25 V), $R = 680\ \Omega$ resistance
$mA \Rightarrow$ milliammeter (0–200 mA)
$A \Rightarrow$ Anode terminal,
$K \Rightarrow$ Cathode terminal, $G \Rightarrow$ Gate

Fig. 3.3

most SCRs J_1 is a low voltage zener diode junction when the device is biased with the anode negative.

To understand the SCR as a three terminal device, it is necessary to consider the SCR as two transistors [Fig. 3.2(a)]. If the SCR of Fig. 3.1 is sliced into two parts and the divided regions are then reconnected with wires, then it can be considered as two transistors Q_1 and Q_2 [Fig. 3.2(b)]. In the configuration the anode current is

$$I_a = I_{C1} + I_{C2} \quad \text{(i)}$$

The collector currents of Q_1 and Q_2 are,

$$I_{C1} = \beta_1(I_{C2} + I_{cbo1}) + I_{cbo1} \quad \text{(ii)}$$

$$I_{C2} = \beta_2(I_{C1} + I_{cbo2}) + I_{cbo2} \quad \text{(iii)}$$

where β_1 and β_2 are the gains of Q_1 and Q_2 and I_{cbo1} and I_{cbo2} are the respective base leakage currents. Solving (i), (ii) and (iii), we get

$$I_a = \frac{(1+\beta_1)(1+\beta_2)(I_{cbo1}+I_{cbo2})}{(1-\beta_1\beta_2)} \quad \text{(iv)}$$

From Eqn. (iv), it is clear that at some specified anode current $\beta_1\beta_2$ will be unity. Consequently if there is a voltage applied to the anode and cathode of the SCR, the leakage current of Fig. 3.3 of the two equivalent transistors creates the first term of the numerator. Gate current then needed to create the amount of anode current that will allow the product $\beta_1\beta_2$ to reach unity and turn on the SCR.

Characteristics of SCR : When a supply voltage is applied through a load resistor between anode and cathode of an SCR and voltage is less than breakover voltage, the rectifier will remain OFF and may be turned ON by the application to the gate of a triggering current or voltage adequate to lower the breakover voltage to less than the applied voltage. The rectifier having been turned ON, it latches and it is found to be impractical to stop the conduction by reverse biasing the gate.

It turns off when the anode voltage is reduced below the holding voltage V_H or the anode current below the holding current I_H.

Experimental Data :

A. Specification of the meter used :

Type of the SCR –TYN 604

Table I

Type	Particulars
Voltmeter (multimeter)	0-2000 mV
Repetitive Peak Reverse Voltage (open gate voltage)	25 V
RMS ON-state Current	5.4 mA
Average Trigger Voltage	15 V
Surge Non-repetitive ON-state Current	7 mA
Milliammeter I	0-200 mA
Milliammeter II	0-20 mA

B. Variation of cathode to anode voltage (V_{AK}) with corresponding current (I_{AK}) for a fixed gate current (I_G) :

Table II

No.	I_G = 0mA		I_G = 7.05 mA		I_G = 7.10 mA		I_G = 7.16 mA		I_G = 7.4 mA	
obs	V_{AK} in V	I_{AK} in mA	V_{AK} in V	I_{AK} in mA	V_{AK} in V	I_{AK} in mA	V_{AK} in V	I_{AK} in mA	V_{AK} in V	I_{AK} in mA
1	1	0.5	1	0.5	1	0.5	1	0.5	0.85	1.5
2	2	0.6	2	0.6	2	0.6	2	0.6	0.88	3.4
3	3	0.7	3	0.7	3	0.7	3	0.7	0.90	5.3
⋮	⋮	⋮	⋮	⋮	⋮	⋮	⋮	⋮		
16	17	6.9	15	5.4	16	9.1	16	14.2		

The characteristics of SCR is drawn in Fig. 3.4.

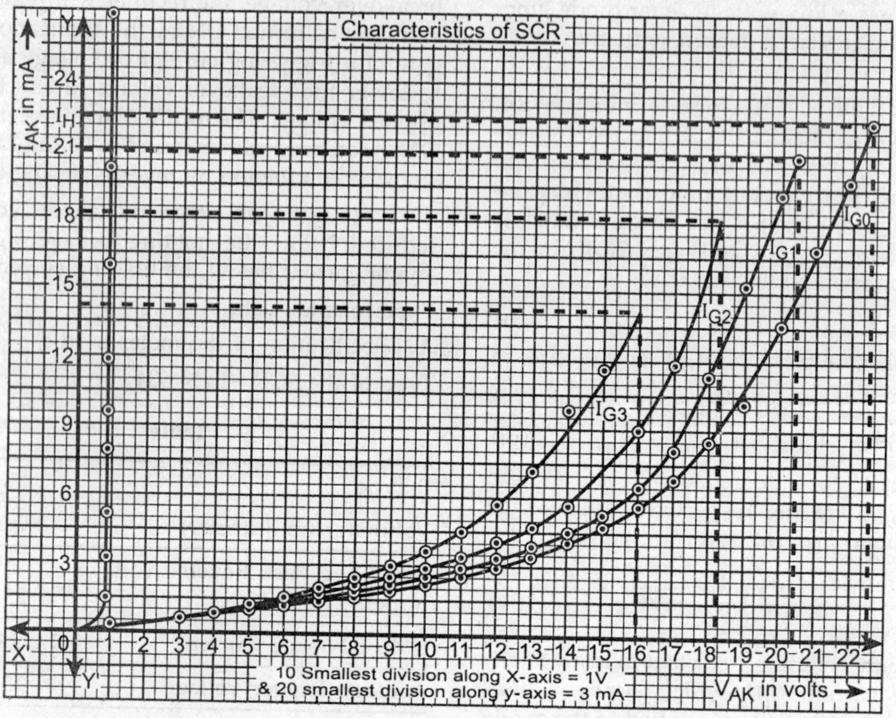

$I_{G0} = 0$ mA, $I_{G1} = 7$ mA, $I_{G2} = 7.10$ mA, $I_{G3} = 7.16$ mA, $I_{G4} = 7.40$ mA

Fig. 3.4

C. Calculation of different parameters from SCR characteristics :

Table III

V_{BO} in volts					IM	VM
$I_G = 0$ mA	$I_G = 7.05$ mA	$I_G = 7.10$ mA	$I_G = 7.16$ mA	$I_G = 7.40$ mA	in mA	in volts
22.60v	20.5v	18.30v	16.00v	0.96v	22.3 mA	0.96 volts

Discussion

(i) In general, the supply voltage is much less than the breakover voltage.

(ii) If the gate current is increased above the required value, the SCR will close at a very reduced supply voltage.

(iii) To turn OFF the SCR from the ON state the anode current should be reduced to holding current.

(iv) By passing an appropriate amount of current the SCR is turned ON and not by break over voltage.

Advantages

(i) Its switching speed is too high upto 10^9 operations/second.

(ii) By a small gate current of few milliamperes it permits control over large current.

(iii) It can provide noiseless operation.

(iv) Its size is small and service is trouble free.

EXPERIMENT 4

TO STUDY UJT CHARACTERISTICS AND RELAXATION OSCILLATOR USING UJT

Theory

The Unijunction Transistor (UJT) is a three terminal semiconductor device with –ve resistance characteristics. It consists of a bar of n-type Si with a small *p*-type emitter near to one of the leads. The two ohmic contacts at the ends of the n-type bar constitute bases 1 and 2 [Fig. 4.1(a)]. The rectifying contact is the emitter.

The n-type bar is lightly doped and acts as a resistor. The effective resistance between base, 1 and 2 is called inter base resistor (R_{BB}). It is also known as double base diode. The standard symbol of a UJT is shown in Fig. 4.1(b).

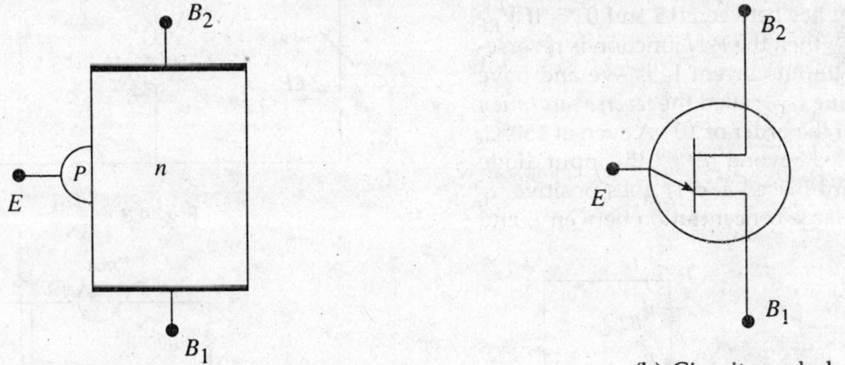

(a) Unijunction transistor (b) Circuit symbol

Fig. 4.1

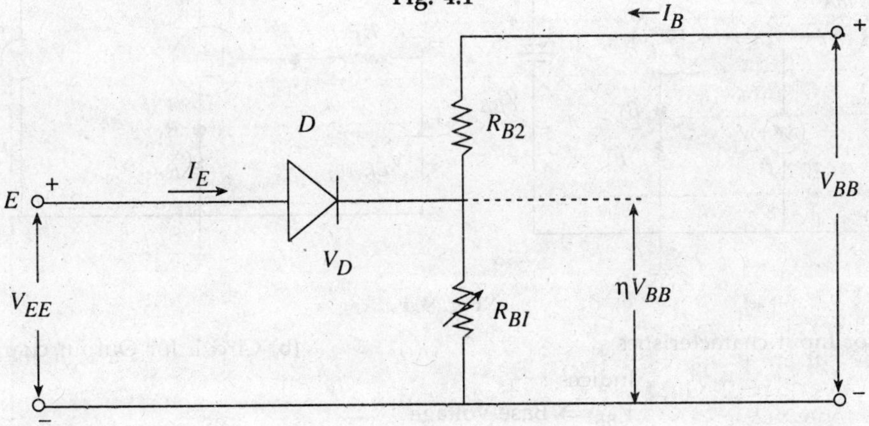

Fig. 4.2

Equivalant circuit of a UJT

B_1 ⇒ Base 1; B_2 ⇒ Base 2; E ⇒ Emitter

V_{EE} ⇒ Emitter voltage; V_D ⇒ Voltage drop a cross diode

V_{BB} = Base voltage; R_{B2} = Base 2 resistance

I_B ⇒ Base current; R_{BI} = Resistance across Base 1

I_E = Emitter current; $\eta\ V_{BB}$ ⇒ Voltage drop at R_{B1}

Usually a fixed inter-base potential V_{BB} is applied between B_1 and B_2 as shown in Fig. 4.2. Most important characteristics of UJT is that of the input diode between E and B_1. If B_2 is open-circuited so that $I_{B2} = 0$, then the input volt-ampere relationship is that of the usual P-N junction. Fig. 4.3 is used for getting UJT characteristics for a fixed value of V_{BB}. If $I_E = 0$, then the silicon bar may be considered as an ohmic resistor R_{BB} between base leads. Usually R_{BB} lies in the range 5-10 kΩ. Between B_1 (or B_2) and the n side of the emitter junction the resistance is R_{B1} (or R_{B2}) so that $R_{BB} = R_{B1} + R_{B2}$. Under this condition of small or zero value of I_E, the voltage on the n side of the emitter junction is ηV_{BB} where

$\eta = \dfrac{R_{B_1}}{R_{BB}}$ and is called the intrinsic Stand-off Ratio. It usually lies between 0.5 and 0.75. If V_{EE} is less than ηV_{BB} then the *P-N* junction is reverse-biased and the input current I_E is –ve and have a maximum value I_{EO}, called the *reverse saturation current* and is of the order of 10 μA even at 150°C. If V_{EE} is increased beyond ηV_{BB}, the input diode becomes forward biased and I_E goes positive. I_E increases the charge concentration betwen E and B_1 because holes are injected in the *N*-type Si. Since conductivity is proportional to charge density, the resisitance R_{B1} decreases with increasing I_E. Hence, for voltages above the threshold value V_t as I_E is increased, the voltages V_E between E and B_1 decreases because of the decrease in the value of the resistance R_{BI}. Fig. 4.4(a) and (b) are used for studying the input and output characteristics of UJT.

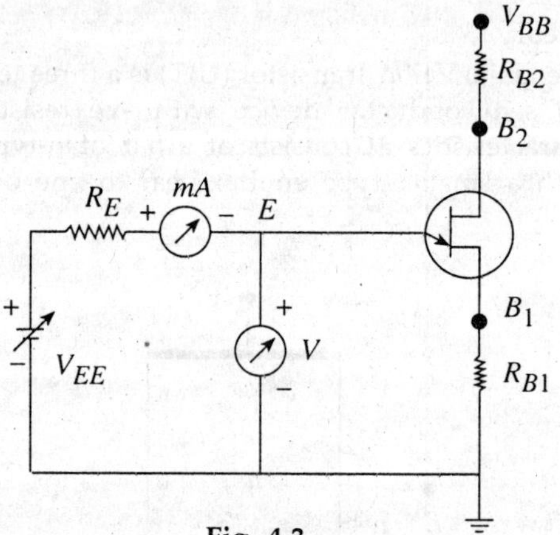

Fig. 4.3

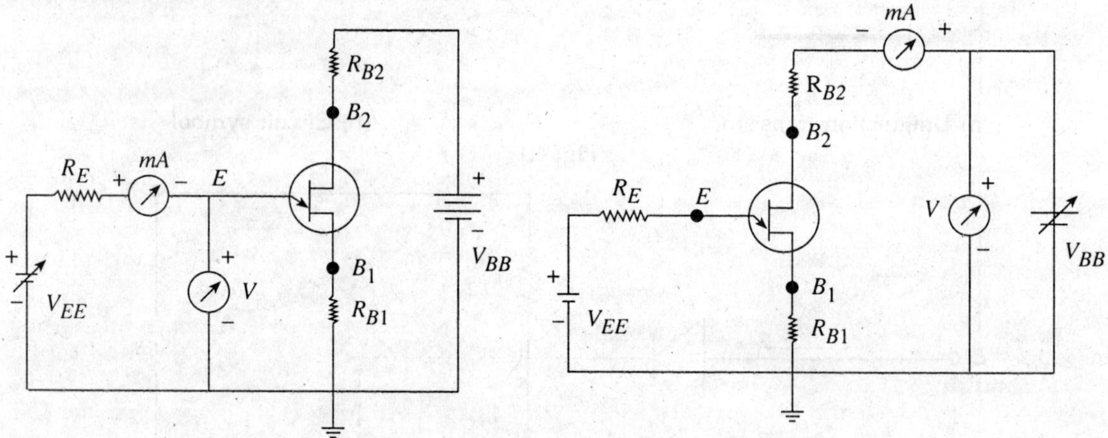

Fig. 4.4

(a) Circuit for Input characteristics.

(b) Circuit for Output characteristics

Indices :

V_{BB} ⇒ Base voltage
V_{EE} ⇒ Emitter voltage
R_E ⇒ Resistance across emitter
R_{B1} ⇒ Resistance across Base 1
R_{B2} ⇒ Resistance across Base 2
mA ⇒ milliammeter (0–20 mA)
V ⇒ Voltmeter (0–15 V)
Type of UJT ⇒ 2N 2646

Indices :

V_{EE} ⇒ Emitter voltage

V_{BB} ⇒ Base voltage

R_E ⇒ Emitter resistance

C_E ⇒ Capacitor between emitter and ground

R_{B1} ⇒ Resistance across Base 1

R_{B2} ⇒ Resistance across Base 2

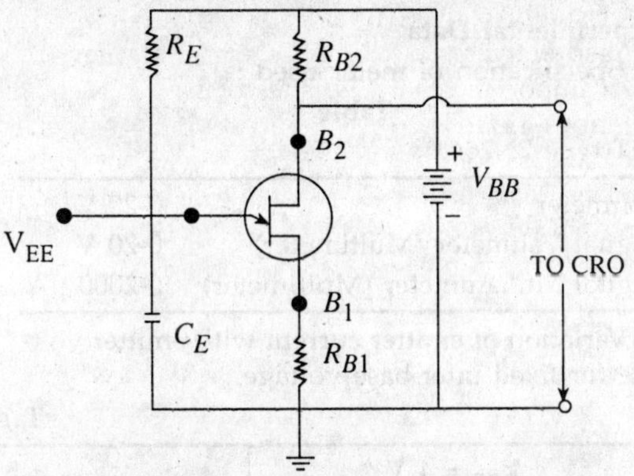

Relaxation oscillator circuit employing a UJT

Fig. 4.5

After the emitter current has become very large value than I_{B2}, we can neglect it. Hence for very large I_E the input characteristic asymptotically approaches the curve for $I_{B2} = 0$. This results in a minimum valley point where the resistance changes from –ve to +ve. Above the valley point (the current I_V) we obtain the saturation region. The voltage at $I_E = 50$ mA is called saturation voltage V_E and is of the order of 3 V. Region left to the peak voltage V_P ($I_P \sim 25$ μA) is called cut off region. The peak voltage is given by,

$$V_P = \eta V_{BB} + V_t \,. \quad \text{(i)}$$

Since the temperature coefficient of R_{BI} is the same as that of R_{B2}, then $\eta = \frac{R_{B1}}{R_{B1}+R_{B2}}$ should be independent of temperature. Experimentally temp. coefficient of η is called Intrinsic stand-off ratio which is less than 0.01 per cent per 1°C and may be either +ve or –ve. For input characteristics the peak voltage increases linearly with V_{BB} and the valley is very broad. So it is difficult to give an exact value of I_V but V_V can be determined accurately.

In the output characteristics for $I_E = 0$, a straight line curve is obtained, i.e., with emitter open circuited the *n*-type Si bar is essentially ohmic. Reciprocal slope of this line gives R_{BB}.

UJT Relaxation Oscillator : The circuit is shown in Fig. 4.5. The resistors R_{B1} and R_{B2} are not essential to the circuit but are included because the voltages developed across these resistors may prove useful. The capacitor charges to the peak voltage V_P. The device turns ON and the capacitor discharges to the valley voltage V_V and the cycle repeats. The charging time T_1 is given

by, $$T_1 = R_E C_E \frac{\eta(V-V_V)}{V-V_P} \quad \text{(ii)}$$

Pulses develop across R_{B1} and R_{B2} during the interval when the device is ON and the capacitor discharges. The pulses at B_1 has an abrupt leading edge, but the abrupt drop at the training edge may not be easily apparent if the holding current is small in comparison to the current at the moment of breakdown. The B_2 current is smaller, being the B_1 current minus the emitter current, but is more nearly constant during the capacitor discharges.

The period of oscillation and thus the frequency is determined primarily by the values of R_E and C_E. It is most practical to start design by selecting a trial value for R_E rather than for C_E because R_E must meets certain conditions for the oscillator to operate. If R_E is too large, the UJT will never fire. If R_E is too small, the UJT will not turn off. If R_E is within the range, the period of oscillation is given by

$$T = R_L C_L \frac{\eta(V-V_V)}{V-V_P} \quad \text{(iii)}$$

where V is the supply voltage.

Experimental Data

A. Specification of meter used :

Table I

UJT type-2N2646

Voltmeter	0-20 V
Digital Voltmeter (Multimeter)	0-20 V
Digital Milliammeter (Multimeter)	0-2000 μA

B. Variation of emitter current with emitter voltage for fixed inter-base voltage :

Table II

V_{BB} = 4 V		V_{BB} = 5 V		V_{BB} = 6 V	
Emitter Vol. V_{EE} in volts	Emitter. Cur. I_E in μA	Emitter Vol. V_{EE} in volts	Emitter Cur. I_E in μA	Emitter Vol. V_{EE} in volts	Emitter Cur. I_E in μA
0.90	1	1.00	1	1.00	1
2.00	2	2.00	1.5	2.90	2
2.20	3	2.40	2	3.20	3
2.30	4	2.70	3	3.30	4
⋮	⋮	⋮	⋮		
1.50	95	1.70	286	1.81	250
1.40	180	1.71	315	1.83	290
1.43	195	1.72	336	1.87	305
1.50	262	1.73	348	1.90	326
1.55	367	1.75	355	1.91	348

The variational pattern is drawn in Fig. 4.6 for V_{BB} = 4 V. Similar patterns can be drawn for other V_{BB} values.

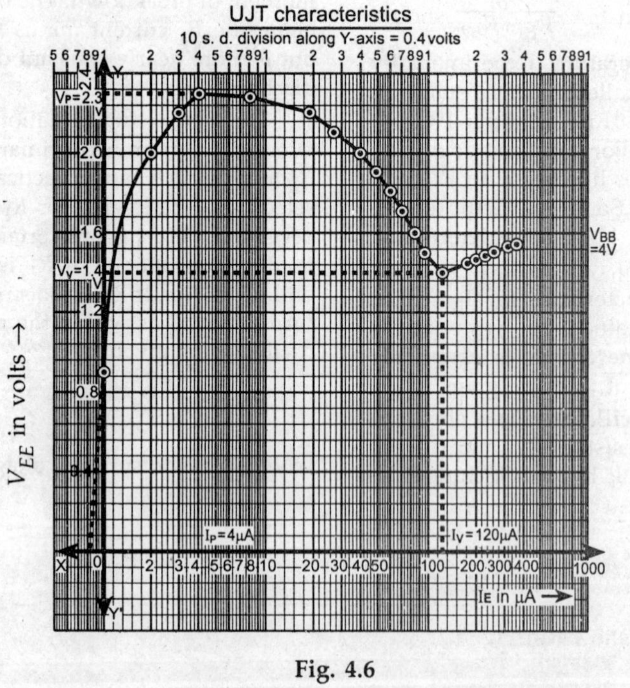

Fig. 4.6

C. To study relaxation oscillator using UJT :

Table III

Set No.	Amplitude of V_V in: volt/div (n)	No of scale div. bet. peak to peak (d)	Peak to peak value (nd) in volts	Amplitude of (V_P-V_V) in: volt/div (n')	No. of scale div. betn. peak to peak(d')	Peak to peak value (n'd') in volts	Amplitude of T_1 in: Time/div (t)	No. of scale div(s)	Total value (ts) in ms	Amplitude of T_2 in: Time/div (t')	No of scale div (s')	Total value (t's') in ms
I	0.1	14	1.4	0.1	22	2.2	1 ms	21	21 ms	1 ms	7	7 ms
II	0.1	16	1.6	0.1	28	2.8	1 ms	22.5	22.5 ms	1 ms	7.5	7.5 ms
III	0.1	18	1.8	0.1	32	3.2	1 ms	24	24 ms	1 ms	8	8 ms

D. Calculation of theoretical and practical frequencies :

We have $f = \frac{1}{T}$, where

$$T = R_E C_E \frac{\eta(V-V_v)}{(V-V_P)} \text{ and}$$

$V_P = \eta V_{BB} + V_t$ (V_t is the voltage drop across emitter diode which is 0.6 V typical)

$$\text{or, } \eta = \frac{V_p - V_t}{V_{BB}}$$

Table IV

Value of R_E in kΩ	Value of C_E in μF	Value of η	Value of V_v in volt	Value of (V_P-V_v) in volt	T in ms	Th f in Hz	Practical frequency
		0.425	1.4	2.2	20.25 ms	49.38 Hz	
33 kΩ	0.1 μF	0.440	1.6	2.8	22.44 ms	44.56 Hz	48.23 Hz
		0.450	1.8	3.2	22.28 ms	44.88 Hz	

E. Study of the input characteristics :

Table V

V_{BB} = 4 V: V_{EE} in volts	I_E in μA	V_{BB} = 5 V: V_{EE} in volts	I_E in μA	V_{BB} = 6 V: V_{EE} in volts	I_E in μA
2.30	4	2.80	4	3.30	10
2.28	8	2.70	20	3.00	20
2.20	19	2.60	27	2.90	30
⋮	⋮	⋮	⋮	⋮	⋮
1.40	180	1.80	120	2.10	125

The characteristics are shown in Fig. 4.7.

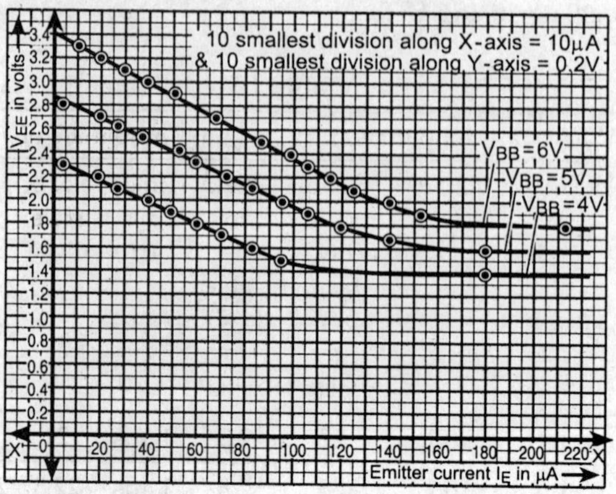

Fig. 4.7

F. Study of the output characteristics :

Table VI

I_E = 10 mA		I_E = 20 mA		I_E = 30 mA	
V_{BB} in volts	I_{B2} in μA	V_{BB} in volts	I_{B2} in mA	V_{BB} in volts	I_{B2} in mA
1.01	0	1.70	0	1.70	0
1.25	1.26	1.75	2.10	1.75	2.10
1.50	3.32	2.00	6.06	2.00	6.06
⋮	⋮	⋮	⋮	⋮	⋮
3.25	7.98	3.75	11.94	3.75	13.38

The characteristics are shown in Fig. 4.8.

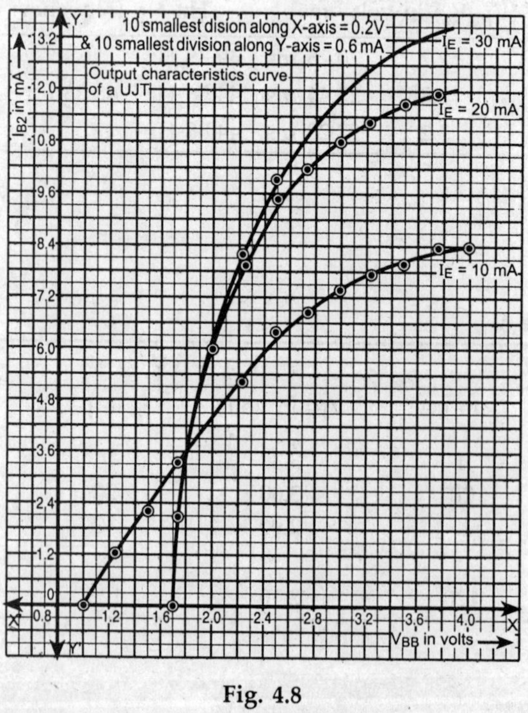

Fig. 4.8

Discussion

(i) It is very difficult to get the cut-off region in the UJT characteristics curve.

(ii) Sometimes the UJT may be heated up very rapidly and then reading should not be taken.

(iii) η may be of different values for different temperatures.

(iv) The negative terminal of CRO should be grounded properly to get the waveform pattern and proper measurements.

EXPERIMENT 5

CONSTRUCTION OF A SIMPLE POWER SUPPLY : TO STUDY LINE AND LOAD REGULATION

Theory

A regulated power supply gives a d.c. output voltage that remains substantially constant even when the a.c. supply voltage or the d.c. load current changes. The system contains a rectifier with filter and a voltage regulator. The rectifier transforms the input alternating voltage into a d.c. voltage (or current). The output of the rectifier contains the fluctuating component of current, called the ripple. The filter, following the rectifier stage, smoothes out the ripples.

The output from the filter can vary with the changes in the a.c. supply voltage or the load resistance. The voltage regulation tends to limit this fluctuation. The ripple factor ν is obtained from

$$\nu = \frac{\pi}{2\sqrt{3}\omega CR} = 0.01 \text{ where } \omega = 2\pi f \quad \text{(i)}$$

$$\text{and } R = \frac{E_{dc}}{I_{max}} = \frac{V_m}{I_{max}} \quad \text{(ii)}$$

Fig. 5.1 is a full wave rectifier circuit with capacitor filter . The output voltage obtained here is not fully regulated. In the next stage we add a zener diode Z with suitable resistance R_s. as shown in Fig. 5.2 Here R_s is obtained by the relation, $V_{dc} - V_z = I_{max} R_s$ (iii)

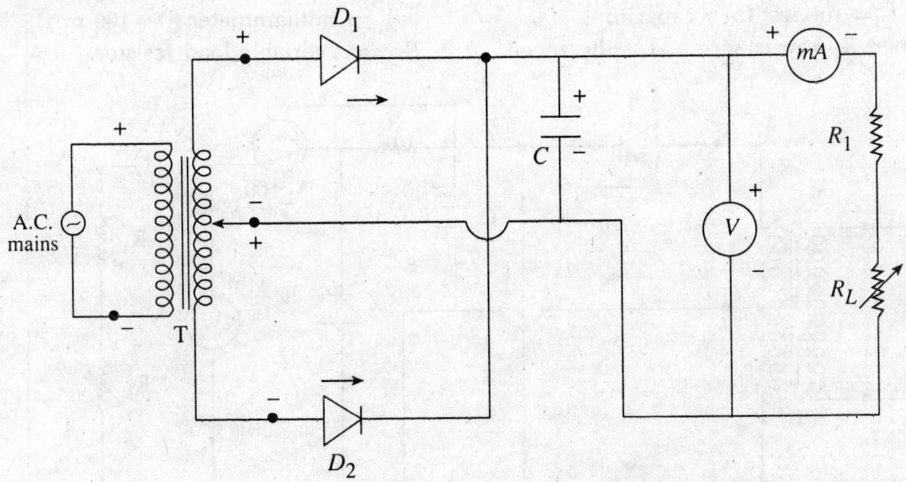

Fig. 5.1 : Power supply circuit without zener diode

a.c. mains ⇒ variable a.c. voltmeter
T ⇒ a stepdown transformer
D_1 and D_2 ⇒ diodes
C ⇒ capacitor
V ⇒ voltmeter (0–100V)
mA ⇒ milliammeter (0–100 mA)
R_L ⇒ a variable resistor.

where V_{dc} is the output d.c. voltage of the unregulated supply at 100 mA load current. In this situation we achieve a regulated power supply in which the output voltage remains fixed. The maximum load current available is equal to the current through the zener when R_L is infinite. This current is limited by the dissipation rating of the zener. So to achieve a power supply where the load current is greater than the current allowed through the zener, emitter follower circuit of Fig. 5.3 is used. Here,

$$V_{output} = (V_z - 0.6) \text{ V}$$

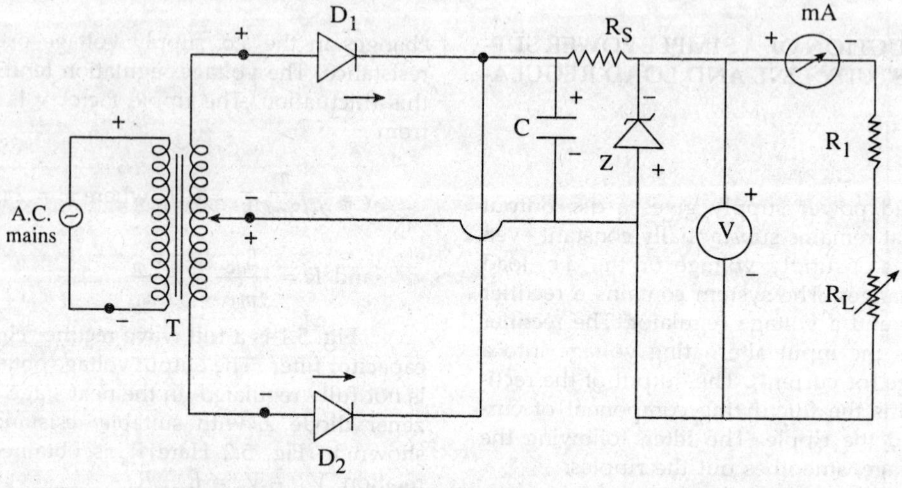

Fig. 5.2 : Power supply circuit employing a zener diode

T ⇒ a stepdown transformer
D_1 and D_2 ⇒ diodes; C ⇒ capacitor
Z ⇒ zener; R_s = resistance across the zener
V ⇒ voltmeter (0 – 100V)
mA ⇒ milliammeter (0 – 100 mA)
R_L ⇒ variable load resistor.

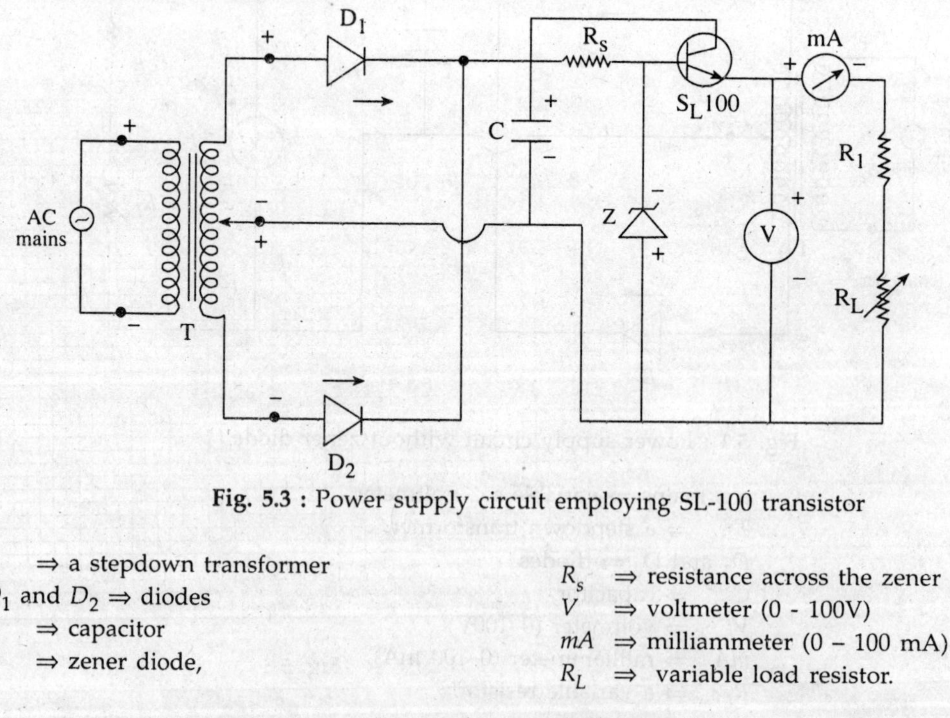

Fig. 5.3 : Power supply circuit employing SL-100 transistor

T ⇒ a stepdown transformer
D_1 and D_2 ⇒ diodes
C ⇒ capacitor
Z ⇒ zener diode,
R_s ⇒ resistance across the zener
V ⇒ voltmeter (0 - 100V)
mA ⇒ milliammeter (0 – 100 mA)
R_L ⇒ variable load resistor.

0.6 V refers to the base-emitter voltage in the forward biased condition.

Experimental Data

A. Calculation of the resistance R_1 :

Value of the capacitor C = 200 μF

Input voltage = 220 volt A.C.

Output voltage = 11.5 volt D.C.

I_{max} = 100 mA

$\therefore$ peak output voltage $V_m = 11.2 \times \sqrt{2}$

$= 15.84$ volt

$\therefore$ value of the resistance, $R_1 = \dfrac{E_{dc}}{I_{max}}$

$= \dfrac{V_m}{I_{max}}$

$= \dfrac{15.84 \times 10^3}{100}\ \Omega$

$= 158.4\ \Omega$

Available value of the resistance $R_1 = 188\ \Omega$.

B. Data for load regulation :

Table I

No. of obs.	D.C. output voltage in V	Load current in mA
1	18 V	2 mA
2	18 V	5 mA
3	18 V	15 mA
⋮	⋮	⋮
11	17.0 V	80 mA

Load regulation curve without zener is shown in Fig. 5.4.

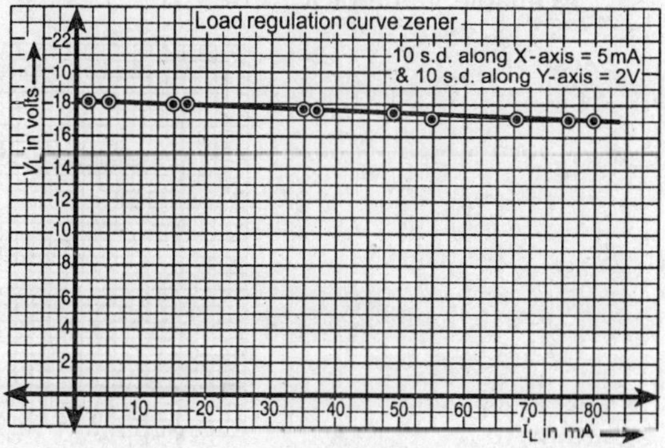

Fig. 5.4

C. Data for input regulation :
Load current fixed at 10 mA.

Table II

No. of obs.	A.C. input voltage in V	D.C. Output voltage in V
1	100	7.5
2	110	8
3	120	9
⋮	⋮	⋮
12	210	16

Input regulation curve without zener is drawn in Fig. 5.5.

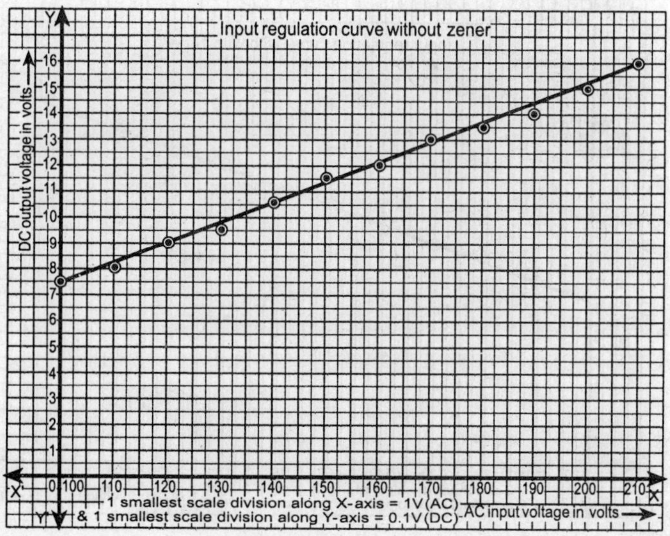

Fig. 5.5

D. Data with zener diode :

Zener used : EC IZ 10A

V_{dc} = 15.84 V

Now we have $(V_{dc} - V_z) = I_{max} R_s$

$$\text{or, } R_s = \frac{(V_{dc} - V_z)}{I_{max}}$$

$$= \frac{15.84 - 9.1}{100 \times 10^{-3}} \Omega$$

$$= 67.4 \ \Omega$$

∴ calculated value of R_s = 67.4 Ω

Available value of R_s = 67.2 Ω

E. Data for load regulation with zener diode :

Table III

No. of obs.	D.C. output voltage in V	Load current in mA
1	13.5 V	1 mA
2	13.5 V	3 mA
3	13.5 V	5 mA
⋮	⋮	⋮
14	13.5 V	62 mA

Load regulation curve with zener diode is drawn in Fig. 5.6.

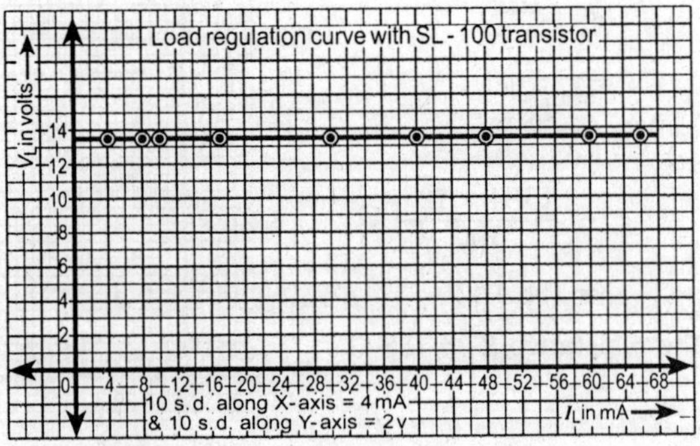

Fig. 5.6

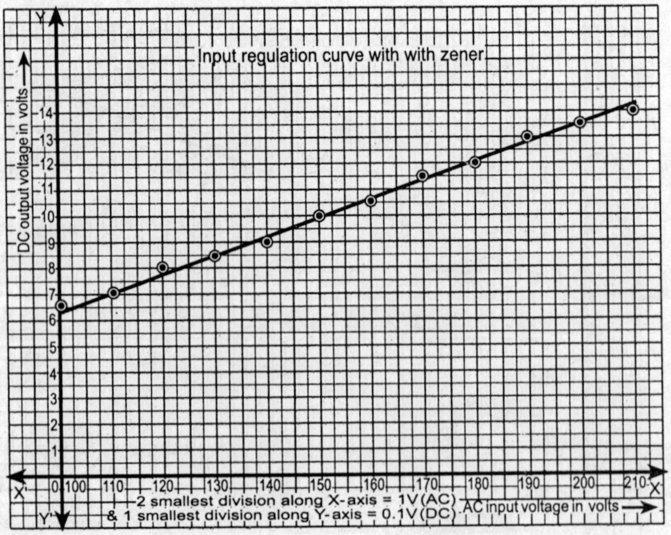

Fig. 5.7

F. Data for input regulation with zener diode :

Load current fixed at 10 mA.

Table IV

No. of obs.	A.C. input voltage in V	DC output voltage in V
1	100	6.5
2	110	7
3	120	8
⋮	⋮	⋮
12	210	14

Input regulation curve with zener is shown in Fig. 5.7

G. Data for SL-100 transistor (Load reguation) :

Table V

No. of obs.	D.C. output voltage in V	Load current in mA
1	13.5 V	1 mA
2	13.5 V	8 mA
3	13.5 V	10 mA
⋮	⋮	⋮
8	13.5 V	66 mA

Load regulation curve with SL-100 transistor is shown in Fig. 5.8.

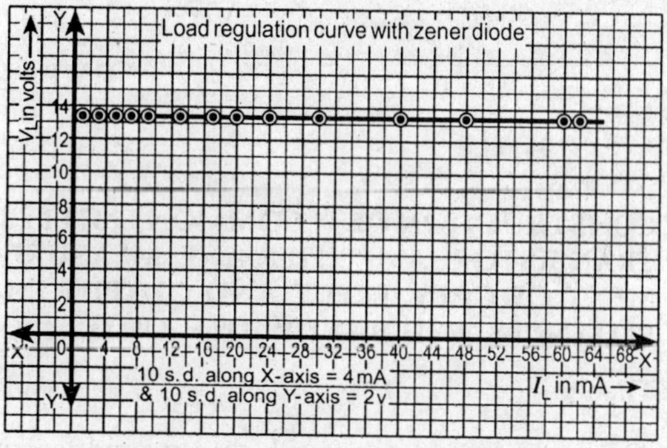

Fig. 5.8

H. Data for input regulation with SL-100 transistor :

Load current fixed at 10 mA.

Table VI

No. of obs.	A.C. input voltage in V	D.C. output voltage in V
1	100	6
2	110	7
3	120	7.5
⋮	⋮	⋮
12	210	14

Input regulation curve with SL-100 is drawn in Fig. 5.9.

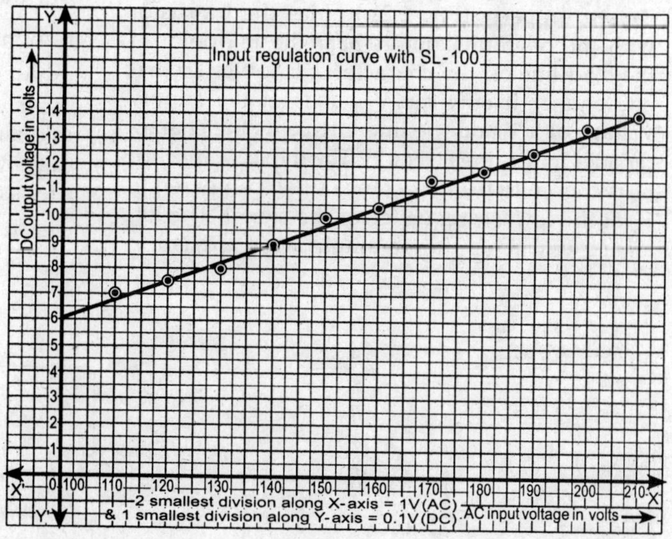

Fig. 5.9

Discussion

(i) Before switching on the A.C. mains be sure that the load R_L is high

(ii) While varying the load R_L, care should be taken that I_L does not exceed the maximum permissible current.

(iii) The output voltage may change with I_L as well as with the input A.C. voltage. As we are studying the load regulation, the constancy of the input A.C. voltage should be maintained.

(iv) The smaller the value of percentage load regulation better is the regulation.

EXPERIMENT 6

CONSTRUCTION OF A COMMON EMITTER TRANSISTOR AMPLIFIER AND MEASUREMENT OF GAIN, INPUT AND OUTPUT IMPEDANCES AND BANDWIDTH OF IT

Theory

An amplifier is a device which can produce an enlarged but exact replica of the input signal wave-form. Fig. 6.1 shows the circuit diagram of a single stage transistor amplifier. The power suply V_{CC} and the resistances R_1, R_2, R_c, R_e are chosen properly (Fig. 6.2) to bias the transistor with the operating points near the midpoint of the load line on the characteristic curves so that one end of it cuts the V_{CE} axis at the supply voltage of 9 V and the line intercepts the characteristics in the linear region. The bypass capacitor C_s is chosen to have reactance small compared to R_e at the lowest frequency of the signal. The a.c. signal voltage (V_i) to be amplified is applied to the base of the transistor via a blocking capacitor C_i. Now a small fluctuation in base current causes a large fluctuation in collector current which on flowing through the load resistance R produces an amplified output voltage (Vo). The Vo is 180° out of phase with V_i. The voltage gain A_v of the amplifier is $|A_v| = \frac{V_o}{V_i}$ (i)

The variation of gain $|A_v|$ with the frequency f of the input signal is known as the frequency response of the amplifier. The gain $|A_v|$ is maximum and remains constant over a certain frequency range, known as midband. Below the midband the gain drops off with decreasing frequency and above the midband the gain also drops off with increasing frequency. The frequencies at which the voltage gain is $1/\sqrt{2}$ times the

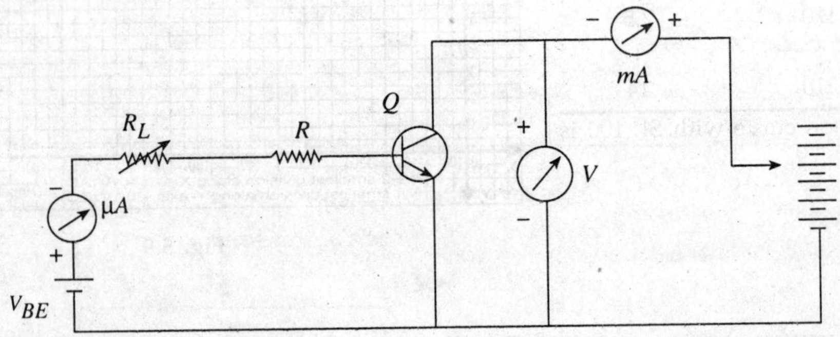

Fig. 6.1

Q ⇒ SL 100 transistor (N–P–N)
V_{BE} ⇒ supply voltage applied at the base-emitter junction
R_L ⇒ a variable resistor
R ⇒ a resistance of value 1 kΩ
μA ⇒ microammeter (0-100 μA)
V ⇒ voltmeter (0–10 V)
mA ⇒ milliammeter (0–50 mA)
V_{CE} ⇒ supply voltage applied at the collector-emitter junction.

maximum value are called half power frequencies. The seperation between these two half power frequencies is known as the bandwidth Δf of the amplifier such that

$$\Delta f = (f_2 - f_1) \qquad \text{(ii)}$$

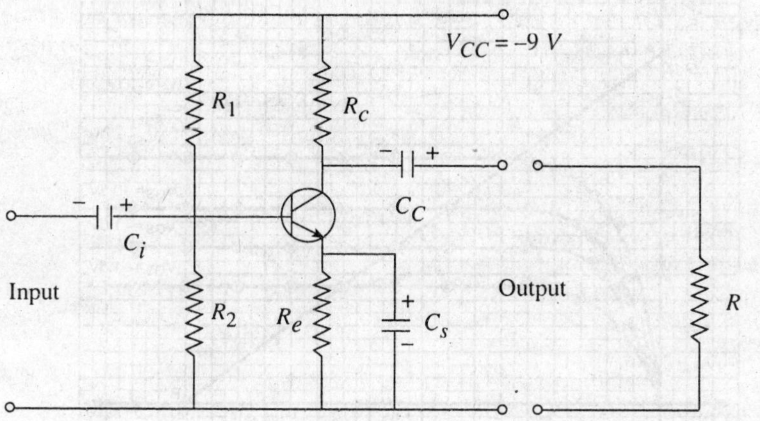

Fig. 6.2

Input ⇒ *a* variable frequency generator

C_i ⇒ blocking capacitor (4.7 μF)

R_1 ⇒ resistance applied at the base terminal (22 kΩ)

R_2 ⇒ other res. applied at the base (22 kΩ)

R_c ⇒ Resistance applied at the callector terminal (378 Ω)

R_e ⇒ resistance applied at the emitter terminal (119 Ω)

C_s ⇒ bypass capacitor (150 μF)

C_c ⇒ capacitor connected at the collector (47 μF)

V_{cc} ⇒ *d.c. S*upply voltage (–9 V)

R ⇒ load resistance connected at the output (470 Ω)

Output ⇒ this terminal is connected to the CRO

Experimental Data

A. Specification of the meters used :

Table I

Meter used	Range
Microammeter	(0–100 μA)
Milliammeter	(0–50 mA)
Voltmeter 1	(0–3 V)
Voltmeter 2	(0–50 V)

B. Data for drawing the collector characteristics of the transistor :

Table II

No.	I_{B1}=20 μA		I_{B2}=40 μA		I_{B3}=60 μA		I_{B4}=80 μA	
of obs.	V_{CE} in V	I_C in mA	V_{CE} in V	I_C in mA	V_{CE} in V	I_C in mA	V_{CE} in V	I_C in mA
1	0	0	0	0	0	0	0	0
2	0.02	0.5	0.02	5.0	0.02	7.0	0.02	10.0
3	0.04	2.5	0.04	6.0	0.04	9.0	0.04	12.0
⋮	⋮	⋮	⋮	⋮	⋮	⋮	⋮	⋮
18	10.00	4.0	10.00	8.5	10.00	13.0	10.00	18.0

C. Calculation of resistances and capacitances from the characteristic curve :

The characteristics are drawn in Fig. 6.3.

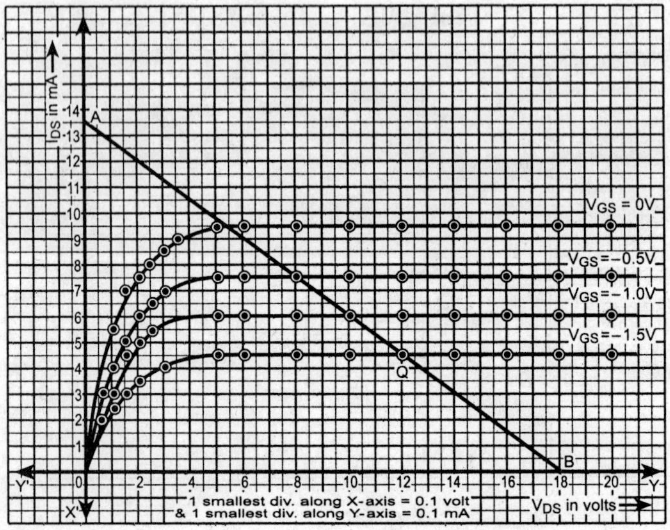

Fig. 6.3

From the characteristic curve we have the load resistance,

$$R_L = \frac{BO}{AO} = \frac{9}{18} \text{ k}\Omega = 500\ \Omega$$

First we set, $R_e + R_c = R_L$ (iii)

Now let us divide R_C and R_e in the ratio 3 : 1

i.e., $R_c : R_e = 3 : 1$

Thus from Eqn. (iii),

$$R_e + 3R_e = 500\ \Omega$$

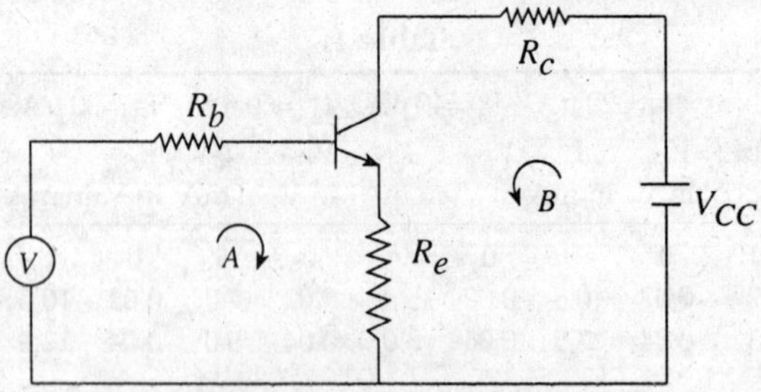

Fig. 6.4

or, $4\ R_e = 500\ \Omega$

$$\therefore \quad R_e = \frac{500}{4} = 125\ \Omega$$

$$\therefore \quad R_c = R_L - R_e = (500 - 125)\ \Omega = 375\ \Omega$$

Now, we have $V_{BE} = 0.3$ volt

$V_{CC} = -9$ V

From graph, V_{CE} = V ...

I_B = A ...

I_C = A ...

we take $R_2 = 22\ \text{k}\Omega = 22 \times 10^3\ \Omega$

and so $R_1 = 22\ \text{k}\Omega$

(applying Kirchhoff's law in the loops A and B of Fig. 6.4).

From Fig. 6.4, we have

$$R_b = \text{resistance at the base} = \frac{R_1 R_2}{R_1 + R_2} = 11\ \text{k}\Omega$$

$$V = \frac{V_{cc} R_2}{R_1 + R_2} = 4.5\ \text{V}.$$

Let $C_1 = 5\ \mu F$ (given).

C_C and C_S are calculated as—

$$X_{CS} = \frac{1}{\omega C_S} << R_e \left(\sim \frac{Re}{10}\right)$$

$$\therefore C_S = \frac{10}{Re\omega} = \frac{10}{125 \times 2\pi f} = \frac{10}{125 \times 2\pi \times 80}$$

[Let lower cut off freq. f = 80 Hz]

= 159.23 μF.

$$X_{CC} = \frac{1}{\omega C_c} << R = R_L \left(\sim \frac{R_L}{10}\right)$$

$$\therefore C_C = \frac{10}{R_L \omega} = \frac{10}{375 \times 2\pi f}$$

$$= \frac{10}{375 \times 2\pi \times 80} = 53.07\ \mu F$$

Available values of the resistances and capacitances :

R_L = 470 Ω

R_e = 119 Ω

R_C = 378 Ω

R_2 = 22 × 10^3 Ω

R_1 = 22 × 10^3 Ω

C_i = 4.7 μF

C_s = 150 μF

C_c = 47 μF

D. Data for drawing the frequency response curve after achieving the amplifier circuit :

Table III

Input voltage fixed at 20 mV

No. of obs.	Frequency in Hz	Output voltage in mV	Vol. gain $A_v = \frac{V_o}{V_i}$	No. of obs.	Frequency in Hz	Output voltage in mV	Vol. gain $A_v = \frac{V_o}{V_i}$
1	200 Hz	200	10.0	22	60 kHz	1100	55.0
2	300 Hz	350	17.5	23	60 kHz	1100	55.0
3	400 Hz	425	21.25	24	70 kHz	1100	55.0
⋮	⋮	⋮	⋮	⋮	⋮	⋮	⋮
21	40 kHz	1100	55.0	37	1100 kHz	520	26.0

Voltage gain vs. frequency curve is drawn in Fig. 6.5

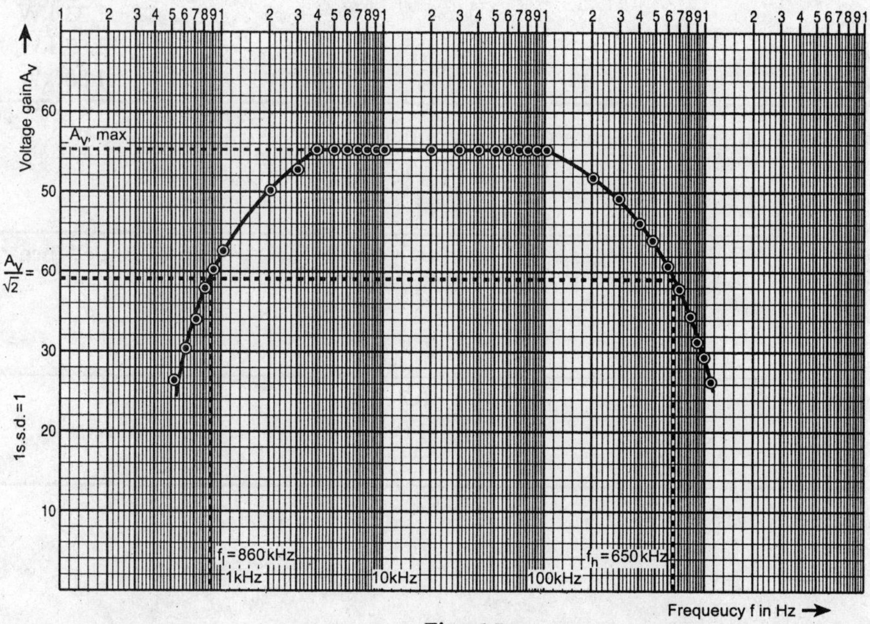

Fig. 6.5

Calculation of bandwidth from the frequency response curve :

Table IV

Midband gain (A_o)	Gain at half power freq. ($A_o/\sqrt{2}$)	Half power freq. from the graph Lower (f_l)	Upper(f_h)	Bandwidth $\Delta f = (f_h - f_l)$
55.0	38.9	860 Hz	650 kHz	649.14 kHz

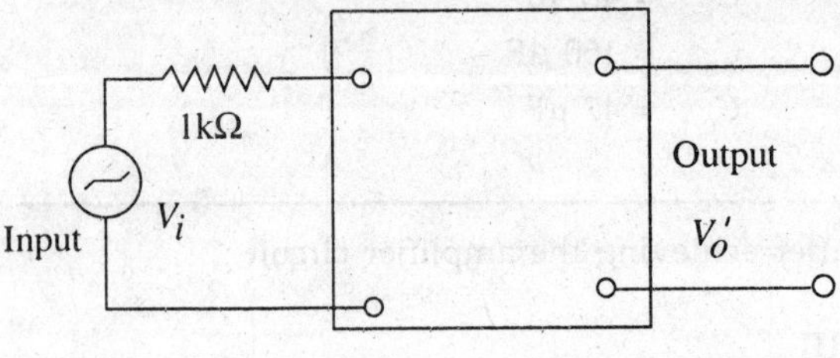

Fig. 6.6

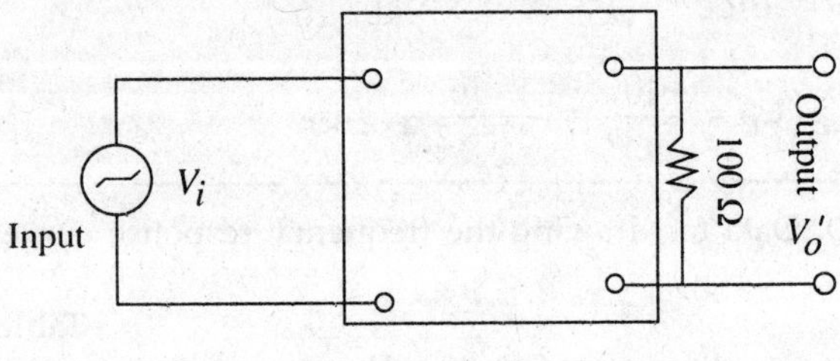

Fig. 6.7

E. Measurement of the input impedance of the amplifier :
(Using Fig. 6.6)

Table V

Frequency in kHz	Output voltage without res. (V_o)	Output voltage with res. (V_o')	Input impedance $Z_i = \left(\frac{V_o' r}{V_o - V_o'}\right)$
1	1.75	1	1.33 kW
10	1.2	0.75	1.67 kW
100	1	0.65	1.86 kW

F. Measurement of the output impedance of the amplifier :
(Using Fig. 6.7)

Table VI

Frequency in kHz	Output voltage without res. (V_o)	Output voltage with res. (V_o)	Output impedance $\frac{(V_o - V_o')r}{V_o'}$
1	1.2	0.7	71 Ω
10	1.4	1.0	40 Ω
100	1.6	1.2	33 Ω

Discussion

(i) After drawing the d.c. load line the Q point should be properly adjusted and the resistances and capacitances needed to make the standard amplifier circuit should be calculated to get proper amplification.

(ii) The d.c. load line should be drawn in such a way that it passes through the linear region of the characteristics.

(iii) The amplitude of V_i should be small such that the Q point is not driven either to cut-off or saturation region and also should be maintained at a constant value throughout the whole observation.

EXPERIMENT 7

CONSTRUCTION OF A FET AMPLIFIER AND MEASUREMENT OF GAIN AND BAND-WIDTH OF THE AMPLIFIER

Theory

An amplifier is a device which can produce an enlarged but exact replica of the input signal waveform. Fig. 7.1 shows the circuit diagram of FET amplifier. The power supply V_{DD}, the resistances R_g, R_s, R_D, and capacitances C_i, C_c, C_s are chosen properly for biasing (Fig. 7.2). The Q point is chosen near the mid-point of the load line drawn at the FET drain characteristics curve.

Amplifier with a very high input inpedance uses field effect-transistors. The frequency response of a FET is very broad and the power amplification is also large. The gate terminal of FET taken in this case is of P-type material. For a FET in proper condition there exists a P-N junction between the gate and the source and also between the gate and the drain. The bypass

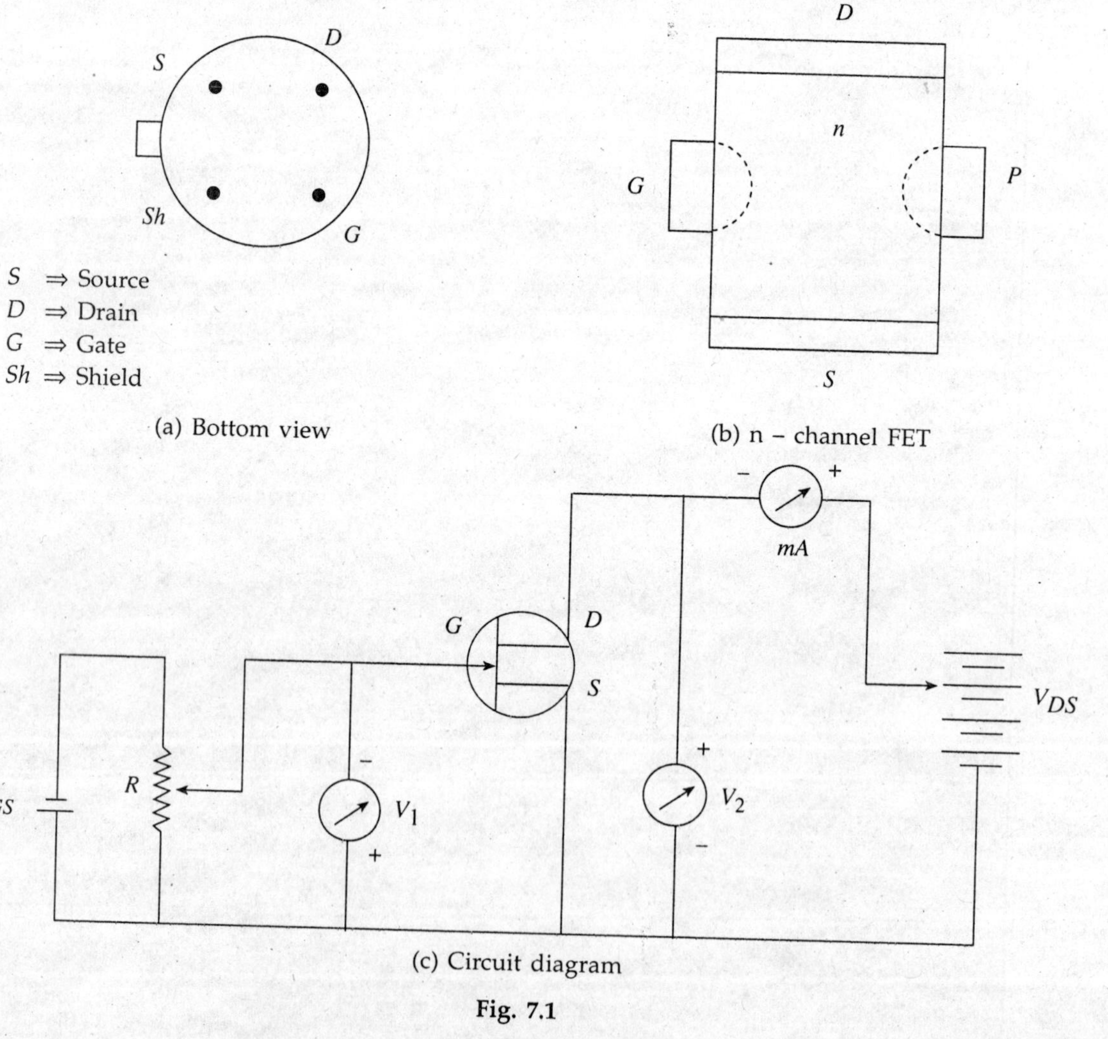

(a) Bottom view

(b) n – channel FET

(c) Circuit diagram

Fig. 7.1

Indices of Fig. 7.1(c) are given below :

V_{GS} ⇒ Supply Voltage applied at the gate to source junction
R ⇒ Variable load resistance
V_1 ⇒ Voltmeter connected across the gate and source (0 – 3 V)
G ⇒ Gate
D ⇒ Drain
S ⇒ Source
V_2 ⇒ Voltmeter connected across the drain and source (0 – 50 V)
mA ⇒ milliammeter (0 – 50 mA)
V_{DS} ⇒ Supply voltage connected across the drain and source

capacitor C_s is chosen to have a small reactance compared to R_s. The A.C. signal voltage (V_i) to be amplified is to be applied to the gate terminal of the FET, via a blocking capacitor C_i. A small fluctuation in the gate current on flowing through the load resistance R produces an amplified output voltage (V_o). This is 180° out of phase with V_i. The voltage gain A_v of the amplifier is

$$|A_v| = \left|\frac{Vo}{Vi}\right| \quad \text{(i)}$$

Variation of $|A_v|$ with the frequency of the input signal is known as the frequency response of the amplifier. The gain $|A_v|$ is maximum and remains constant over a certain frequency range, known as the midband. Below it the gain drops off with decreasing frequency and above also it drops off with increasing frequency. The frequencies at which the voltage gain is $\frac{1}{\sqrt{2}}$ times the maximum value are called the *half power frequencies*. The seperation between these two is known as the band-width Δf of the amplifier such that

$$\Delta f = (f_2 - f_1) \quad \text{(ii)}$$

Fig. 7.2 is the standard amplifier circuit using a FET. Here R_s acts as a self-biasing resistance and the voltage drop across R_s due to the drain current must be equal to the gate bias required.

$$V_{GS} = I_D R_s \quad \text{(iii)}$$

$$\text{and} \quad R_D + R_s = R_L \quad \text{(iv)}$$

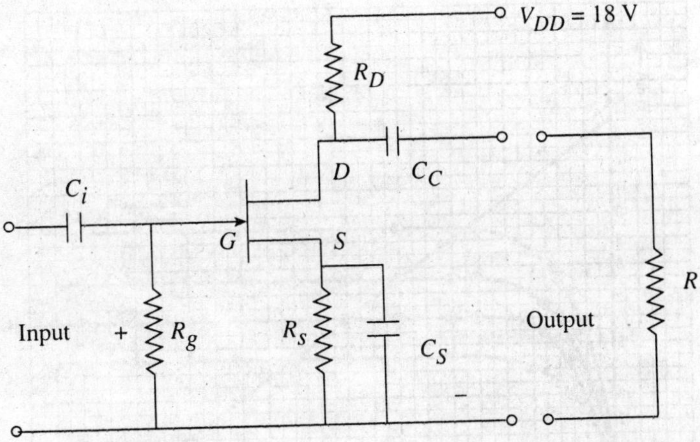

Fig. 7.2

Input ⇒ A variable frequency generator
C_i ⇒ Blocking capacitor (4.7 µF)
R_g ⇒ resistance connected at the gate (1 MΩ)
G ⇒ Gate
D ⇒ Drain
S ⇒ Source
R_D ⇒ Resistance connected at the drain terminal (860 Ω)
C_C ⇒ Capacitance connected from drain to output (10 µF)
C_S = Capacitance connected across R_s (47 µF)
R_S = Resistance connected at the source terminal (380 Ω)
V_{DD} ⇒ Supply voltage (18 V)
R ⇒ Load resistance connected at the output
Output ⇒ This terminal is connected to the CRO.

Using (iii) and (iv) R_D and R_s can be calculated. C_S and C_C are determined as,

$$XC_s = \frac{1}{C_C\omega} << R_s \sim \frac{R_s}{10}$$

and $XC_C = \dfrac{1}{C_c\omega} << \text{R} \sim \dfrac{\text{R}}{10}$; where $R = R_L$

and $\omega = 2\pi f$, f being the lower cut-off frequency.

Experimental Data

A. Specification of the meters used :

Table I

Type of the FET : BFW-10H_2

Meter used	Range
Voltmeter 1	(0–3 V)
Voltmeter 2	(0–50 V)
Milliammeter	(0–50 mA)

B. Data for the FET drain characteristics :

Table II

No. of obs.	V_{GS} = 0 V: V_{DS} in volts	I_D in mA	V_{GS} = –0.5 V: V_{DS} in volts	I_D in mA	V_{GS} = –1.0 V: V_{DS} in volts	I_D in mA	V_{GS} = –1.5 V: V_{DS} in volts	I_D in mA
1	0.5	3.0	0.5	2.0	0.5	1.5	0.5	1.5
2	1.0	5.5	1.0	4.0	1.0	3.0	1.0	2.5
3	1.5	7.0	1.5	5.0	1.5	4.5	1.5	3.0
⋮	⋮	⋮	⋮	⋮	⋮	⋮	⋮	⋮
17	20.0	9.5	20.0	7.5	20.0	6.0	20.0	4.5

FET drain characteristics are drawn in Fig. 7.3.

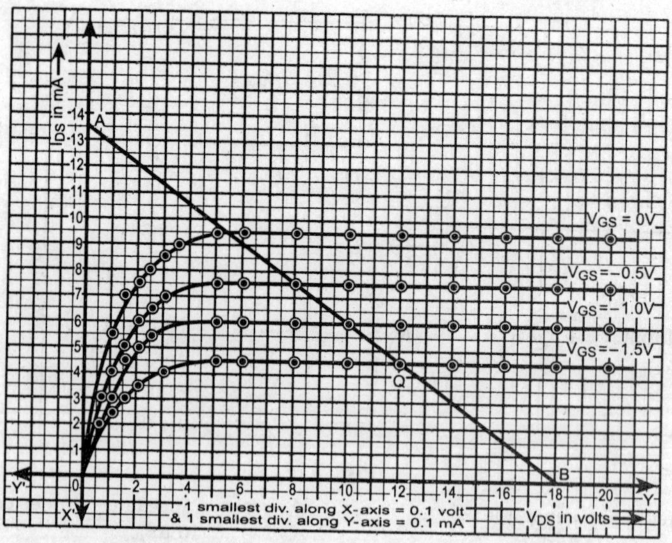

Fig. 7.3

C. Calculation of resistances and capacitances from the characteristic curves :

From the characteristics we have the load resistance,

$$R_L = \frac{BO}{AO} = \frac{18}{13.5} = 1.34 \text{ k}\Omega$$

Now $V_{GS} = I_D R_S$.

$$\therefore R_s = \frac{V_{GS}}{I_D} = \frac{1.5}{4.5} = 0.333 \text{ k}\Omega = 333\ \Omega$$

Now, $R_D + R_S = R_L$

$\therefore R_D = R_L - R_S$

$= 1.34 - 0.333 = 1 \text{ k}\Omega$

Let $R_g = 1$ MΩ (given).

and $C_i = 5$ μF (given).

$$\text{Now, } X_{C_S} = \frac{1}{C_S\omega} << R_s \sim \frac{R_s}{10}.$$

$$\therefore C_S = \frac{10}{\omega R_S}.$$

$$= \frac{10 \times 10^6}{2\pi \times 100 \times 333} \text{ μF}$$

$$= 47.8 \text{ μF}$$

$$\text{and } X_{C_C} = \frac{1}{C_c\omega} << R \sim \frac{R}{10} = \frac{R_L}{10}$$

$$\therefore C_c = \frac{10}{\omega R_L} = \frac{10 \times 10^6}{2\pi \times 100 \times 1333} = 11.94 \text{ μF}$$

Available values of the resistances and capacitances :

$R_L = 1.54$ kΩ

$R_s = 380$ Ω

$R_g = 1$ MΩ

$R_D = 860$ Ω

$C_i = 4.7$ μF

$C_s = 47$ μF

$C_c = 10$ μF.

D. Data for drawing the frequency response curve :

Table III

Input voltage fixed at 20 mV.

No. of obs.	Frequency f in Hza	Output voltage in mV	Vol. gain $Av = \frac{V_o}{V_i}$	No. of obs.	Frequency f in Hza	Output voltage in mV	Vol. gain $A_v = \frac{V_o}{V_i}$
1	10 Hz	120	6	22	30 kHz	240	12
2	20 Hz	160	8	23	40 kHz	240	12
3	30 Hz	190	9.5	24	50 kHz	240	12
⋮	⋮	⋮	⋮	⋮	⋮	⋮	⋮
21	20 kHz	240	12	39	1100 kHz	115	5.75

Frequency response curve is shown in Fig. 7.4.

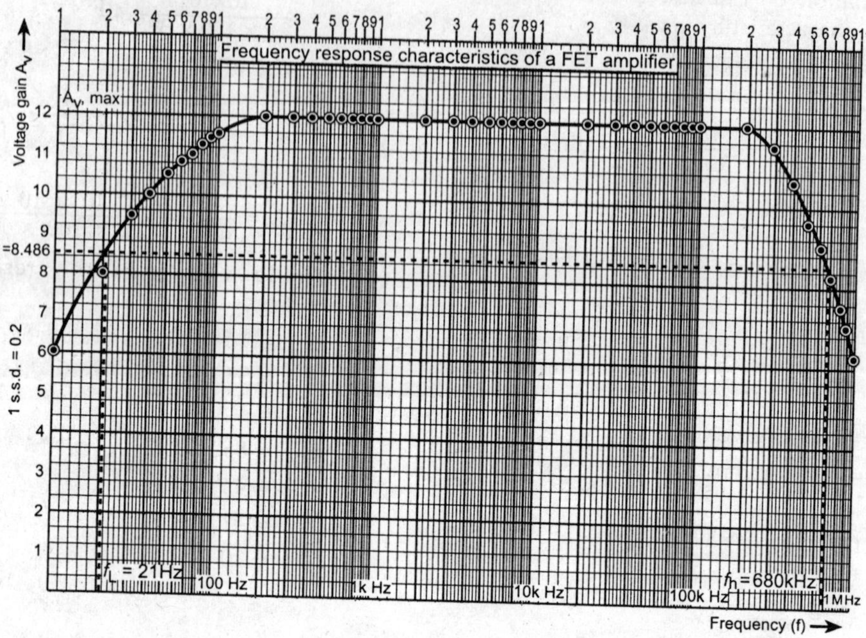

Fig. 7.4

E. Calculation of the bandwidth from the frequency response curve :

Table IV

Mid band gain (A_o)	Gain at half power freq. ($A_o/\sqrt{2}$)	Half power freq.		Bandwidth
		Upper (f_h)	Lower (f_l)	$\Delta f = f_h - f_l$
12.00	8.48	680 kHz	21Hz	679.979 kHz

Discussion

(i) After drawing the d.c. load line the Q point should be properly adjusted and the resistances and capacitances needed to make the standard amplifier circuit should be properly calculated to get proper amplification.

(ii) The d.c. load line should be drawn in such a way that it passes through the linear part of the FET drain characteristics.

(iii) The amplitude of V_i should be small such that the Q point is not driven either to cut-off or saturation region.

(iv) The input voltage V_i should be maintained at a constant value throughout the whole experiment.

EXPERIMENT 8

DESIGN AND CONSTRUCTION OF A MULTIVIBRATOR CIRCUIT AND TO STUDY ITS PERFORMANCE AND CHARACTERISTICS

Theory

The multivibrator is a relaxation oscillator generating sinusoidal waveforms. Basically the multivibrator is a two-stage amplifier or oscillator operating in two modes or states. Each amplifier stage feeds back the other such that the active element of one stage is driven to saturation and the other to cut-off. A new set of actions, producing the opposite effects, then follows. Thus the saturated stage becomes cut-off and cut-off stage saturates. The operation of the multivibrator is based on the fact that no two active elments have exactly identical characteristics. The circuit diagram for finding the D.C. gain of a transistor is shown in Fig. 8.1 The multivibrator circuit is basically two symmetrical CE amplifier stages, each supplying a feedback signal to the other (Fig. 8.2). The transistor Q_1 is forward-biased by the supply voltage V_{CC} and the resistance R_{BI} while Q_2 is forward-biased by V_{CC} and the resistance R_{B2}. The collector-emitter voltages of Q_1 and Q_2 are determined respectively by the load resistances R_{C1} and R_{C2} together with V_{CC}. The capacitor C_2 couples the output of Q_1 to the input of Q_2 whereas C_1 couples the output of Q_2 to the input of Q_1. In practice $R_{BI} = R_{B2}$, $R_{C1} = R_{C2}$ and $C_1 = C_2$.

We can now differentiate the pulses to examine the performance of a differentiator circuit using Figs. 8.3, 8.4 and 8.5. Here the input is applied to a series combination of a resistor and capacitor and the output is taken from the two ends of the resistor. The output is a train of sharp positive and negative pulses when the input is a square wave.

Experimental Data

A. Calculation of D.C. gain for different transistors :

Base supply voltage kept at 2 V.

Table I

Transistor I			Transistor II			Transistor III			Transistor IV		
I_b in μA	I_c in mA	gain β	I_b in μA	I_c in mA	gain β	I_b in μA	I_c in mA	gain β	I_b in μA	I_c in mA	gain β
20	7.0	350	20	2.0	100	20	2.0	100	20	0.6	30
40	14.2	355	40	4.7	117.5	40	5.0	125	40	1.5	37.5
60	21.0	350	60	7.3	121.6	60	7.7	128.3	60	3.2	53.3
80	27.4	342.5	80	10.2	127.5	80	10.6	132.5	80	5.2	65

Collector supply voltage kept at 9 V.

Since the D.C. gain of transistors II and III are nearly equal they are employed to make the multivibrator circuit. Average gain of the two transistors are $\beta = 119.05$.

Let us take $C_1 = 0.01\ \mu F$

$C_2 = 0.01\ \mu F$

$$R_B = \frac{1}{1.4fC}$$

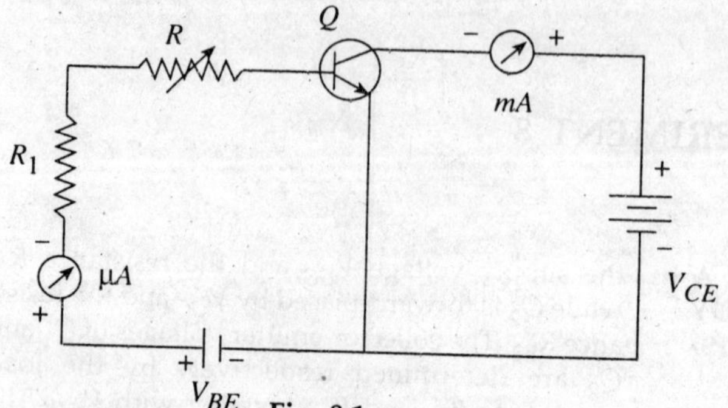

Fig. 8.1

V_{BE} ⇒ Supply voltage connected at the base-emitter junction

V_{CE} ⇒ Supply voltage connected at the collector-emitter junction

R_1 ⇒ Resistance (1.5 kΩ)

R ⇒ A variable resistance

mA ⇒ Milliammeter,
μA = Microammeter

Q ⇒ transistor (BC – 107, N–P–N)

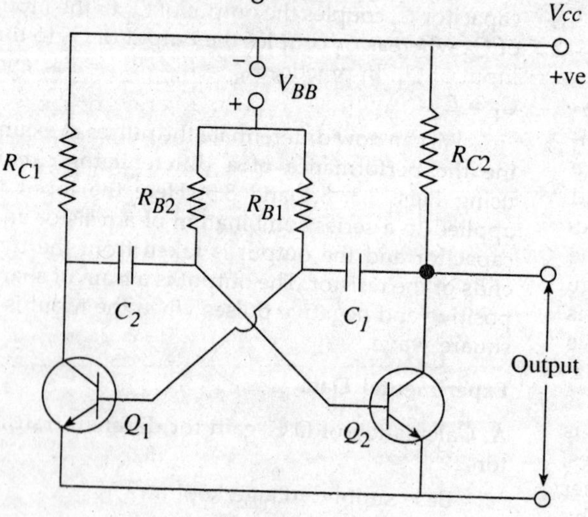

Fig. 8.2

V_{BB} ⇒ Base supply voltage

V_{CC} ⇒ Collector supply voltage

Q_1 and Q_2 ⇒ Two transistors (BC–107, N–P–N)

C_1 and C_2 ⇒ Capacitors connected to the collector of Q_1 and base of Q_2 and vice versa (0.01 μF)

R_{C1} and R_{C2} ⇒ Res. corrected at the collectors of Q_1 and Q_2 (1.56 kΩ)

R_{B1} and R_{B2} ⇒ Res. connected at the bases of Q_1 and Q_2 (68kΩ)

Output ⇒ Connected at the CRO

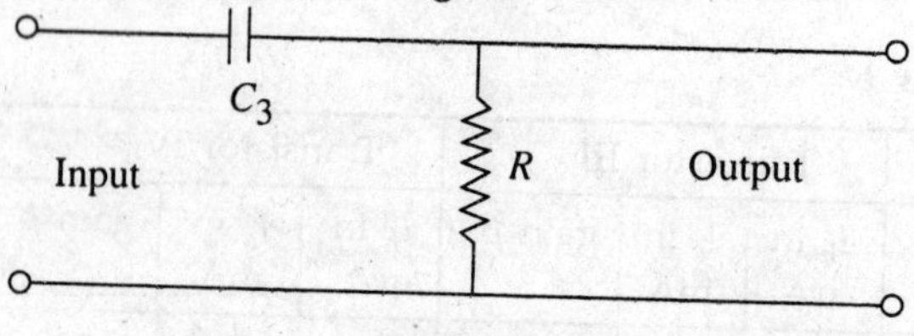

Fig. 8.3

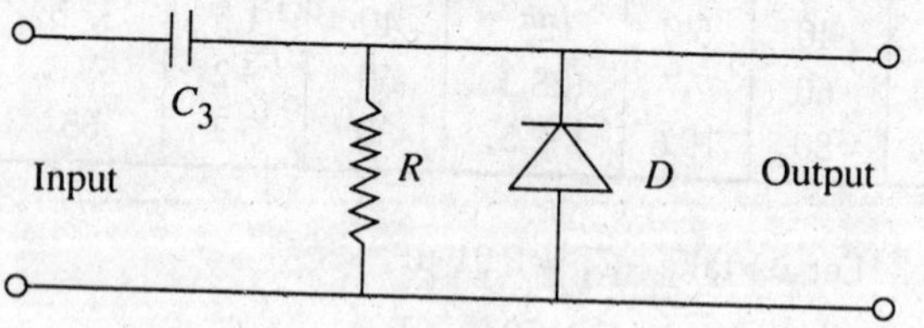

Fig. 8.4

C_3 ⇒ Capacitor (0.01 μF)

R ⇒ Resistances (1 kΩ, 10 kΩ, 100 kΩ, 1 MΩ)

D ⇒ Diode (Ge)

Input ⇒ Connected at the output of Fig. 8.2

Output ⇒ Connected at the CRO.

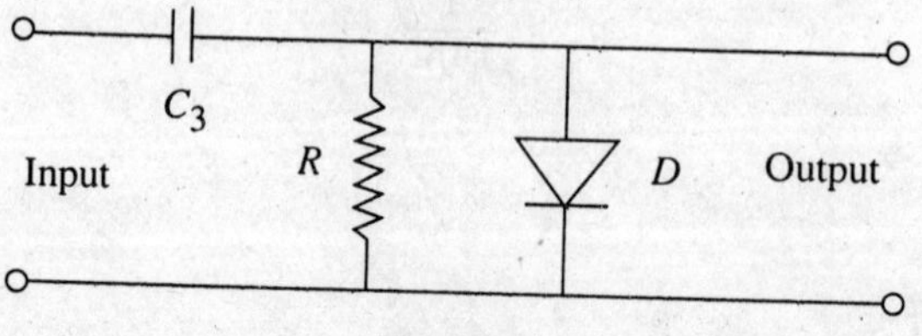

Fig. 8.5

Let us take the frequency of the oscillator as

$$f = 800 \text{ Hz.}$$

$$\therefore R_B = \frac{1}{1.4 \times 400 \times 0.01 \times 10^{-6}} \Omega$$

$$= \frac{10^6}{1.4 \times 400 \times 0.01} \Omega$$

$$= 89.3 \text{ k}\Omega$$

$$\therefore R_{B1} = R_{B2} = 89.3 \text{ k}\Omega.$$

$$R_c = 2\left(\frac{R_B}{\beta}\right)$$

$$= 2 \times \frac{89.3 \times 10^3}{119.05} \Omega$$

$$= 1.5 \text{ k}\Omega$$

Available values of $R_{B1} = R_{B2} = 68 \text{ k}\Omega$

$R_C = 1.56 \text{ k}\Omega$

B. **Measurement of the time period of the output voltage waveform :**

Table II

Collector supply voltage V_{CC} in volts	Base supply vol. V_B in volts	Height of the output voltage waveform in volts	Time period of the output voltage waveform in ms
	1	2.4	10.8
	2	2.4	8.4
2	3	2.4	6.8
	4	2.4	6.0
	5	2.4	4.8
	1	4.6	12.0
	2	4.6	10.4
4	3	4.6	9.6
	4	4.6	8.2
	5	4.6	7.2
	1	68	12.2
	2	6.8	10.8
6	3	6.8	9.8
	4	6.8	9.0
	5	6.8	8.4
	1	8.8	12.4
	2	8.8	11.2
8	3	8.8	10.4
	4	8.8	9.6
	5	8.8	9.0

The time period of the output voltage waveform is drawn in Fig. 8.6.

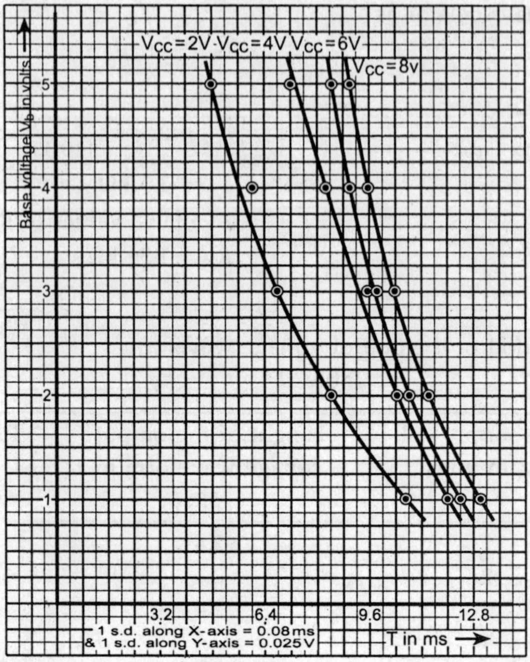

Fig. 8.6

To study the output voltage waveform with the square wave input for the differentiator circuit and with the presence of the diode.

Let us take four RC combinations.

$R_1C_1 = 0.01$ ms, $R_2C_2 = 0.1$ ms, $R_3C_3 = 1$ ms and $R_4C_4 = 10$ ms with $C_1 = C_2 = C_3 = C_4 = 0.01\ \mu F$.

$$\therefore \quad R_1 = \frac{0.01 \times 10^{-3}}{0.01 \times 10^{-6}}$$

$$= 10^3\ \Omega$$

$$= 1\ k\Omega$$

$$R_2 = \frac{0.100 \times 10^{-3}}{0.01 \times 10^{-6}}$$

$$= 10^4\ \Omega$$

$$= 10\ k\Omega$$

$$R_3 = \frac{1 \times 10^{-3}}{0.01 \times 10^{-6}}$$

$$= 100\ k\Omega$$

$$R_4 = \frac{10 \times 10^{-3}}{0.01 \times 10^{-6}}$$

$$= 1\ M\Omega$$

C. Measurement of the pulse height when a differentiator circuit is connected at the output :

Table III

Base supply voltage fixed at	Collector supply voltage fixed at	Res. connected at the output	Pulse height of the output vol. waveform
		1 kΩ	2.4 volts
6 V	5 V	10 kΩ	6.6 volts
		100 kΩ	10.2 volts
		1 MΩ	10.4 volts

Performance of the differentiator circuit for different values of R. (Viz., R = 1 kΩ, 10 kΩ, 100 kΩ and 1 MΩ) is shown in Fig. 8.7(a) and 8.7(b).

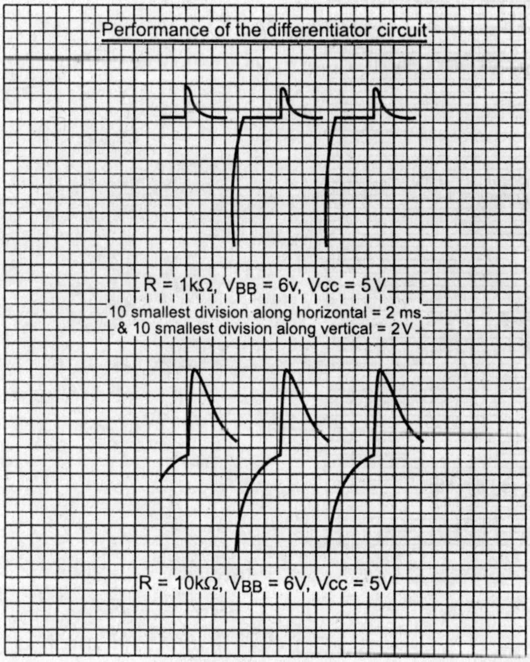

Fig. 8.7(a)

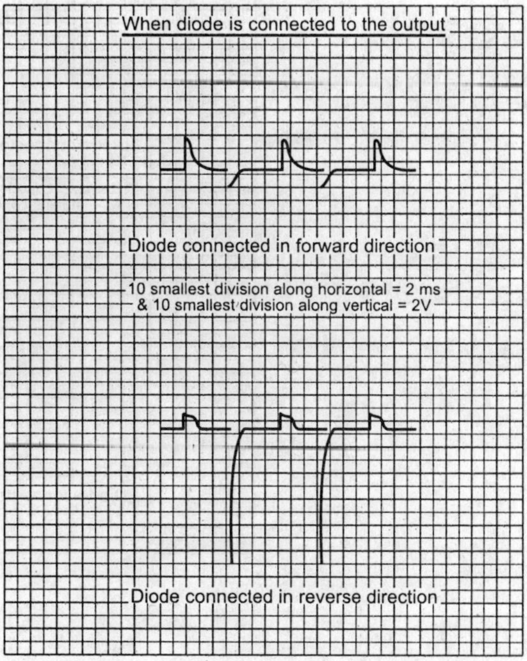

Fig. 8.7(b)

D. Measurement of the pulse height when a diode is connected at the output : Value of $R = 1$ kΩ

Table IV

Base supply vol. fixed at	Collector Supply vol. fixed at	Pulse height when the diode is forward biased	Pulse height when the diode is reverse biased
6 V	5 V	2.4 volts	10.4 volts

When a diode is connected to the output in forward and reversed directions we get the variation as shown in Fig. 8.8.

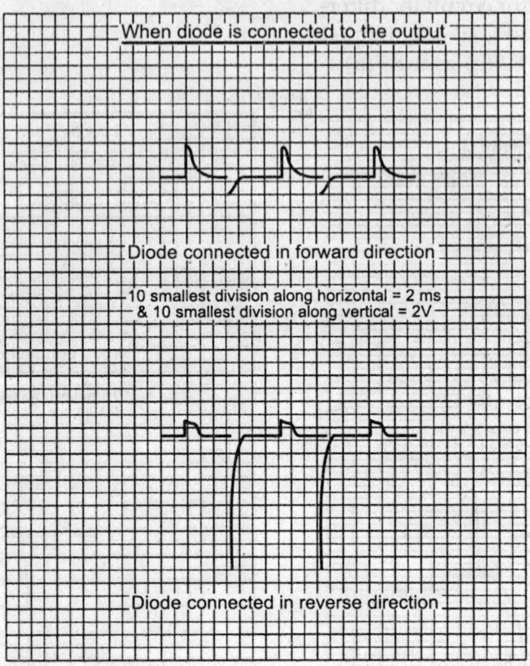

Fig. 8.8

Discussion

(i) Care should be taken while choosing the transistors Q_1 and Q_2. They should be almost identical with the d.c. gain values.

(ii) If Q_1 and Q_2 are not almost identical then difficulties may arise while studying the performance and characteristics of the multivibrator circuit.

(iii) The collector supply voltage should be fixed and readings should be taken for different base supply voltages.

(iv) The values of C_1 and C_2 should be equal with $R_{B1} = R_{B2}$ and $R_{C1} = R_{C2}$ to achieve the multivibrator circuit.

EXPERIMENT 9

VERIFICATION AND CONSTRUCTION OF 'NOT', 'NAND' AND 'XOR' GATES

Theory

NOT gate : The NOT operation is alternatively known as the complementation operation. The NOT operation on a variable A is represented by $\overline{A}$. It means that the variable A is complemented to the alternative value by this operation. The NOT operatioin is represented in Boolean algebra by the equation $Y = \overline{A}$ reads as Y equals NOT A. The truth table for a NOT gate having only one input A is shown below. The pin connections of IC 7404 and IC 7400 are shown in Figs. 9.1 and 9.2 respectively. The circuit symbol of the NOT gate is shown in Fig. 9.3.

Truth Table

Input	Output
A	$Y = \overline{A}$
0	1
1	0

NAND gate : De Margan's second theorem $\overline{A.B} = \overline{A} + \overline{B}$ shows that an AND followed by a NOT gate performs identically as an OR gate whose inputs are the outputs of two NOT circuits. The logic circuit in which a NOT gate follows an AND gate is referred to as a NOT-AND or NAND gate. The logic symbol of two input NAND gate is depicted in Fig. 9.4. The Boolean expression of a two input NAND gate is $Y = \overline{A.B}$. The truth table is shown below.

Truth Table

Input		Output
A	B	$Y = \overline{A} \cdot \overline{B}$
0	0	1
0	1	1
1	0	1
1	1	0

XOR gate : The exclusive OR gate, also termed as XOR gate or EOR gate, is a two-input gate, the

$+V_{CC} \Rightarrow$ Connected to the positive of the supply voltage (+5V)

$-V_{CC} \Rightarrow$ Connected to negative of the supply voltage (0v)

1, 3, 5, 9, 11 and 13 ⇒ Input terminals of six NOT gates fabricated in IC 7404

2, 4, 5, 8, 10 and 12 ⇒ Output terminals of them respectively.

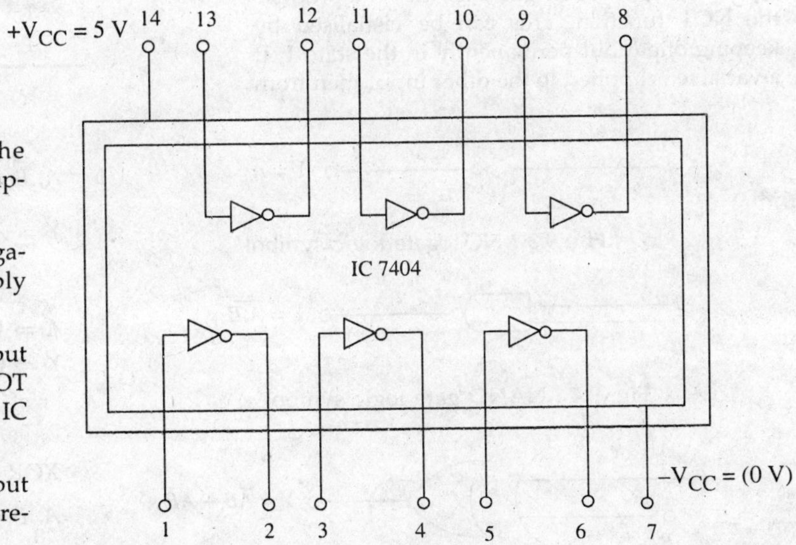

Fig. 9.1 : IC 7404

$+V_{CC} \Rightarrow$ Connected to the positive of the supply voltage (+5 V)

$-V_{CC} \Rightarrow$ Connected to the negative of the supply voltage (0 V)

1, 2; 4, 5; 9, 10 and 12, 13 $\Rightarrow$ Input terminals of the four NAND gates fabricated in IC 7400

3, 6, 8 and 11 are the respective output terminals.

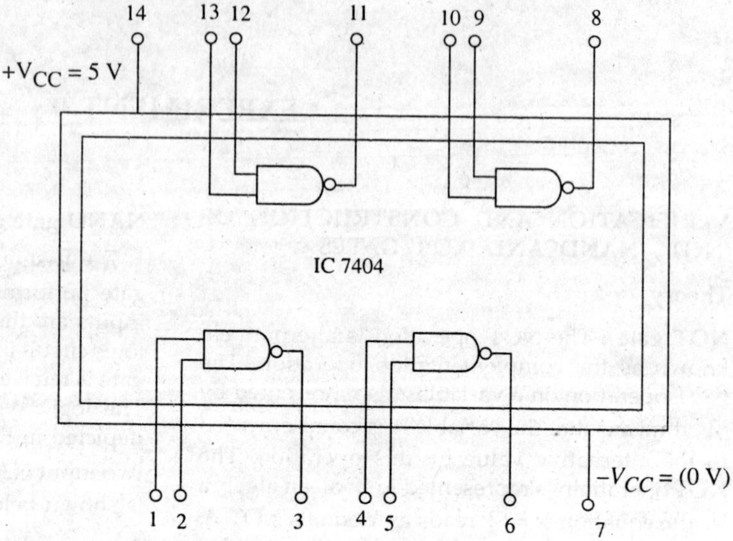

Fig. 9.2 : IC 7400

logic symbol of which is depicted in Fig. 9.5. The output Y of the XOR gate takes the state 1 if either A or B but not both, assumes the state 1. The Boolean expression is : $Y = \bar{A}B + \bar{B}A$.

The truth table of the XOR gate is shown below. The difference from the OR gate is that for the OR gate $Y = 1$, when $A = B = 1$. So the OR gate is termed as the INCLUSIVE-OR gate. An XOR gate can be realised by combining AND, OR and NOT gates.

XOR gates can invert signal, i.e., they can produce the NOT function. This can be visualised by keeping one input permanently in the state 1. If a variable is applied to the other input then from the truth table of XOR gate we find that the output is similar to that of a NOT gate having that variable input. This is a universal logic element.

Truth Table

Input		Output
A	**B**	$Y = \bar{A}B + \bar{B}A$
0	0	0
0	1	1
1	0	1
1	1	0

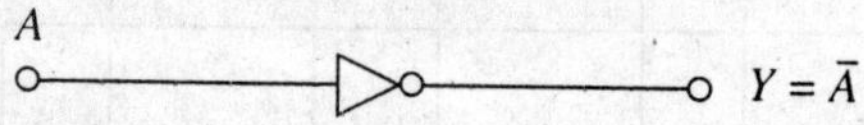

Fig. 9.3 : NOT gate logic symbol

$A \Rightarrow$ Input

$Y \Rightarrow$ Output ($\bar{A}$)

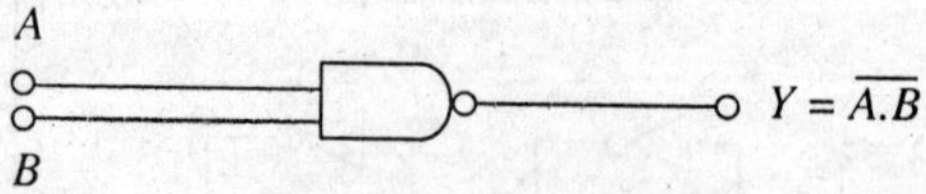

Fig. 9.4 : NAND gate logic symbol

$A, B \Rightarrow$ The two-input terminals

$Y = \overline{A.B} \Rightarrow$ Output

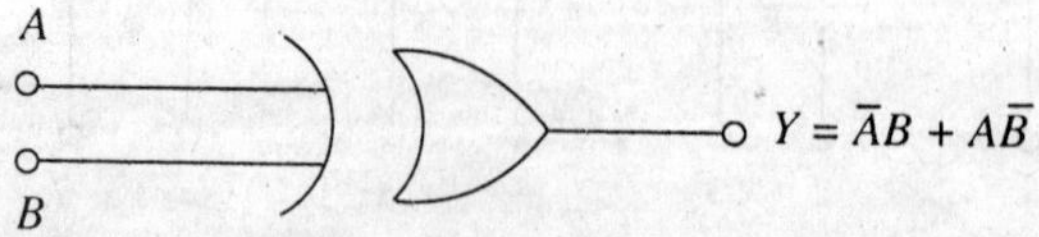

Fig. 9.5 : XOR gate logic symbol

XOR gate logic symbol

$A, B \Rightarrow$ Two input terminals

$Y = \bar{A}B + A\bar{B} =$ Output

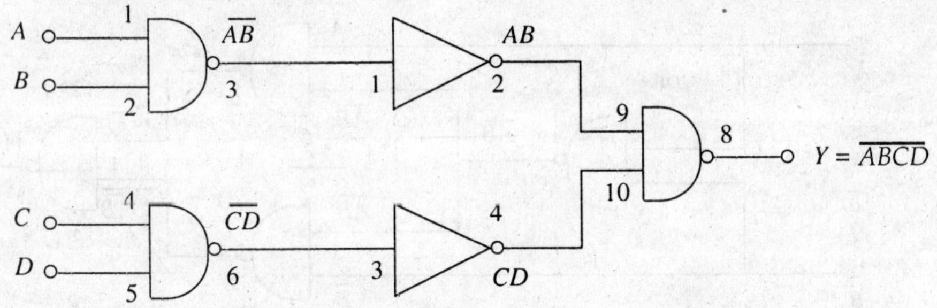

Fig. 9.6 : Construction of a four input 'NAND' gate

A, B, C, D	⇒	Input terminals at 1, 2, 4, 5 of IC 7400
$\overline{AB}$, $\overline{CD}$	⇒	Connected from 3 and 6 of IC 7400 to 1 and 3 of IC 7404
AB, CD	⇒	Connected from 2 and 4 of IC 7404 to 9 and 10 of IC 7400
$Y = \overline{ABCD}$	⇒	Output terminal measured at 8 of IC 7400.

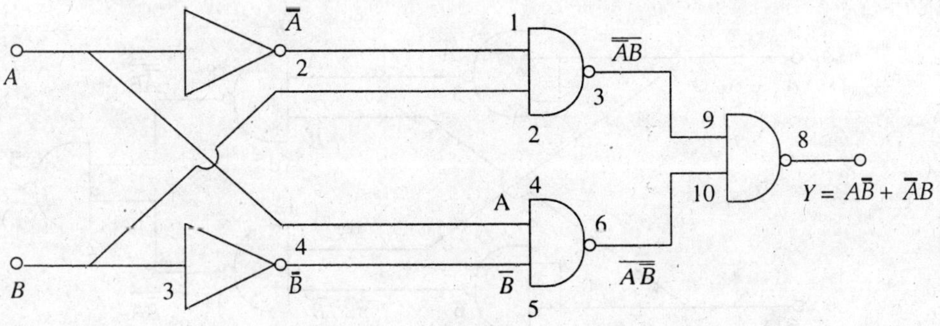

Fig. 9.7 : Construction of 'XOR' gate (Method I)

A, B ⇒	Input terminals at 1 and 2 of IC 7404
$\overline{A}$, $\overline{B}$ ⇒	Connected from 2 and 4 of IC 7404 to 1 and 5 of IC 7400
$\overline{AB}$, $A\overline{B}$ ⇒	Connected from 3 and 6 of IC 7400 to 9 and 10 of IC 7400
$Y = A\overline{B} + \overline{A}B$ ⇒	Output terminal measured at 8 of IC 7400.

Now-a-days these gates are compacted in the IC chips taking small places in integrated circuits and doing good performances.

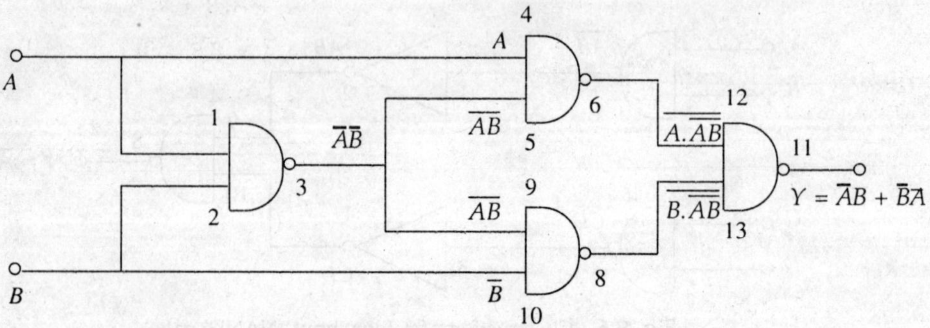

Fig. 9.8 : Construction of XOR gate (Method II)

A, B ⇒ Input terminals at the junction of 1 and 4 & 2 and 10 of IC 7400

$\overline{AB}$ ⇒ Connected from 3 to the junction of 5 and 9 of IC 7400

$\overline{A\overline{AB}}$, $\overline{B\overline{AB}}$ ⇒ Connected from 6 and 8 to 12 and 13 of IC 7400

$Y = \overline{A}B + \overline{B}A$ ⇒ Output measured at 11 of IC 7400.

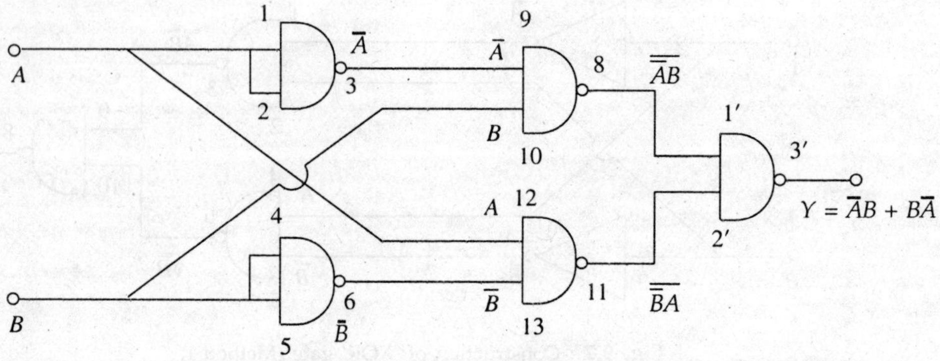

Fig 9.9 : Construction of XOR gate (Method III)

A, B ⇒ Input terminals at the junction of 1 and 2 & 4 and 5 of IC 7400

$\overline{A}$, $\overline{B}$ ⇒ Connected from 3 and 6 to 9 and 13 of IC 7400

$\overline{\overline{A}B}$, $\overline{\overline{B}A}$ ⇒ Connected from 8 and 11 of one IC 7400 to 1′ and 2′ of another.

Experimental Data

A. Verification of NOT gate truth table :

IC used is 7404.

Boolean expression for NOT gate is if input = A output = $Y = \overline{A}$.

Truth Table

Inputs	Output
A	$\overline{A}$
1	0
0	1

In this way all the NOT gates are verified.

B. Verification of NAND gate truth table :

IC used is 7400.

Boolean expression for NAND gate if A and B are two inputs; output $Y = \overline{A.B}$

Truth Table

Inputs		Output
A	B	$Y = \overline{A.B}$
0	0	1
0	1	1
1	0	1
1	1	0

In this way all the NAND gates of IC 7400 are verified.

C. Verification of a four input NAND gate

IC used 7400 and 7404 :

The construction of a four-input NAND gate is shown in Fig. 9.6.

Boolean expression for a four-input NAND gate having inputs *A*,*B*,*C*,*D* is,

Output $Y = \overline{ABCD}$

Truth Table

Inputs				Output
A	*B*	*C*	*D*	$Y = \overline{ABCD}$
0	0	0	0	1
0	0	0	1	1
0	0	1	1	1
0	1	1	1	1
1	1	1	1	0

D. Verification of XOR gate

Boolean expression for XOR gate having inputs *A* and *B* is $Y = \bar{A}B + A\bar{B}$

7486 is an XOR gate. We can construct XOR gate by NAND and NOT gates using IC 7400 and 7404.

Method I : Circuit connection for this is shown in Fig. 9.7.

Truth Table

Inputs		Output
A	*B*	$Y = \bar{A}B + A\bar{B}$
0	0	0
0	1	1
1	0	1
1	1	0

Method II : A single IC 7400 can also be used to get an XOR gate. The circuit used is shown in Fig. 9.8. A simplified output is obtained using Boolean *algebra and De* Morgan's theorem.

$$Y = \overline{\overline{A.\overline{AB}}\;\overline{B.\overline{AB}}}$$

$= \overline{\overline{A.\overline{AB}}} + \overline{\overline{B.\overline{AB}}}$ (using De Morgan's theorem)

$= A.\overline{AB} + B.\overline{AB}$ (using Boolean identity)

$= A.(\bar{A} + \bar{B}) + B.(\bar{A} + \bar{B})$ (using De Morgan's theorem)

$= A.\bar{A} + A.\bar{B} + B.\bar{A} + B.\bar{B}$

$[\because A\bar{A} = B\bar{B} = 0]$

$= A.\bar{B} + B.\bar{A}$

Truth Table

Inputs		$A\bar{B}$	$B\bar{A}$	Output
A	*B*			$Y = A\bar{B} + B\bar{A}$
0	0	0	0	0
0	1	0	1	1
1	0	1	0	1
1	1	0	0	0

The above truth table is verified experimentally and so the above conbination of gates represents an XOR gate.

Method III : The chips of IC 7400 can be used to get an XOR gate. The circuit used is shown in Fig. 9.9. A simpified output is obtained using Boolean algebra and De Morgan's theorem.

Let *A* and *B* are the two inputs and *Y* is the output. Then

$$Y = \overline{\overline{\bar{A}.B}\,.,\,\overline{\bar{B}.A}} = \overline{\overline{\bar{A}B}} + \overline{\overline{\bar{B}A}} = \bar{A}B + \bar{B}A$$

Truth Table

Inputs		$\bar{A}$	$\bar{B}$	$\bar{A}.B$	$\bar{B}.A$	Output
A	*B*					$Y = \bar{A}B + \bar{B}A$
0	0	1	1	0	0	0
0	1	1	0	1	0	1
1	0	0	1	0	1	1
1	1	0	0	0	0	0

The above truth table is verified experimentally and so the above combination of gates represents an XOR gate.

Discussion

(i) For flexibility, connections can be made preferably on a bread board by using single strand wire.

(ii) Unlike analog electronics, here absolute values of input and output voltages are not important. The only requirement is that, these voltages must be either low or high.

(iii) Instead of implementing digital gates by using discrete circuit components, their implementation by using IC is more popular. This is due to low cost, small size, low power requirement and improved performance of ICs.

EXPERIMENT 10

VERIFICATION AND CONSTRUCTION OF 'XNOR' GATE AND HALF ADDER

Theory

XNOR gate : The EXCLUSIVE-NOR gate or the XNOR or ENOR gate is a two-input gate, complement to the XOR gate (Fig. 10.1 and 10.2). The truth table of XNOR gate is given below. The output Y is in the state 1 when both A and B are either 0 or 1. Thus the gate serves as a coincidence checker. The expression for the output of the

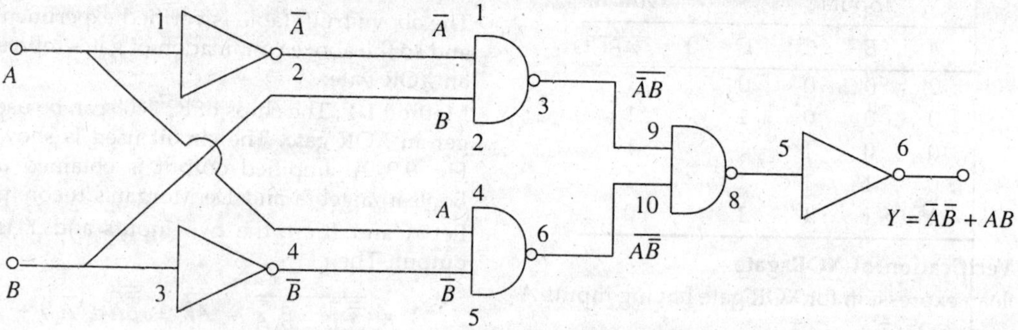

Fig. 10.1 : Construction of a XNOR gate (METHOD I)

A, B ⇒ Input terminals connected at 1 and 3 of IC 7404

$\overline{A}\ \overline{B}$ ⇒ Connected from 2 and 4 of IC 7404 to 1 and 5 of IC 7400

$\overline{\overline{A}B}, \overline{A\overline{B}}$ ⇒ Connected from 3 and 6 to 9 and 10 of IC 7400

$Y = \overline{A}\overline{B} + AB$ ⇒ Output terminal connected from 6 of IC 7404.

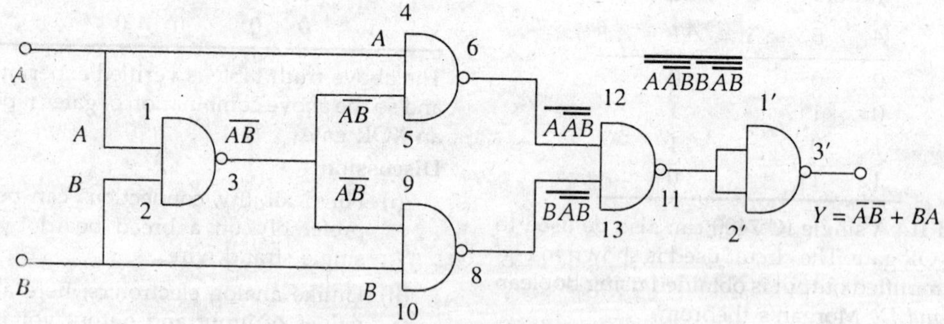

Fig. 10.2 : Construction of a XNOR gate (METHOD II)

A, B ⇒ Input terminals connected at the junction of 1 and 4 and 2 and 10 of IC 7400

$\overline{AB}$ ⇒ Connected from 3 to the junction of 5 and 9 of IC 7400

$\overline{A.\overline{AB}}, \overline{B.\overline{AB}}$ ⇒ Connected both 6 and 12 & 8 and 13 of IC 7400

$\overline{\overline{A.\overline{AB}}.\overline{B\overline{AB}}}$ = Connect from 11 of the IC 7400 to the junction of 1′, 2′ of another

$Y = \overline{A}\overline{B} + BA$ ⇒ Output measured at 3′ of IC 7400.

XNOR gate is :

$$Y = AB + \overline{A}\,\overline{B}.$$

The truth table for the XNOR gate is shown below :

Truth Table

Inputs		Output
A	B	$Y = AB + \overline{A}\,\overline{B}$
0	0	1
0	1	0
1	0	0
1	1	1 .

the sum of A and B is represented by a single digit and D gives this sum. In the last row the sum of A and **B** is 1 + 1 = 10. Here D = 0 and the carry 1 must be taken to be the position of next higher significance. Therefore, C = 1. An inspection of the truth table shows that D can be realized by an XOR gate and C by an AND gate. We can write,

$$D = \overline{A}B + A\overline{B} \text{ and } C = AB$$

Thus the logic circuit of Fig. 10.3 serves as a half-adder.

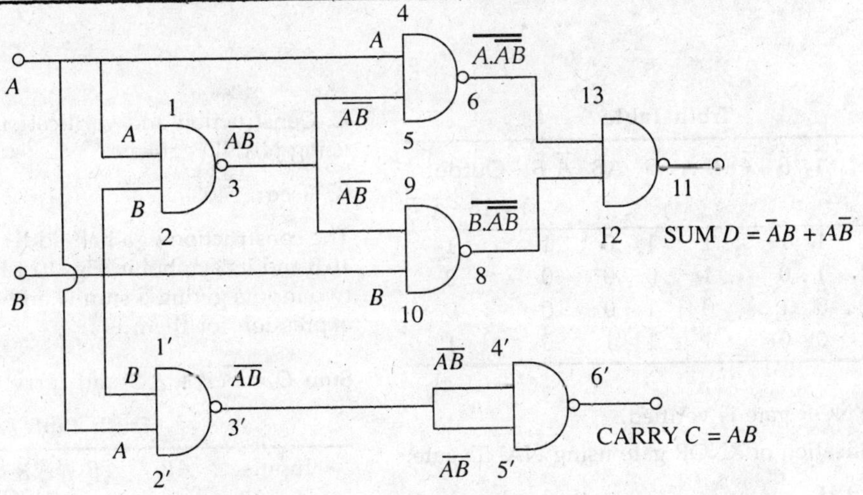

Fig. 10.3 : Construction of a Half-Adder

A ⇒ One input connected at 1 and 2′ of two 7400 ICs

B ⇒ Another input connected at 2 and 1′ of them

$\overline{AB}$ ⇒ Connected from 3 to the junction of 5 and 9 of IC 7400

$\overline{AB}$ ⇒ Connected from 3′ to the junction of 4′ and 5′ of another

$\overline{A\overline{AB}}$, $\overline{B\overline{AB}}$ ⇒ Connected from 6 to 13 and 8 to 12 of IC 7400

SUM D ⇒ $\overline{A}B + A\overline{B}$ = Measured at 11 of IC 7400

CARRY C = AB ⇒ Measured at 6′ of the second IC 7400

Half-Adder : Half-adder has two inputs A and B to represent the bits to be added. There are two outputs D and C the output D gives the sum of A and B when the sum can be represented by a single digit. If the sum is greater than that represented by a single digit, D shows the digit in the sum that has to be the same significance as the individual digits added. The output C gives the carry-bit. In the first three rows of the truth table,

Experimental Data

A. Verification of 'XNOR' gate using 'NAND' and 'NOT' gates :

METHOD I

IC used 7400 and 7404.

The construction of a XNOR gate is shown in Fig. 10.1.

Boolean expression for a XNOR gate. If A and B are the inputs then $Y = \overline{\overline{AB}\;\overline{\overline{A}\,\overline{B}}} = \overline{A}\,\overline{B} + AB$.

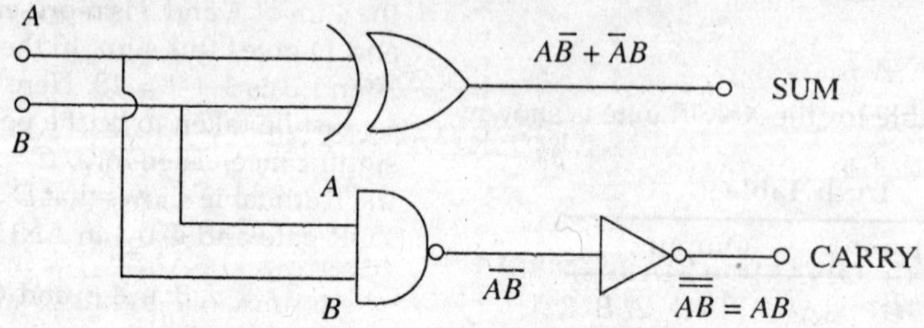

Fig. 10.4 : Symbol of a Half-Adder

Truth Table

Inputs A	B	$\overline{A}$	$\overline{B}$	$\overline{A}$+B	A+$\overline{B}$	AB	$\overline{A}\,\overline{B}$	Output $Y=AB+\overline{A}\,\overline{B}$
0	0	1	1	1	1	0	1	1
0	1	1	0	1	0	0	0	0
1	0	0	1	0	1	0	0	0
1	1	0	0	1	1	1	0	1

Thus XNOR gate is verified.

B. Verification of XNOR gate using NAND gate:

Method II

IC used 7400.

The construction of XNOR gate following this method is shown in Fig. 10.2.

Boolean expression for 'XNOR' gate with A and B as input is :

$$Y = \overline{\overline{(\overline{A.\overline{AB}}.\ \overline{B\overline{AB}})}\ \overline{(\overline{A.\overline{AB}}.\ \overline{B\overline{AB}})}}$$

$$= A.B + \overline{A}\ \overline{B}$$

Truth Table

Inputs A	B	$\overline{A}$	$\overline{B}$	AB	$\overline{A}$.AB	AB.$\overline{B}$	$\overline{A}.\overline{B}$	Output $Y=AB+\overline{A}\ \overline{B}$
0	0	1	1	0	0	0	1	1
0	1	1	0	0	0	0	0	0
1	0	0	1	0	0	0	0	0
1	1	0	0	1	0	0	0	1

Thus XNOR gate is verified.

C. Construction and verification of a half-adder using NAND gates :

IC used : 7400.

The construction of a half-adder is shown in Fig. 10.3 and its symbol in Fig. 10.4 Here A and B are two inputs giving a sum D and carry C. Boolean expression for them is :

Sum $D = \overline{A}\,B + A\,\overline{B}$ and carry $C = AB$

Truth Table

Inputs A	B	AB	$\overline{A}\,B$	$A\,\overline{B}$	Output Sum D	Carry C
0	0	0	0	0	0	0
0	1	0	1	0	1	0
1	0	0	0	1	1	0
1	1	1	0	0	0	1

Thus Half-adder is verified.

Discussion

(i) For flexibility, connections can be made preferably on a bread board using single strand wire.

(ii) Unlike analog electronics, here absolute values of input and output voltages are not important. The only requirement is that, these voltages must be either low or high.

(iii) Instead of implementing digital gates by using discrete circuit components, their implementation by using IC is more popular. This is due to low cost, small size, low power requirement and improved performance of ICs.

EXPERIMENT 11

TO STUDY THE OP-AMP CHARACTERISTICS, OPEN LOOP GAIN AND CMRR

Theory

The operational amplifier is an amplifier with integrated circuit and an ideal OP-AMP has the following characteristics :

(i) Its input resistance is infinite.

(ii) Output resistance is zero.

(iii) Voltage gain is infinite.

(iv) Bandwidth is infinite.

(v) If V_1 and V_2 be the voltages at the inputs, then the output voltage $V_o = 0$, when $V_1 = V_2$.

(vi) Characteristics do not change with temperature.

In practical OP-AMP many factors arise and the above mentioned characteristics are not strictly obeyed.

Inverting OP-AMP : For inverting OP-AMP the +ve terminal is grounded and a signal source is connected in the –ve terminal (Fig. 11.1) The feed-back voltage gain is given by

$$A_{vf} = \frac{V_o}{V_s} = -\frac{R_2}{R_1} \qquad \text{(i)}$$

Non-inverting OP-AMP : The –ve terminal is grounded through a resistance R_1 and a signal source V_s is inserted in the +ve terminal (Fig. 11.2). The voltage gain with feed-back is :

$$A_{vf} = \frac{V_o}{V_s} = 1 + \frac{R_2}{R_1} \qquad \text{(ii)}$$

Open-loop voltage gain : The open loop voltage gain can be measured by the formula,

$$A_v = \frac{V_o}{V_i} = \frac{V_o}{V_j} \times \frac{R_1 + R_2}{R_2} = 221 \frac{V_o}{V_j} \qquad \text{(iii)}$$

(assuming R_1 = 220 kΩ; R_2 = 1kΩ)

2 ⇒ Inverting input
3 ⇒ Non-inverting input
6 ⇒ Output
4 ⇒ –ve supply
7 ⇒ +ve supply

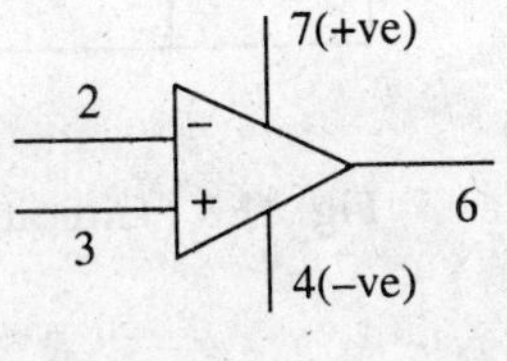

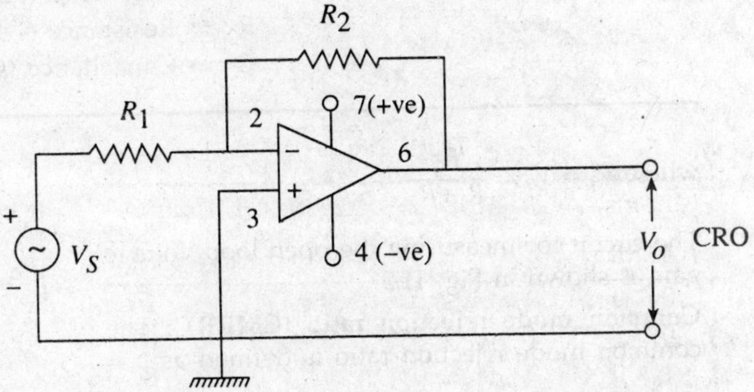

⊝ ⇒ Frequency generator
V_s ⇒ Input supply voltage
R_1 ⇒ Connected from the +ve of the input supply voltage to terminal 2 of OP-AMP (10 kΩ)
3 ⇒ This terminal is grounded
7 ⇒ This is connected to the positive of the D.C. supply
4 ⇒ This is connected to the negative of the D.C. supply
R_2 ⇒ Resistance connected between 2 and 6
V_o ⇒ Output voltage measured at CRO.

Fig. 11.1 : Inverting Amplifier Circuit

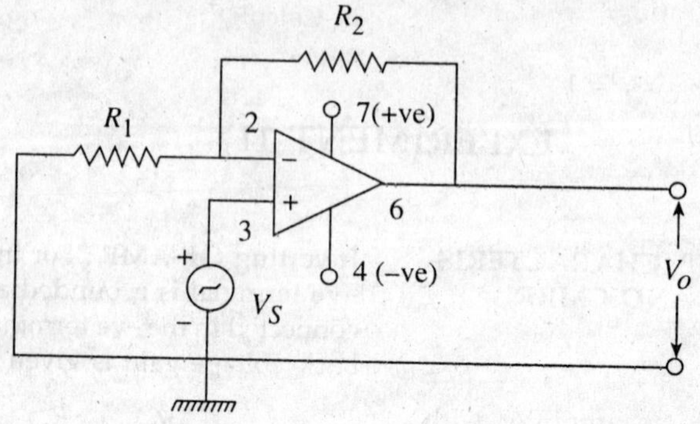

Fig. 11.2 : Non-inverting Amplifier Circuit

V_s ⇒ Input supply voltage
V_o ⇒ Output voltage measured at CRO
R_1 ⇒ Resistance (10 kΩ)
R_2 ⇒ Resistance (100 kΩ)
7 and 4 ⇒ Connected to the positive and negative of the d.c. supply.

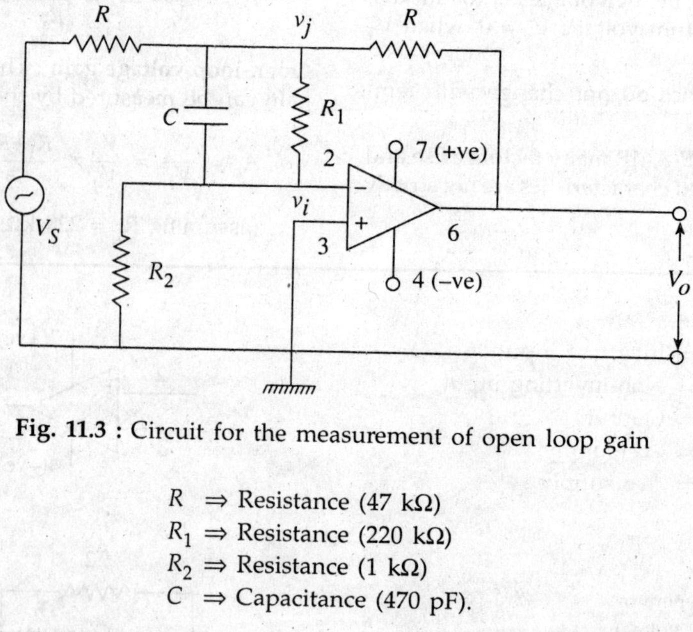

Fig. 11.3 : Circuit for the measurement of open loop gain

R ⇒ Resistance (47 kΩ)
R_1 ⇒ Resistance (220 kΩ)
R_2 ⇒ Resistance (1 kΩ)
C ⇒ Capacitance (470 pF).

where $V_i = V_j \times \dfrac{R_2}{R_1+R_2}$

The circuit for measuring the open loop voltage gain is shown in Fig. 11.3

Common mode rejection ratio (CMRR) : The common mode rejection ratio is defined as $\rho = \left|\dfrac{A_d}{A_c}\right|$ where A_d is the differential voltage gain and A_c is the common mode voltage gain.

It can be shown that CMRR,

$$\rho = \frac{R+R'}{R}\frac{V_s}{V_o} \quad \text{(iv)}$$

The common mode rejection ratio in dB = 20 logρ

Here R = R_1 = 1 kΩ and R′ = R_1' = 1 MΩ

The circuit for measuring CMRR is shown in Fig. 11.4.

Experimental Data :

A. Calculation for the gain of the inverting operational amplifier :

Table I

Frequency in Hz	Input voltage V_s in volts (peak to peak)	Output voltage V_o in volts (peak to peak)	Gain $A_{vf} = \frac{V_o}{V_s}$
100 Hz	0.85 V	11V	12.94
500 Hz	0.82 V	11V	13.41
1000 Hz	0.80 V	11V	13.75
5000 Hz	0.78 V	11V	14.10
10000 Hz	0.75 V	11V	14.66

B. Calculation for the gain of the non-inverting operational amplifier :

Table II

Frequency in Hz	Input voltage V_s in volts (peak to peak)	Output voltage V_o in volts (peak to peak)	Gain $A_{vf} = \frac{V_o}{V_s}$
100 Hz	1 V	12.4 V	12.40
500 Hz	1 V	12.4 V	12.40
1000 Hz	1 V	12.0 V	12.00
5000 Hz	1.1 V	12.8 V	11.64
10000 Hz	1.1 V	12.6 V	11.45

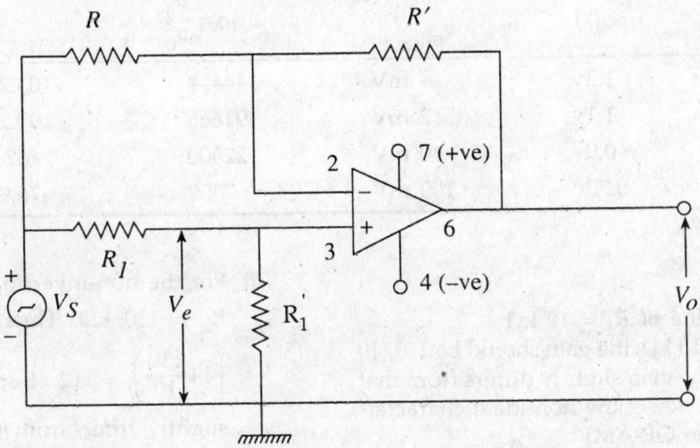

Fig. 11.4 : Circuit for the measurement of CMRR

$R \Rightarrow R_1 \Rightarrow$ Resistance (1 kΩ)
$R' \Rightarrow R_1' \Rightarrow$ Resistance (1 MΩ)
(~) ⇒ Frequency generator
$V_s \Rightarrow$ Input supply voltage
$V_o \Rightarrow$ Output voltage measured at CRO

C. Calculation for the open loop voltage gain :

Table III

Frequency in Hz	V_i in volts (peak to peak)	Output voltage V_o in volts (peak to peak)	Open loop vol gain $A_v = 221\frac{V_o}{V_j}$
100 Hz	5 mV	1.3 V	57460
500 Hz	5 mV	1.3 V	57460
1000 Hz	5 mV	1.3 V	57460
⋮	⋮	⋮	⋮
200000 Hz	220 mV	0.56 V	562
500000 Hz	300 mV	0.50 V	368

D. Data for the common mode rejection ratio (CMRR) :

Table IV

Frequency f in Hz	V_s in volts (peak-to-peak)	V_o in volts (peak-to-peak)	$\rho = \frac{R+R'}{R}\frac{v_s}{v_o}$ vol. gain $= 1001\frac{v_s}{v_o}$	CMRR in dB $= 20 \log \rho$
500 Hz	1.3v	9 mV	144444	103.20
1000 Hz	1.1v	2 mV	91666	99.24
2000 Hz	0.9v	40 mV	22500	87.4
10000 Hz	0.70v	100 mV	7000	76.90

Discussion

(i) As the value of R_1 = 10 kΩ and R_2 = 110 kΩ, the gain should be 110/10 = 11. But the gain slightly differs from that which can detect the non-ideal characteristics of the OP-AMP

(ii) For the non-inverting case R_1 = 10 kΩ, R_2 = 110 kΩ. Thus the gain should be $\left(1+\frac{110}{10}\right) = 12$. But the gain obtained slightly differs from it. It may be due to the tolerance of resistances.

EXPERIMENT 12

TO DESIGN AND CONSTRUCT A PULSE GENERATOR OF VARIABLE FREQUENCY AND WIDTH AND TO STUDY ITS PERFORMANCE.

Theory : Fig. 12.1 is the fundamental block diagram of IC 555. The device consists of two comparators A_1 and A_2, a flip-fliop F, output state Q_1 and transistor switch Q_2. The two comparators A_1 and A_2 are basically OP-AMP, the output of which switches between a +ve saturation voltage and zero, according to its input voltage which is either below or above a reference voltage. The reference voltage is applied to two comparators developed across a voltage divider which consists of three equal resistances of $R = 5\ k\Omega$

The comparator A_1 is known as threshold comparator and A_2 is known as the trigger comparator. A_1 is referenced at $\frac{2}{3}\ V_{cc}$ and A_2 is referenced at $\frac{1}{3}\ V_{cc}$. The outputs of the two comparators control the flip-flop F. In its quiescent state, the output is very high. Consequently, the transistor switch Q_2 remains closed and the output terminal 3 is at low state.

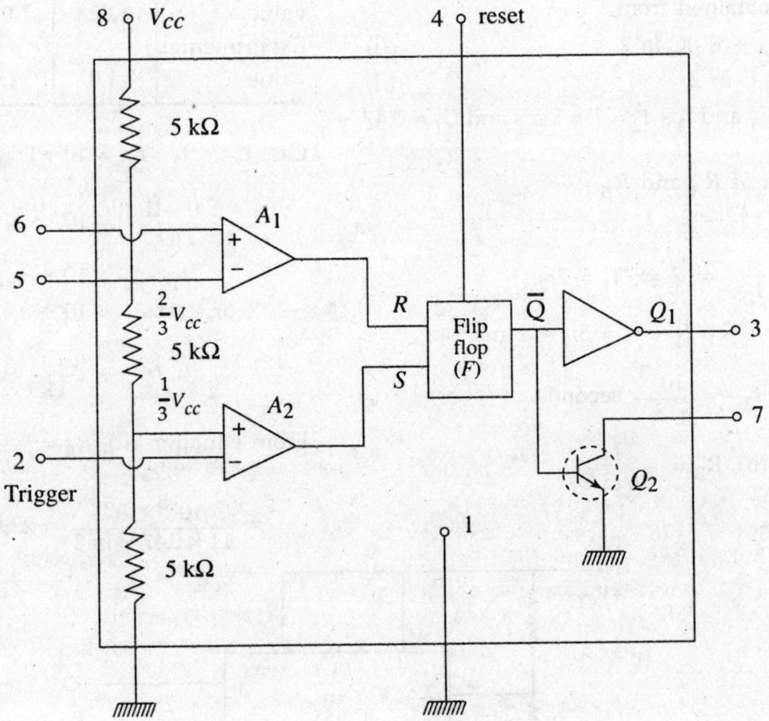

Fig. 12.1 : Block diagram of IC 555

If a –ve pulse of height below $\frac{1}{3}\ V_{cc}$ is applied, the flip-flop output changes state. This will cause the input terminal high (just below V_{cc}) and the transistor switch becomes open. Now to bring the state of the flip-flop to the original condition, a +ve pulse is needed at terminal 2. This is done by raising the pot of terminal 6 above $\frac{2}{3}V_{cc}$. This is how the timer operates. This timer can be used as astable, monostable multivibrator, comparator, etc. when suitable components are connected externally.

AIM 1 : To construct an astable multivibrator with IC 555 for different pulse width.

Theory : The Astable multivibrator has two quasistable state. It switches back and forth from one state to other, remaining in each for a time depending upon the charging of capacitance circuit. It requires no external triggering and is used to generate square wave.

Working formula

The time period is given by $T = t_1 + t_2$, where t_1 and t_2 are obtained from :

$$t_1 = (R_A + R_B)C \ln 2 \qquad \text{(i)}$$

$$t_2 = R_B C \ln 2 \qquad \text{(ii)}$$

where $C = C_T$ and $t_1 + t_2 = T = 1$ ms and $C_T = 0.47$ μF (given).

Calculation of R_A and R_B

Case I : $t_1 : t_2 = 2 : 1$

$$\frac{t_1}{t_2} = 2 \Rightarrow t_1 = 2t_2.$$

$$\therefore T = t_1 + t_2 = 3t_2 = 1 \text{ ms}.$$

$$t_2 = \frac{10^{-3}}{3} \text{ second}.$$

From Eqn. (ii), $R_B = \dfrac{t_2}{0.693\, C_T}$

$$= \frac{10^{-3}}{3 \times 0.693 \times 0.47 \times 10^{-6}} = 1.023 \text{ k}\Omega$$

and $t_1 + t_2 = 10^{-3}$ sec.

$$t_1 + \frac{t_1}{2} = 10^{-3}$$

$$t_1 = \frac{2}{3}10^{-3} \text{ sec.}$$

$$\therefore R_A + R_B = \frac{t_1}{C_T \ln 2}$$

$$= \frac{2}{3}\times 10^{-3} \cdot \frac{1}{0.47 \times 10^{-6} \times \ln 2} = 2.0468 \text{ k}\Omega.$$

$$\therefore R_A = (2.046 - 1.023) \text{ k}\Omega = 1.023 \text{ k}\Omega.$$

Table I

	R_B in kΩ	R_A in kΩ	C_T in μF	C in μF
Calculated value	1.023	1.023	0.47 μF	0.01
Experimental value	1.0	1.0	(given)	given)

Case II : $t_1 : t_2 = 10 : 1$

$$\frac{t_1}{t_2} = 10 \Rightarrow t_1 = 10t_2.$$

$$t_1 + t_2 = 10^{-3}$$

$$\text{or,} \quad 11t_2 = 10^{-3}$$

$$\therefore \quad t_2 = \frac{10^{-3}}{11} \text{ sec.}$$

From equation (ii), $R_B = \dfrac{t_2}{C_T \ln 2}$

$$= \frac{10^{-3} \times 10^{6}}{11 \times 0.47 \times \ln 2} = 279 \ \Omega.$$

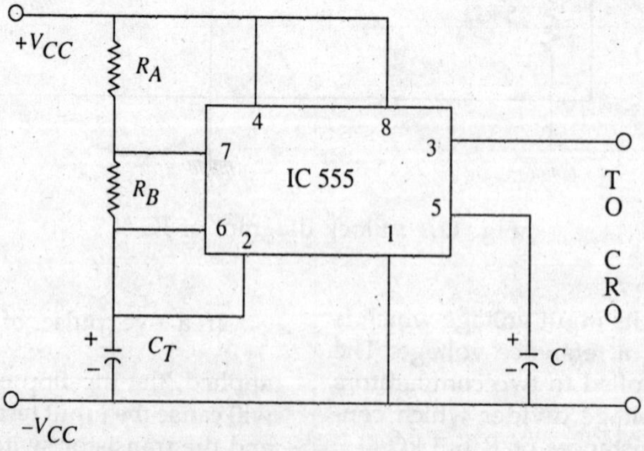

Fig. 12.2 : Circuit diagram of astable multivibrator with IC 555

Now, $t_2 = \frac{t_1}{10}$.

$t_1 + t_2 = 10^{-3}$ sec.

$\frac{11}{10}t_1 = 10^{-3}$.

$\therefore \quad t_1 = \frac{10}{11}\times10^{-3}$ sec.

$\therefore$ from equation (i), $R_A + R_B = \frac{t_1}{C_T \, ln\, 2}$

$$= \frac{10 \times 10^{-3}}{11 \times 0.47 \times 10^{-6} \times ln\, 2} = 2.791 \text{ k}\Omega.$$

$\therefore \quad R_A = (2791 - 279)\ \Omega = 2.512$ kΩ.

Table II

	R_B in kΩ	R_A in kΩ	C_T in μF	C in μF
Calculated value	279	2.512	0.47	0.01
Expt. value	270	2.7	(given)	(given)

Case III. $t_1 : t_2 = 40 : 1$

or, $\frac{t_1}{t_2} = 40$.

$\therefore \quad t_1 = 40\, t_2$

$t_1 + t_2 = 10^{-3}$ sec

$41\, t_2 = 10^{-3}$ sec

$t_2 = \frac{10^{-3}}{41}$ sec

$\therefore$ from equation (ii), $R_B = \frac{t_2}{C_T \, ln\, 2}$

$$= \frac{10^{-3}}{41 \times 0.47 \times 10^{-6} \times ln\, 2} = 74.86\ \Omega.$$

Now $t_2 = \frac{t_1}{40}$.

$t_1 + t_2 = 10^{-3}$ sec

or, $\frac{41}{40}t_1 = 10^{-3}$ sec

$t_1 = \frac{40}{41}\times10^{-3}$ sec.

$\therefore$ from equation (i), $R_A + R_B = \frac{t_1}{C_T \, ln\, 2}$

$$= \frac{40 \times 10^{-3}}{41 \times 0.47 \times 10^{-6} \times ln\, 2} = 2994.69\Omega$$

$\therefore \quad R_A = (2994.69 - 74.86)\Omega = 2.92$ kΩ

Table III

	R_B in Ω	R_A in kΩ	C_T in μF	C in μF
Calculated value	74.86	2.92	0.47	0.01
Actual value used	68	2.7	(given)	(given)

EXPERIMENTAL DATA

Verification of the time period of pulse as obtained from the astable multivibrator.

Table IV

Calculated values					Observed values in CRO				
$\frac{t_1}{t_2}$	R_B in Ω	R_A in kΩ	t_1 from Eq. (i) in ms	t_2 from Eq. (ii) in ms	Voltage V_{cc} in V	t_1 in ms	t_2 in ms	Ratio t_1/t_2	Pulse height in V
2	1.023	1.023	0.66	0.33	9	0.66	0.34	1.94	8.8
					⋮	⋮	⋮	⋮	⋮
					5	0.65	0.34	1.91	4.5
10	279	2.512	0.9	0.09	9	0.96	0.1	9.6	8.6
					⋮	⋮	⋮	⋮	⋮
					5	0.96	0.1	9.6	4.6
40	74.86	2.92	0.97	0.02	9	0.9	0.03	30	8.6
					⋮	⋮	⋮	⋮	⋮
					5	0.9	0.04	22.5	4.5

AIM 2 : To construct monostable multivibrator circuit with IC-555.

Theory

This circuit (Fig. 12.3) has only one stable state and one quasistable state. If an input triggering pulse is applied, the multivibrator switches to quasi-stable state. However, no external signals required to induce its transition back to the stable state. Basic application of this circuit is to establish a fixed time interval, the begining and end of which are worked by an abrupt discontinuity in voltage waveform.

Working Formula

The time period of the pulse

$T = t_1 = 1.1\ C_T\ R_T$

where $C_T = 0.47\ \mu F$.

$R_T = 220\ \Omega + 1\ k\Omega$ + pot. resistance.

In the astable circuit $R_A = 2.7\ k\Omega$, $R_B = 68\Omega$. and we take the ratio of $t_1 : t_2 = 40 : 1$.

Table V

Value of R_T in Ω	Value of C_T in μF	Calculated value of $t_1 = 1.1\ C_TR_T$ (ms)	V_{CC} in volt	Value of t_1 in (ms)	Value of t_2 in ms	Pulse height in volt
220+ $(1+1.2)\times10^3$ = 2420	0.47	1.25	9 ⋮ 3	1.2 ⋮ 1.2	0.65 ⋮ 0.80	8.4 ⋮ 2.4
220 + $(1+2.6)10^3$ = 3820	0.47	1.97	9 ⋮ 3	1.9 ⋮ 1.3	0.9 ⋮ 0.8	8.4 ⋮ 2.4
220 + $(1+4.8)10^3$ = 6020	0.47	3.11	9 ⋮ 3	3.0 ⋮ 3.3	0.8 ⋮ 0.9	8.4 ⋮ 2.4
220 + $(1+6.2)10^3$ = 7420	0.47	3.83	9 ⋮ 3	3.9 ⋮ 3.9	0.6 ⋮ 0.8	8.4 ⋮ 2.5

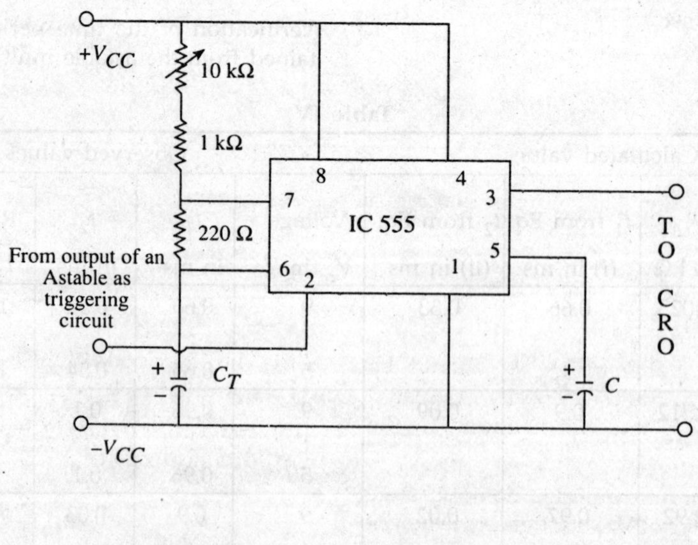

Fig. 12.3 : Circuit diagram of monostable multivibrator with IC 555

AIM 3 : Construction of pulse generator whose time period and pulse width can be varied.

Theory : It is constructed by connecting the astable and monostable multivibrator circuit as shown in Fig. 12.4. Here R_A is replaced by a potentiometer. Thus if R_T is varied keeping R_A fixed, the pulse width changes while the time period is constant. When R_A varies keeping R_T fixed, the time changes while pulse width remains the same.

Experimental Data

Table VI

Position of R_A (kΩ)	Position of R_T (kΩ)	Observed t_1 in ms	Observed t_2 in ms	Time period $T = t_1 + t_2$ (ms)	Width of the pulse in ms	Remark on the Expt.
Fixed (high)	Varied	$6 \times 0.1 = 0.6$	$24.5 \times 0.1 = 2.45$	3.05	0.6	Here time period is fixed, but pulse width is varied
		⋮	⋮		⋮	
		$27.5 \times 0.1 = 2.75$	$3 \times 0.1 = 0.3$	3.05	2.75	
Varied	Fixed (low)	$5.5 \times 0.1 = 0.55$	$2.5 \times 0.1 = 0.25$	0.80	0.55	Width of the pulse is const., but time period is varied.
		⋮	⋮	⋮	⋮	
		$6 \times 0.1 = 0.6$	$24 \times 0.1 = 2.4$	3.0	0.6	

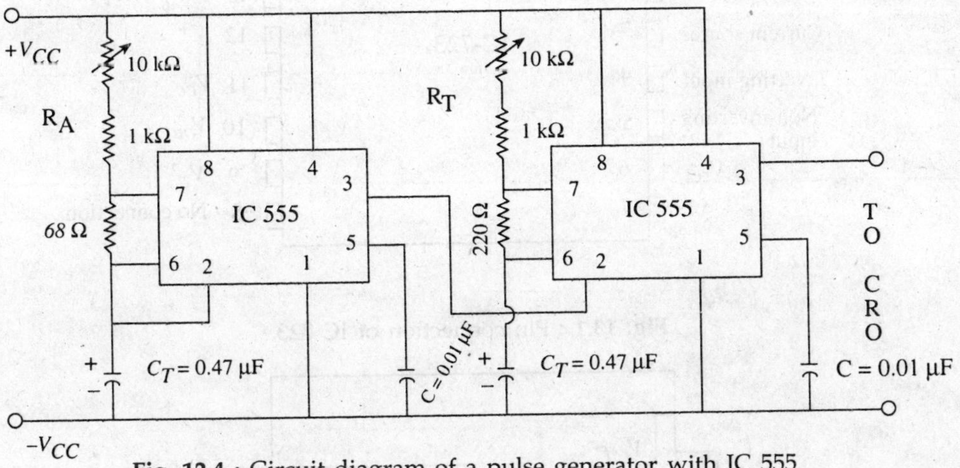

Fig. 12.4 : Circuit diagram of a pulse generator with IC 555

Discussion

(i) Experimental value of $t_1 : t_2$ should be nearly equal to the theoretically calculated value but there is some variation in the result.

(ii) Pulse width and time period do not change with the change of V_{CC} though pulse height changes.

(iii) The time period of the trigger pulse is greater than that of the monostable multivibrator.

(iv) Pin-5 must be connected with capacitance to diminish noise.

EXPERIMENT 13

TO CONSTRUCT A REGULATED POWER SUPPLY WITH CONTINUOUSLY VARIABLE OUTPUT USING IC 723 AND TRANSISTORS SL 100 AND 2N 3055.

Theory

The aim of power supply unit is to produce a steady direct output voltage and current from a source of alternating voltage. The rectifier O/P contains a ripple or an A.C. component. Superposed rectified O/P is fed to the input of IC 723 to get a continuously variable regulated output.

The pin connection of IC 723 is shown in Fig. 13.1 while the essential of the internal circuit of IC 723 is shown in Fig. 13.2. The circuit diagram of unregulated power supply is given in Fig. 13.3.

Construction : *To construct the unregulated power supply.* Output voltage measured across the capacitor is given as $V_{d.c} = 18.68$ volt.

Max. load current = 1000 mA = 1 A.

$$\text{Thus } R_L = \frac{V_{d.c}}{I_{max}} = \frac{18.68}{1} = 18.68\ \Omega.$$

The capacitor C = 2200 μF (given).

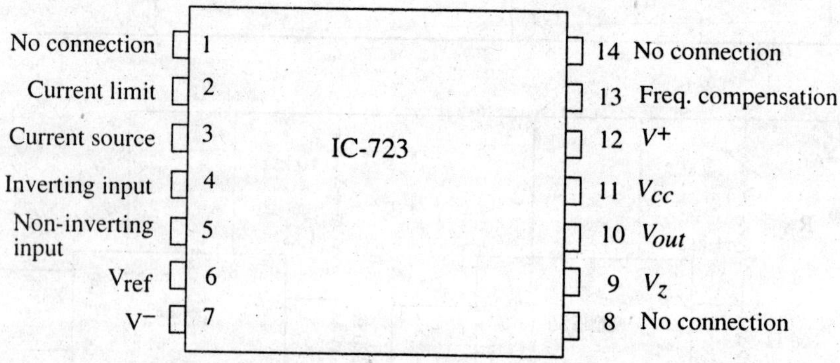

Fig. 13.1 : Pin connection of IC 723

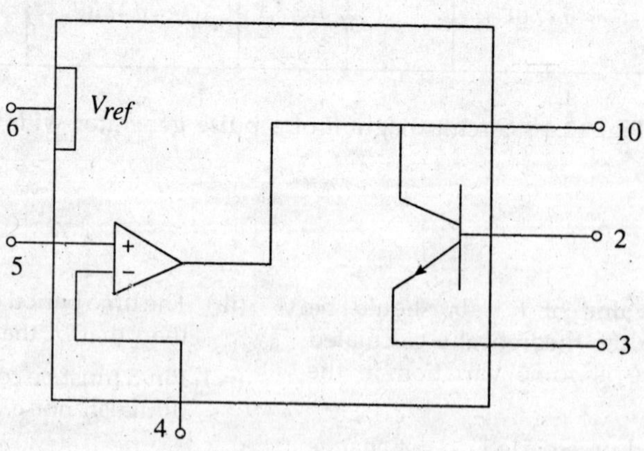

Fig. 13.2 : The essential of the internal circuit of IC 723

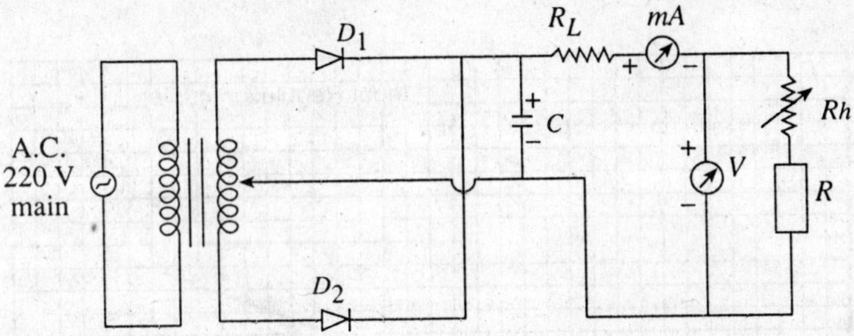

Fig. 13.3 : Unregulated power supply

Experimental Data

1. Unregulated power supply :

(a) *Data for load regulation* :

Max. load current obtained = 700 mA.

Input A.C. voltage = 220 volt.

Table I

Load current in mA	Output voltage in volt
0	16.34
100	14.42
200	12.40
⋮	⋮
700	4.67

The load regulation curve is drawn in Fig. 13.4.

(b) *Data for input regulation* :

Load current is fixed at 200 mA.

Table II

Input A.C. voltage in volts	Output D.C. voltage in volts
80	2.02
90	2.67
100	3.21
⋮	⋮
220	12.38

The input regulation curve is drawn in Fig. 13.5.

Fig. 13.4

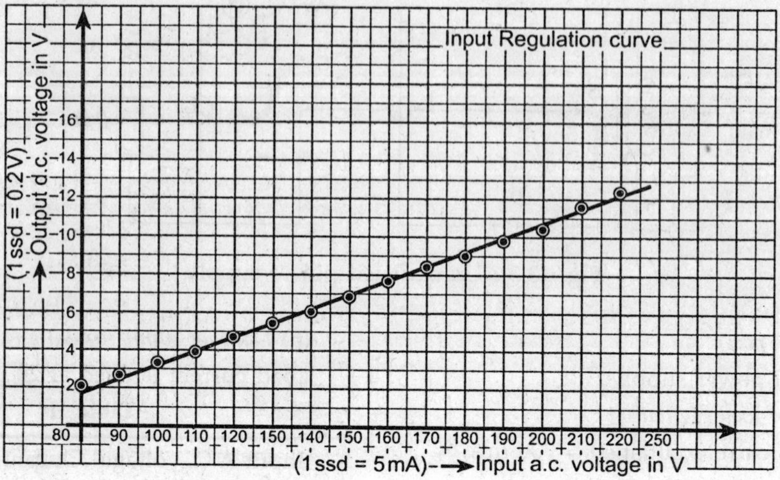

Fig. 13.5

To construct the regulated power supply :

These calculations are to find the proper values of the resistance R_1, R_2, R_A, R_B and R_K for the regulated power supply shown in Fig. 13.6.

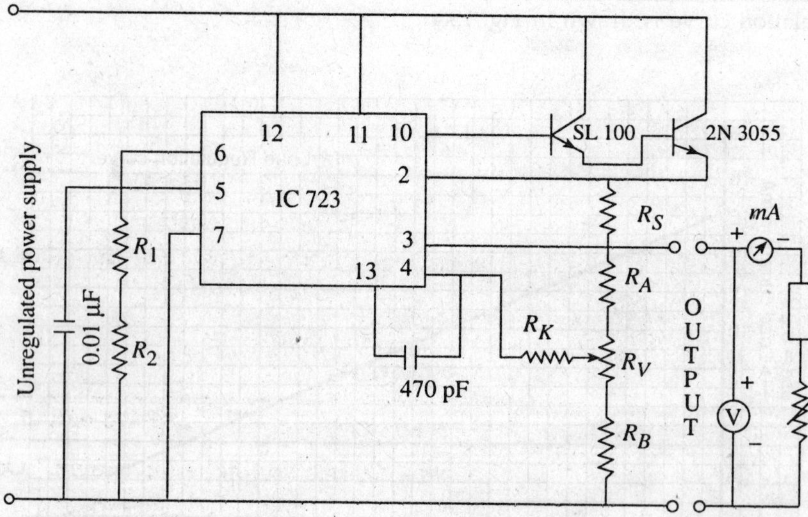

Fig. 13.6 : A regulated power supply with IC 723

For this purpose the given eqns. are

$$\frac{R_2 \times 7}{R_1+R_2} = 2 \text{ volt.} \quad (1)$$

and $$\frac{7}{R_1+R_2} = 5 \times 10^{-3} \text{A}. \quad (2)$$

[Here, Ref. voltage : 2 volt.
Voltage range : 3 V → 9 V.
Max. load current = 500 mA]

Solving eqns. (1) and (2), we get

$$R_2 = \frac{2}{5 \times 10^{-3}} = 400\ \Omega$$

$$R_1 = \frac{5R_2}{2} = 1000\ \Omega.$$

Since the range of the regulated power supply is 3 V to 9 V, we have the following equations :

$$V_{max} \times \frac{R_B}{R_A+R_V+R_B} = V_{ref} \quad (3)$$

and $$V_{min} \times \frac{R_B+R_V}{R_A+R_V+R_B} = V_{ref} \quad (4)$$

Since $V_{max} = 9$ V, $V_{min} = 3$ V and $V_{ref} = 2$ V so eqn. (3) becomes

$$\frac{R_B}{R_A+R_B+R_V} \times 9 = 2$$

$$\Rightarrow 2(R_A + R_B + R_V) = 9 R_B. \quad (5)$$

Eqn. (4) becomes $$\frac{R_B+R_V}{R_A+R_B+R_V} \times 3 = 2$$

$$\Rightarrow 2(R_A + R_B + R_V) = 3(R_B + R_V) \quad (6)$$

From equations (5) and (6)

$$9R_B = 3R_B + 3R_V$$

$$R_B = \frac{3}{6} R_V = \frac{R_V}{2} \ [\because R_V = 4.7 \text{ k}\Omega] \text{ (chosen)}$$

$$= \frac{4.7}{2} \text{k}\Omega = 2.35 \text{ k}\Omega.$$

From Equation (6), $2R_A = 3R_B + 3R_V - 2R_B - 2R_V = R_B + R_V$

$$\therefore R_A = \frac{R_B+R_V}{2}$$

$$= \frac{2.35 + 4.7}{2} \text{k}\Omega = 3.52 \text{ k}\Omega.$$

$$\text{Again } R_K = \frac{R_1R_2}{R_1+R_2} = \frac{1000 \times 400}{1000 + 400} = 285.7\ \Omega.$$

Table III

Calculated and used values of rsistances

Resistances	R_1	R_2	R_A	R_B	R_K
Calculated value	1 kΩ	400 Ω	3.52 k Ω	2.35 k Ω	285.7 Ω
Used value	1 kΩ	330 Ω	3.3 kΩ	2.2 kΩ	330 Ω

Calculation of R_S for three sets

Set I : For current limit

$$I_{limit} = 1000 \text{ mA} = 1 \text{ A}$$

$$\therefore R_S = \frac{0.6}{1} = 0.6\Omega$$

Set II : For current limit

$$I_{limit} = 750 \text{ mA}$$

$$\therefore R_S = \frac{0.6}{750 \times 10^{-3}} = 0.8\ \Omega.$$

Set III : For current limit

$$I_{limit} = 500 \text{ mA}$$

$$\therefore R_S = \frac{0.6}{500 \times 10^{-3}} = 1.2\ \Omega.$$

Experimental data

(i) *Data for load regulation* : (For set - 1)
$R_S = 0.6\ \Omega$.

Table IV

Fixed output voltage = 3 V		Fixed output voltage = 4 V		Fixed output voltage = 5 V	
Load current in mA	Output voltage in volt	Load current in mA	Output voltage in volt	Load current in mA	Outut voltage in volt
0	3.00	0	4.0	0	5.0
100	3.00	100	3.99	100	4.99
200	2.99	200	3.99	200	4.98
⋮	⋮	⋮	⋮	⋮	⋮
900	2.71	900	3.79	900	4.77
920	2.71	920	2.71	920	4.42
920	0.18	920	0.18	920	0.18

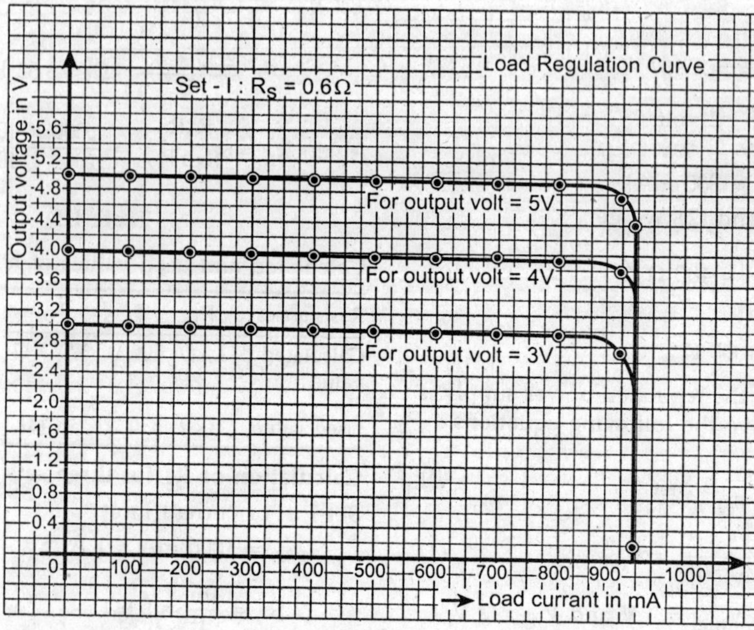

Fig. 13.7

The load regulation curve is drawn in Fig. 13.7 Repeat this observation for two other values of R_S.

(ii) *Data for input regulation* :

Current fixed at 300 mA.

Table V

Set I	R_S = 0.6 Ω
Input A.C. voltage in volt	Output D.C. voltage in volt
80	1.25
90	1.47
100	1.74
⋮	⋮
210	2.99
220	2.99

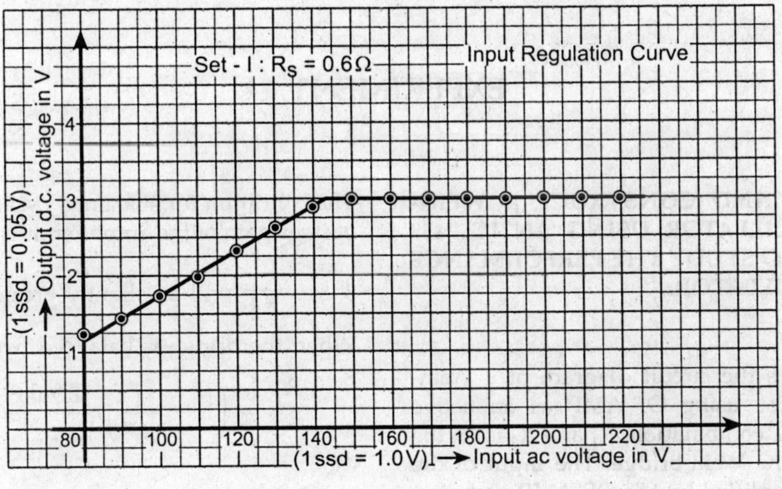

Fig. 13.8

The input regulation curve is drawn in Fig. 13.8. Repeat this observation for another values of R_S.

Discussion

(i) All circuit connections must be soldered properly and before soldering, the circuit should be well designed so that there is no mixing up. Further no connections should be left floating.

(ii) Before giving power to the circuit it must be checked so that the IC used is not damaged.

EXPERIMENT 14

TO DESIGN AND CONSTRUCT A WEIN-BRIDGE OSCILLATOR USING AN IC 7458 AND ALSO TO STUDY THE PERFORMANCE OF THE OSCILLATOR.

Theory

Fig. 14.1 shows the circuit diagram of a Wein-bridge oscillator using OP-AMP as an active device. The *R*-*C* combination, R_1 and R_2 form the four arms of the Wein-bridge. The input of the bridge V_i is amplified by the OP-AMP such that it becomes the bridge supply (Fig. 14.2) V_o. The circuit uses +ve feedback through the RC combination and a –ve fedback through the R_2. To get oscillation, Barkhausion criteria must be satisfied (i.e., the loop gain must be unity and the phase-shift around the loop must be zero). The frequency of oscillation in obtained from

$$f = \frac{1}{2\pi RC}$$

The condition for sustained oscillation is that the voltage gain of the amplifier must be greater than 3. So. $\frac{R_1}{R_2}$ will be slightly greater than 2.

When the bridge is balanced, we have

$$\frac{R_1}{R_2} = \left(R + \frac{1}{j\omega c}\right) / \frac{R/j\omega c}{R+\frac{1}{j\omega c}}$$

$$= \frac{\left(R + \frac{1}{j\omega c}\right)}{R/j\omega c}$$

$$= 2 + j\frac{R^2 - \frac{1}{\omega^2 C^2}}{\frac{R}{\omega C}}$$

Equating real and imaginary parts. We have

$$\frac{R_1}{R_2} = 2$$

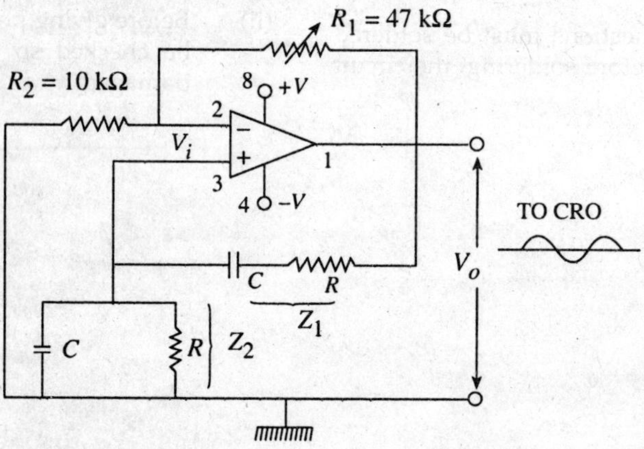

Fig. 14.1 : Circuit diagram of a Wein-bridge oscillator with IC 7458

and $R^2 - \frac{1}{\omega^2 C^2} = 0$

$$\omega = \frac{1}{CR}$$

$$2\pi f = \frac{1}{CR}$$

$$f = \frac{1}{2\pi RC} \quad \text{(i)}$$

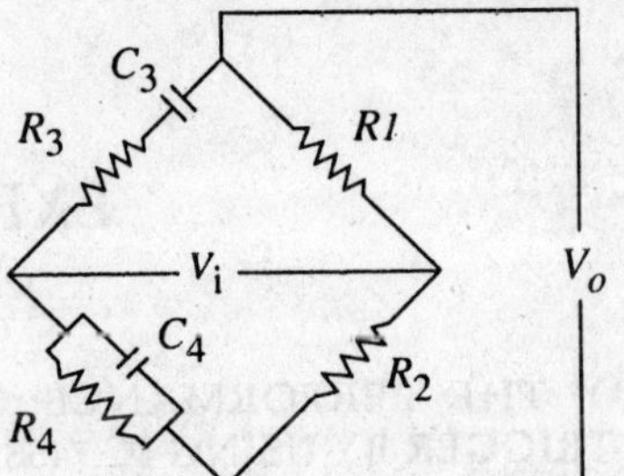

Experimental Data

Table I

Experimental and calculated time period

Value of capacitance C in μF	Value of resistance R in kΩ	Time period T as obtained from CRO in ms	Calculated time period $T = 2\pi RC$ in ms
	10	6.3	6.28
0.1	15	9.8	9.42
	18	11.4	11.30
	10	0.6	0.63
0.01	15	0.92	0.94
	18	1.06	1.13

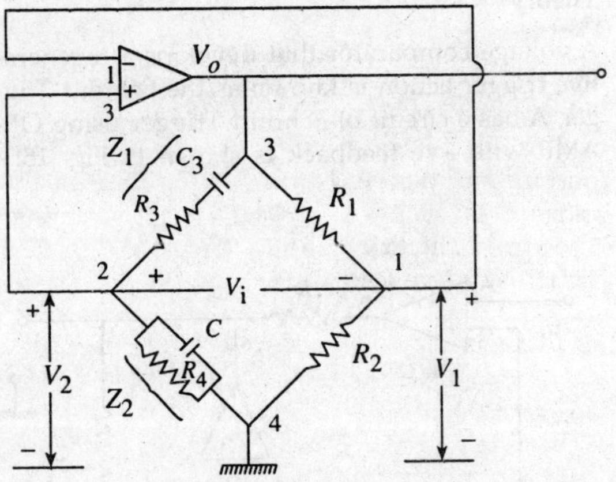

Fig. 14.2 : Basic diagram of a Wein-bridge oscillator

Discussion

(i) If we vary R_1 from zero to maximum, then for $R_1 < 2R_2$ we get normal D.C. line in the CRO output.

(ii) For $R_1 = 2R_2$, we get perfect sine wave as output.

(iii) For $R_1 > 2R_2$, sine wave gets distorted and distortion increases with increase of R_1.

EXPERIMENT 15

TO STUDY THE PERFORMANCE OF A SCHMITT TRIGGER BY USING IC 7458 AND USE OF THE VOLTAGE COMPARATOR

Theory

A voltage comparator that develops a regenerative trigger action is known as the Schmitt Trigger. A basic circuit of Schmitt Trigger using OP-AMP with +ve feedback is shown in Fig. 15.1.

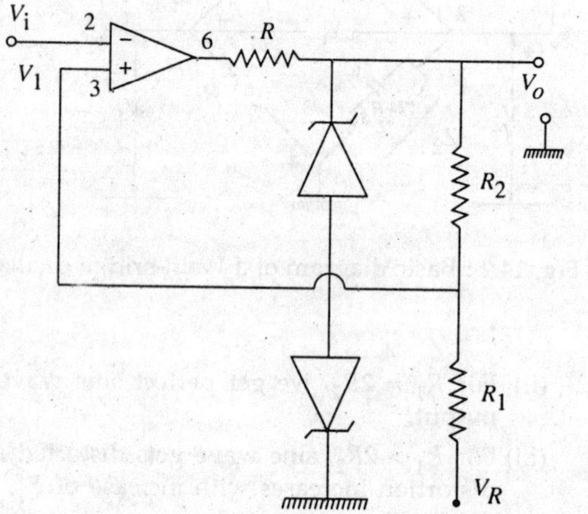

Fig. 15.1 : Basic circuit of a Schmitt Trigger using IC 7458

When $V_i < V_1$, the output voltage V_o is at +ve saturation level ($+V_{sat}$) and under this condition,

$$V_1 = V_R + \frac{(V_{sat} - V_R)}{R_1 + R_2} . R_1$$

$$= \frac{R_2 V_R}{R_1 + R_2} + \frac{R_1 V_{sat}}{R_1 + R_2} \qquad \text{(i)}$$

If V_i is now increased then the output remains constant at $+V_{sat}$ and $V_i \approx V_1$ untill $V_i = V_1$. This threshold or triggering voltage of the output regeneratively switches to $-V_{sat}$ and remains at this value as long as $V_i < V_1$.

The voltage at the non-inverting terminal for $V_i < V_1$ is given by

$$\frac{R_2 V_R}{R_1 + R_2} - \frac{R_1 V_{sat}}{R_1 + R_2} = V_2 \qquad \text{(ii)}$$

If the voltage V_i is now decreased, the output remans at $-V_{sat}$ untill $V_i = V_2$. At this voltage, the output regeneratively switches to $+V_{sat}$. The complete characteristics is shown in Fig 15.2. The circuit triggers at different voltages depending on whether the signal is increasing or deceasing. Thus the circuit shown in Fig 15.3 exhibits hysteresis property. We have the hysteresis voltage,

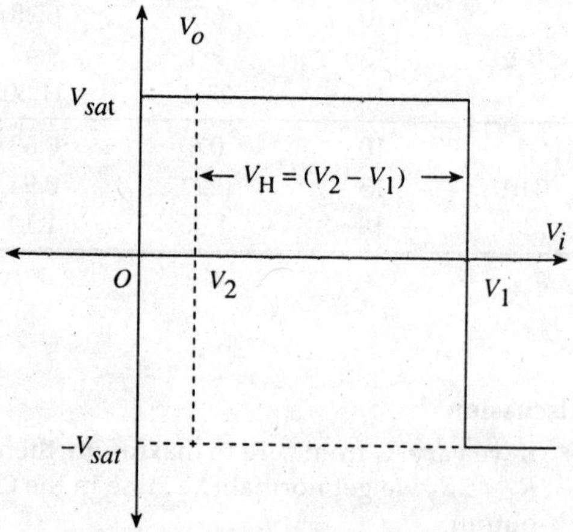

Fig. 15.2 : Time-voltage graph of Schmitt Trigger

$$V_H = V_1 - V_2 = \frac{2R_1}{R_1 + R_2} V_{sat} \qquad \text{(iii)}$$

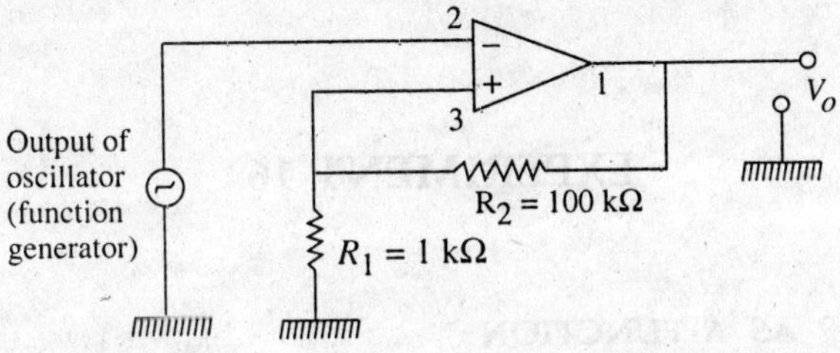

Fig. 15.3 : Experimental circuit of schmitt-trigger with IC 7458

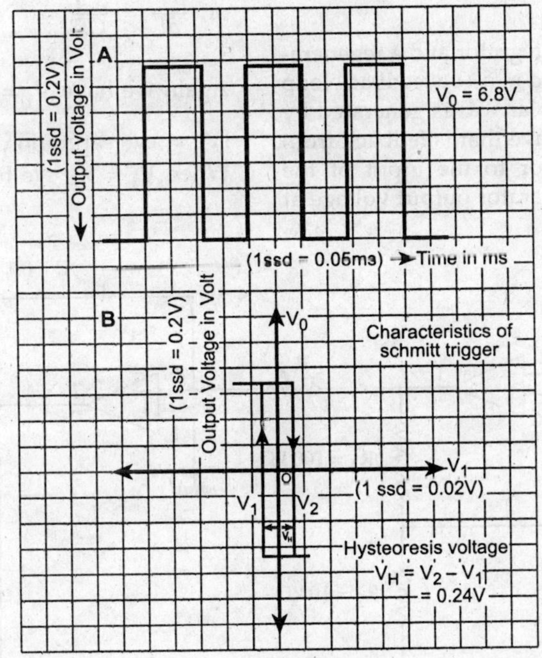

Fig. 15.4

Experimental Data

Data for hysteresis voltage (V_H)

R_1 = 1kΩ; R_2 = 100 kΩ

In Fig 15.4(A) we have plotted output voltage in volt against time in ms while in Fig 15.4(B) output voltage V_o is plotted against V_i.

Table I

The output voltage V_o in volt from graph	Hysteresis V_H in volts from theory $V_H = \frac{2R_1V_0}{R_1+R_2}$ volts	Hysteresis V_H in volts from experiment
6.8	$\frac{2\times10^3\times6.8}{(1+100)\times10^3}$ = 0.134	0.24

Discussion

(i) Distorted hysteresis takes place for high amp. of the output of the oscillator.

(ii) The input V_i should be varied beyond the limits of the hysteresis range V_H.

(iii) With more efficient OP-AMP, the experiment can be done more accurately.

(iv) Since the gain of the OP-AMP is not infinite, the two inputs are not equal and thus output is variable.

EXPERIMENT 16

TO STUDY AN OP-AMP AS A FUNCTION TRIANGULAR GENERATOR

Theory

The circuit contains an integrator and a regenerative comparator connected in a +ve feedback loop (Fig. 16.1). Linear triangular waves generated by integrating the square wave that is fedback from output of the comparator to the input of the integrator with the comparator output voltage at

$$dV = \frac{dQ}{C} = \frac{1}{C}\, idt = \frac{V_o}{CR}\, dt$$

or, $$\frac{dV}{dt} = \frac{Vo}{CR}$$

Again we have $V_1 = \dfrac{V_o R_2}{R_1+R_2} + \dfrac{V_t R_1}{R_1+R_2}$

[V_o = +ve saturation voltage of the comparator]

When $V_1 = V_R$, we have

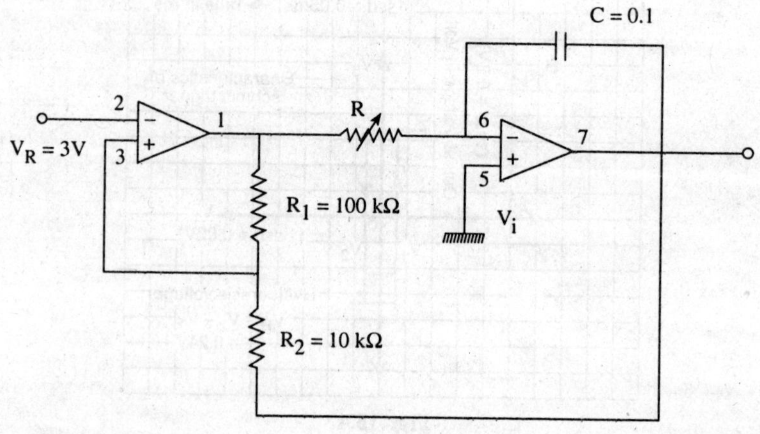

Fig. 16.1 : Circuit diagram for triangular wave generator with IC 7458

its positive saturation value V_o. The integrator output voltage (V+) decreases linearly at the rate V_o/CR till it reaches the lower tips voltage of the comparator. The output voltage of the comparator then switches rapidly to its –ve saturation value $-V_o$ and the integrator output voltage increases linearly at the rate V_o/CR. As the integrator output voltage attains the trip voltage of the comparator, the output of the comparator switches rapidly to V_o and the process is repeated. We can write,

$$V_R = \frac{V_o R_2}{R_1+R_2} + \frac{V_{min} R_1}{R_1+R_2} \qquad \text{(i)}$$

Also, $$V_1 = -\frac{V_o R_2}{R_1+R_2} + \frac{V(t) R_1}{R_1 + R_2}$$

$$V_R = -\frac{V_o R_2}{R_1+R_2} + \frac{V_{max} R_1}{R_1+R_2} \qquad \text{(ii)}$$

From equations (i) and (ii), we have

$$\frac{2V_o R_2}{R_1+R_2} = \frac{V_{max} R_1}{R_1+R_2} - \frac{V_{min} R_1}{R_1+R_2}$$

or, $V_{max} - V_{min} = 2V_o \frac{R_2}{R_1}$

Hence $T_1 = 2V_o \frac{R_2}{R_1} \frac{CR}{V_0} = 2RC \frac{R_2}{R_1}$

$\therefore \quad T = 2T_1$

$= 4RC \frac{R_2}{R_1} \qquad \text{(iii)}$

Equation (iii) is the required time period.

Experimental Data

A. When V_i is connected to ground

$V_i = 0$ volt.

$R_2 = 10\ k\Omega$; $R_1 = 100\ k\Omega$; $V_R = 3.0$ V

Table I

Value of capacitor C in μF	Value of Resistance R in kΩ	Time period (T) obtained from CRO in ms	Calculated time period (T) where $T = 4RC\frac{R_2}{R_1}$	Peak-to-peak voltage in volt
	10	0.5	0.4	2.1
	15	0.7	0.6	2.0
0.1	⋮	⋮	⋮	⋮
	33	1.45	1.32	1.9
	⋮	⋮	⋮	⋮

A typical time – voltage graph for R = 15 kΩ is shown in Fig. 16.2.

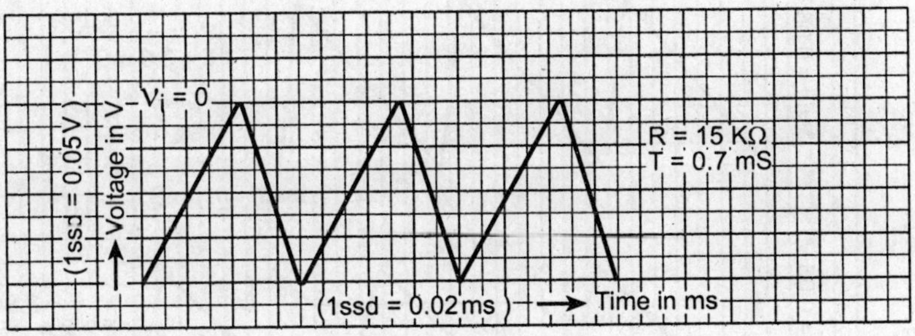

Fig. 16.2

B. When V_i is not zero

$R_2 = 10\ k\Omega$; $R_1 = 10\ k\Omega$; $V_R = 3.0$ V; R = 39 kΩ.

Table II

Value of input voltage V_i in volt	T_1 in ms	T_2 in ms	Total time period (T) where $T = T_1+T_2$ ms	Peak-to-peak voltage in volt
+6.75	0.5	1.3	1.8	1.85
+2.84	0.75	0.75	1.5	1.85
– 4.34	3.2	0.4	3.6	1.90
⋮	⋮	⋮	⋮	⋮

As a typical value we have shown variation of voltage in V with time in ms in Fig. 16.3.

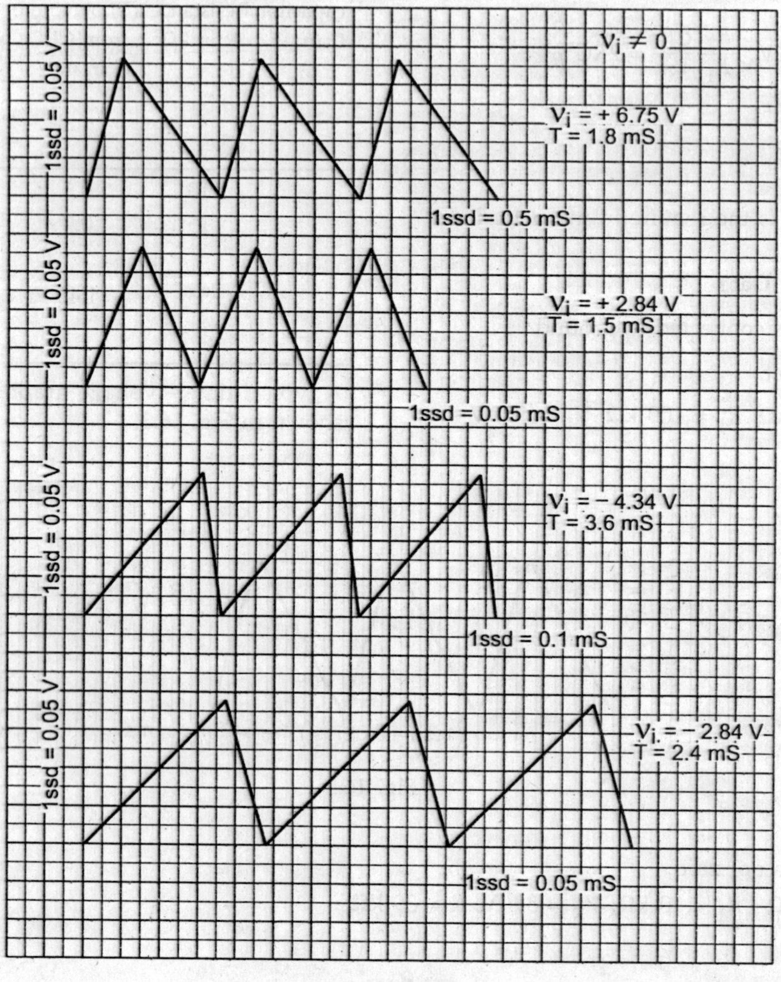

Fig. 16.3

Discussion

(i) According to the circuit diagram the time period of the square wave and triangular wave is the same.

(ii) If V_i with –ve voltage increases, the symmetrical triangular wave changes in sawtooth wave.

EXPERIMENT 17

TO STUDY THE OFFSET VOLTAGE AND SLEW RATE OF AN OPAMP AND TO CONSTRUCT A PULSE GENERATOR USING OP AMP

1. To measure input off-set voltage

The input off set voltage is the d.c. voltage required to provide zero output voltage with no input signal or source resistance (Fig. 17.1).

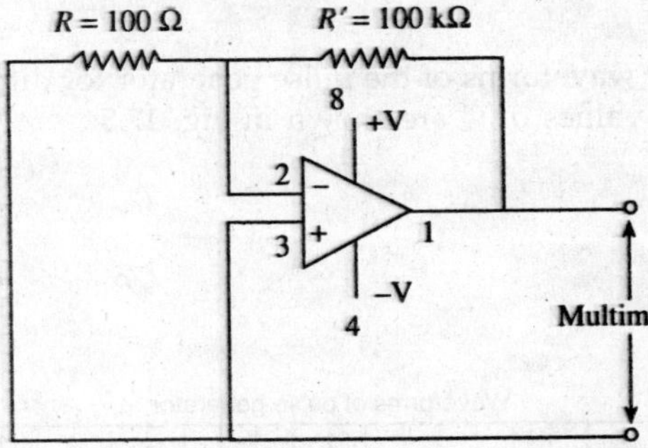

Fig. 17.1 : To measure the input offset voltage

As it depends very much on temperature and supply voltage, the measurement is made at room temperature.

The output voltage may be written as

$$E_o = -\frac{R+R'}{R}V_{io}$$

where V_{io} is the input off-set voltage.

Thus, $V_{io} = -\dfrac{RE_o}{R+R'}$ (i)

Experimental Data

Table I

Value of R in Ω	Value of R′ in kΩ	E_o as measured in volt	$V_{io} = -\dfrac{RE_o}{R+R'}$ volt
100	100	-7.5×10^{-3}	7.49×10^{-6}

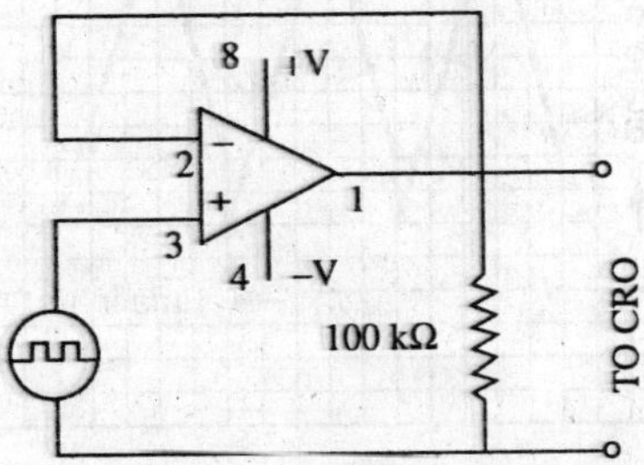

Fig. 17.2 : To measure the slew-rate of an OP-AMP

2. Measurement of slew-rate

This is the maximum rate of change of output voltage with supply (Fig 17.2). Output slewing is measured in the gain follower circuit. The output is connected to the oscilloscope and slopes of the +ve slewing and –ve slewing are measured.

$$\text{Slew-rate, } S_r = \frac{\Delta V_o}{\Delta t} \text{ volt/sec} \quad \text{(ii)}$$

Table II

For +ve slewing			For −ve slewing		
Δt (in μs)	ΔV_o (in volt)	slew rate $S_r = \frac{\Delta V_o}{\Delta t}$ V/μs	Δt (in μs)	ΔV_o (in volt)	slew rate $S_r = \frac{\Delta V_o}{\Delta t}$ V/μs
10	3	$\frac{3}{10} = 0.3$	12	3	$\frac{3}{12} = 0.25$

By plotting voltage against time we have drawn the slew-rate in Fig. 17.3.

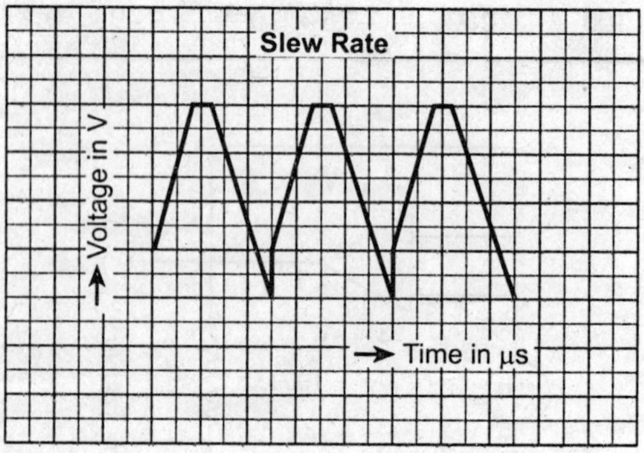

Fig. 17.3

3. Use of an OP-AMP as a pulse generator

The circuit diagram of a pulse generator using IC 7458 is shown in Fig. 17.4.

The time period of a pulse generator is :

$$T = 2RC\, l_n \left(1 + \frac{2R_2}{R_1}\right). \qquad \text{(iii)}$$

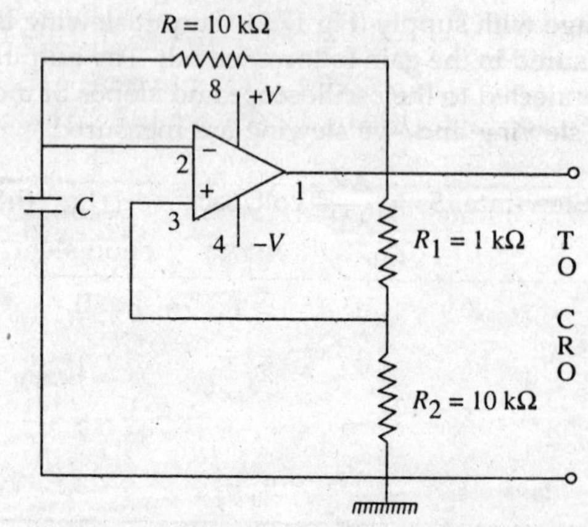

Fig. 17.4 : Circuit of a Pulse Generator using IC 7458

Table III

$R = R_2 = 10$ kΩ, $R_1 = 1$ kΩ

Different values of capacitor C in μF	Time period of (T) as obtained from the CRO in ms	Calculated time period (T) $T = 2RC\, l_n \left(1 + \frac{2R_2}{R_1}\right)$ in ms
0.01	0.6	0.608
0.10	5.8	6.08
0.22	14.0	13.39
0.47	29.0	28.6

The waveforms of the pulse generator for different values of C are shown in Fig. 17.5.

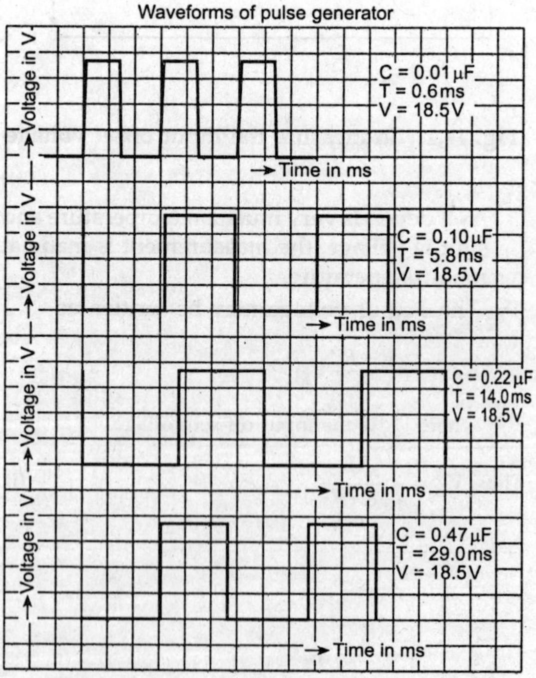

Fig. 17.5

EXPERIMENT 18

CONSTRUCTION OF ENCODER, DECODER AND S-R FLIP-FLOP

[A] Construction of an encoder

An encoder converts an active input signal into a coded output signal. Fig. 18.1 gives the general idea of an encoder. There are *n*-input lines, only one of which is active. Internal logic within the encoder converts the active input to a

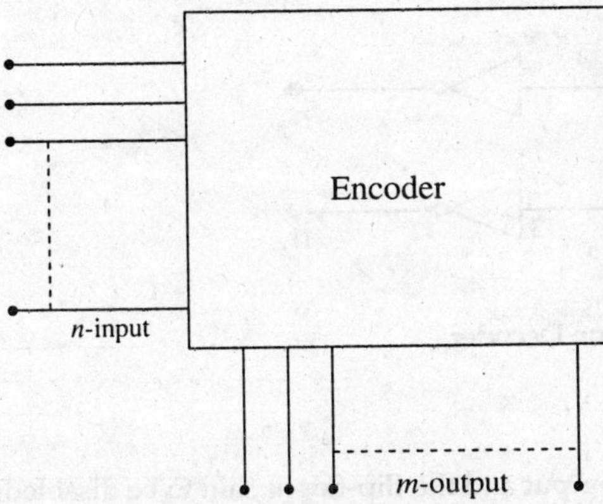

Fig. 18.1

coded binary output with *m*-bits. A decimal to *BCD* encoder is constructed here. Fig. 18.2 shows the required circuit diagram. *A, B, C* and *D* are four inputs used for denoting decimal numbers 0, 1, 2, 3; Y_1 and Y_0 are two binary outputs.

Truth table for encoding decimal nos 0 to 3

Decimal Input	Voltage across *A,B,C,D* four input terminals				$Y_1=\overline{C}\,\overline{D}$ (MSB)	$Y_0=\overline{B}\,\overline{D}$ (LSB)
	A	*B*	*C*	*D*		
0	H	L	L	L	0	0
1	L	H	L	L	0	1
2	L	L	H	L	1	0
3	L	L	L	H	1	1

Here 'H' denotes high voltage rate i.e., converted to + 5 V and L, denotes low voltage rate, i.e., grouded. Therefore, this truth table shows that decimal input is coded to binary output by encoder.

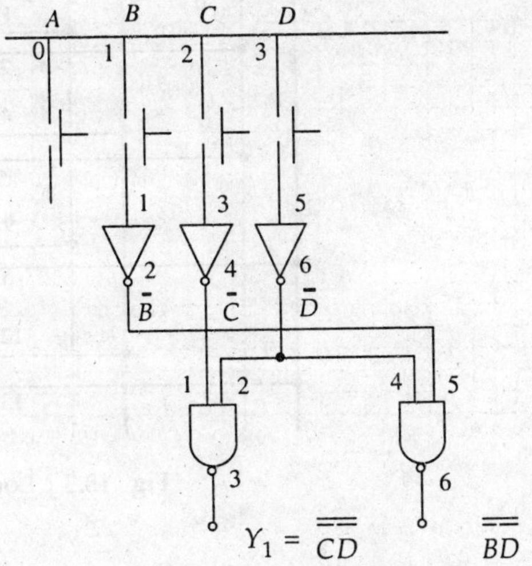

Fig. 18.2 : Circuit diagram of an encoder

[B] Construction of a decoder

A decoder converts a binary number to its decimal equivalent. The logic circuit of a one-to-four decoder is shown in Fig. 18.3. Only one of the 4-input lines will be high at a time. The high output terminal will be decimal equivalent to the input binary number.

Binary inputs B_0	Binary inputs B_1	Terminal with a high output	Decimal equivalent
0	0	$D_0 = \overline{B}_0\,\overline{B}_1$	0
0	1	$D_1 = \overline{B}_0\,B_1$	1
1	0	$D_2 = B_0\,\overline{B}_1$	2
1	1	$D_3 = B_0\,B_1$	3

[C] Constructioin of a clocked S-R flip-flop

The output of the digital circuits considered previously depends entirely on the inputs of those circuits. However, there are requirements for a digital device whose output will remain unchanged once set, even if there is a change in

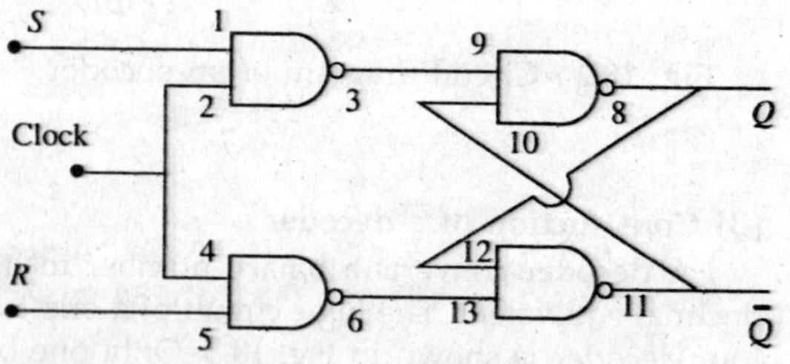

Fig. 18.4 : Circuit diagram of S-R flip flop

the inputs. Such a device could be used to store a binary digit. A flip-flop is one such device. It is a bistable electronic device. Its output is either low or high. It also has a memory. Its output will remain unaltered untill something is done to change it. For this reason a flip-flop can be regarded as a memory device. Ordinary S-R flip-flop are said to be trasparent, i.e., only changed information at R or S is transmitted immediately to the output Q and $\overline{Q}$.

The addition of 2-NAND gates at inputs R and S is shown in Fig. 18.4 and this will result in a flip-flop that can be enabled or disabled. When the clock pulse is low, neither of the input affects

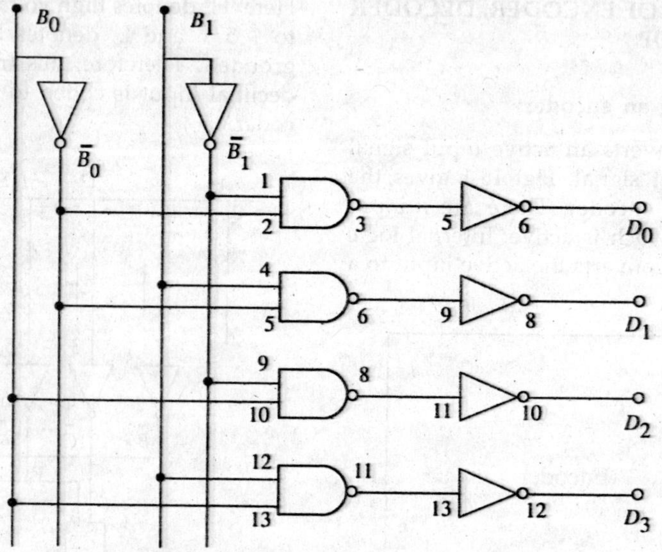

Fig. 18.3 : Logic circuit for Decoder

the output and the flip-flop is said to be disabled. When the clock pulse is high, information at R and S inputs will be directly transmitted to the output. The flip-flop is then said to be enable.

Truth table for S-R flip-flop

Clock pulse	S	R	Q	$\overline{Q}$
L	×	×	No change	No change
H	0	0	No change	No change
H	1	0	1	0
H	0	1	0	1
H	1	1	Not defined	Not defined

Here L = Low ; H = High.

Discussion

(i) During the experiment, pin nos. 14 and 7 must be connected to +5 V and to ground respectively.

EXPERIMENT 19

SIMPLIFICATION OF BOOLEAN EXPRESSION USING KARNAUGH MAP AND CIRCUIT IMPLEMENTATION

Karnaugh map is a simple method to construct a basic digital circuit corresponding to a given truth table. Actually this method provides a simple straight forward procedure for minimizing a Boolean function. It consists of an array of squares or cells which contain all the necessary information in the truth table arranged in such a way that allows quick visual simplification of the logical equation according to some very simple rules.

Problem 1

Construct the basic circuit design corresponding to the truth table 1.

Solution : The Boolean expression of the truth table is

$$Y = \overline{A}\,B\,\overline{C} + A\,B\,\overline{C} + ABC$$

Karnaugh Map :

		C(+)	$\overline{C}$(0)
(0 0)	$\overline{A}\,\overline{B}$	0	0
(0 1)	$\overline{A}\,B$	0	1
(1 1)	$A\,B$	1	1
(1 0)	$A\,\overline{B}$	0	0

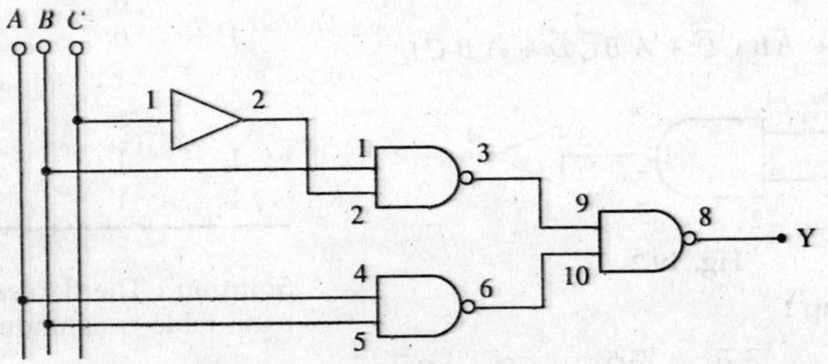

Fig. 19.1 : Basic circuit design

Truth Table 1

A	*B*	*C*	*Y*
0	0	0	0
0	0	1	0
0	1	0	1
0	1	1	0
1	0	0	0
1	0	1	0
1	1	0	1
1	1	1	1

The minimized expression in the sum of product form is

$$Y = AB + B\overline{C}$$

$$= \overline{\overline{AB} + \overline{B\overline{C}}} \text{ (using De-Morgon's theorem).}$$

Therefore, the circuit diagram for the truth table will be as shown in Fig. 19.1 and all the combinations of Truth Table 1 are verified with circuit diagram shown in Fig. 19.1.

Problem 2

Construct the basic circuit design corresponding to the Truth Table 2.

Truth Table 2

A	B	C	D	Y
0	0	0	0	0
0	0	0	1	0
0	0	1	0	0
0	0	1	1	0
0	1	0	0	1
0	1	0	1	0
0	1	1	0	1
0	1	1	1	0
1	0	0	0	0
1	0	0	1	0
1	0	1	0	0
1	0	1	1	0
1	1	0	0	1
1	1	0	1	0
1	1	1	0	1
1	1	1	1	0

Solution : The Boolean expression of this truth table is

$$Y = \bar{A}B\bar{C}\bar{D} + \bar{A}BC\bar{D} + AB\bar{C}\bar{D} + ABC\bar{D}$$

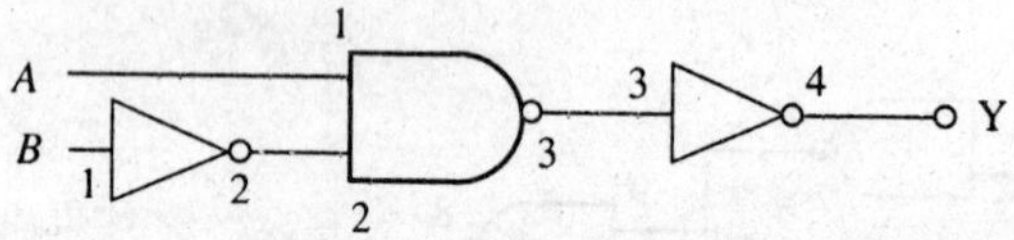

Fig. 19.2

Karnaugh map :

	$\bar{C}\bar{D}$	$\bar{C}D$	CD	$C\bar{D}$
$\bar{A}\bar{B}$	0	0	0	0
$\bar{A}B$	1	0	0	1
AB	1	0	0	1
$A\bar{B}$	0	0	0	0

The minimized expression in the sum of product form is given by

$$Y = \bar{A}B\bar{D} + AB\bar{D} = B\bar{D}$$

The circuit construction for the truth table is given in Fig. 19.2. The truth table is verified using circuit diagram of Fig. 19.2.

Problem 3

Construct the basic circuit design for the Truth Table 3 given bellow :

Truth Table 3

A	*B*	*C*	*D*	*Y*
0	0	0	0	0
0	0	0	1	0
0	0	1	0	1
0	0	1	1	0
0	1	0	0	0
0	1	0	1	0
0	1	1	0	0
1	0	0	0	1
1	0	0	1	1
1	0	1	0	1
1	0	1	1	1
1	1	0	0	1
1	1	0	1	1
1	1	1	0	1
1	1	1	1	1

Solution : The Boolean expression of the above truth table is as under :

$$Y = \bar{A}\bar{B}C\bar{D} + A\bar{B}\bar{C}\bar{D} + A\bar{B}\bar{C}D + A\bar{B}C\bar{D} + A\bar{B}CD + AB\bar{C}\bar{D} + AB\bar{C}D + ABC\bar{D} + ABCD$$

Karnough map :

	$\bar C\bar D$	$\bar C D$	CD	$C\bar D$
$\bar A\bar B$	0	0	0	1
$\bar A B$	0	0	0	0
AB	1	1	1	1
$A\bar B$	1	1	1	1

The minimized expression in the sum of product form is given by

$$Y = AB + A\bar B + \bar B C\bar D = A + \bar B C\bar D$$

$$= \overline{\bar A . \overline{\bar B C\bar D}} \text{ (using De Morgan's theorem)}$$

The circuit construction for the truth table is shown in Fig. 19.3 and the truth table is verified.

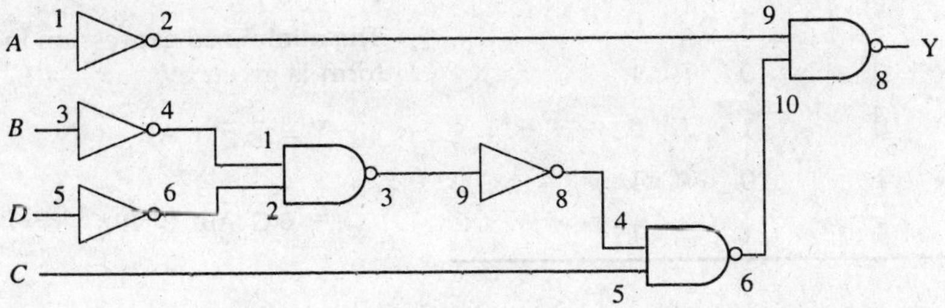

Fig. 19.3

Problem 4

Construct the basic circuit design corresponding to the Truth Table 4.

Truth Table 4

A	B	C	Y
0	0	0	1
0	0	1	1
0	1	0	0
0	1	1	1
1	0	0	1
1	0	1	1
1	1	0	0
1	1	1	0

Solution : The Boolean expression of the above truth table is

$$Y = \bar A\bar B\bar C + \bar A\bar B C + \bar A BC + A\bar B\bar C + A\bar B C$$

Karnaugh map :

	C	$\bar C$
$\bar A\bar B$	1	1
$\bar A B$	1	0
AB	0	0
$A\bar B$	1	1

The minimized expression in the sum of product form is given by,

$$Y = \bar A\bar B + \bar A C + A\bar B$$

$$= \bar B + \bar A C$$

$$= \overline{B . \overline{\bar A C}} \text{ (using De-Morgan's theorem)}$$

Therefore the circuit diagram for the truth table is given by Fig. 19.4 and all the combinations of truth table are verified with this circuit diagram.

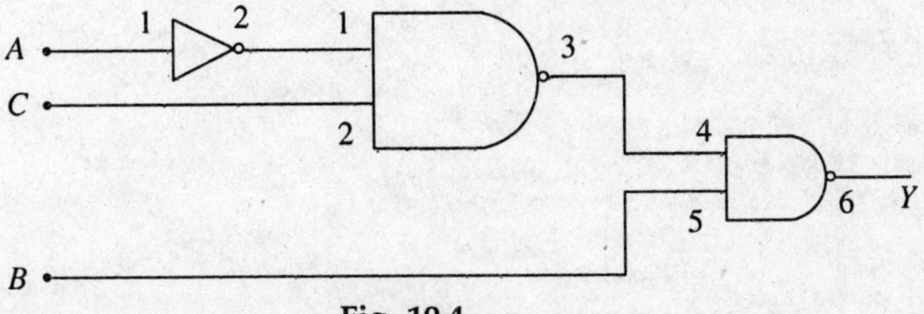

Fig. 19.4

Problem 5

Construct the basic circuit design corresponding to the Truth Table 5.

Truth Table 5

A	B	C	Y
0	0	0	1
0	0	1	0
0	1	0	0
0	1	1	0
1	0	0	1
1	0	1	0
1	1	0	1
1	1	1	1

Solution : The Boolean expression of the above truth table is :

$$Y = \bar{A}\,\bar{B}\,\bar{C} + A\bar{B}\,\bar{C} + AB\bar{C} + ABC$$

Karnaugh map :

	$\bar{C}$	C
$\bar{A}\,\bar{B}$	1	0
$\bar{A}B$	0	0
AB	1	1
$A\bar{B}$	1	0

The minimized expression in the sum of product form is given by

$$Y = \bar{B}\,\bar{C} + AB$$

$$= \overline{\overline{\bar{B}\,\bar{C}}\,.\,\overline{AB}} \quad \text{(using De–Morgan's theorem)}$$

Therefore, the circuit diagram for the truth table is given by Fig. 19.5 and all the combination of the truth table is verified with this circuit diagram.

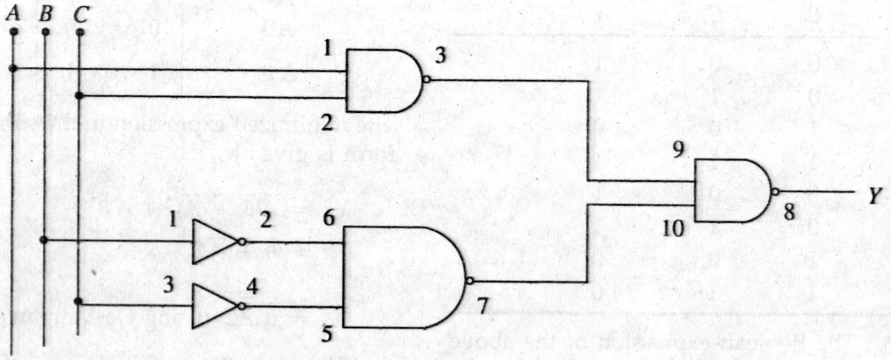

Fig. 19.5

Problem 6 : Construct the basic circuit design correspondiing to the Truth Table 6.

Truth Table 6

A	B	C	D	Y
0	0	0	0	1
0	0	0	1	1
0	0	1	0	1
0	0	1	1	1
0	1	0	0	1
0	1	0	1	0
0	1	1	0	1
0	1	1	1	0
1	0	0	0	1
1	0	0	1	1
1	0	1	0	1
1	0	1	1	1
1	1	0	0	1
1	1	0	1	0
1	1	1	0	1
1	1	1	1	0

Solution : The Boolean expression of the above truth table is given by

$$Y = \bar{A}\,\bar{B}\,\bar{C}\,\bar{D} + \bar{A}\,\bar{B}\,\bar{C}D + \bar{A}\,\bar{B}C\bar{D} + \bar{A}\,\bar{B}CD + \bar{A}B\bar{C}\,\bar{D} + \bar{A}B\bar{C}\,\bar{D} + A\bar{B}\,\bar{C}\,\bar{D} + A\bar{B}C\bar{D} + A\bar{B}\,\bar{C}D + AB\bar{C}\,\bar{D} + ABC\bar{D} + A\bar{B}\,\bar{C}D$$

Kanraugh map :

	$\bar{C}\,\bar{D}$	$\bar{C}D$	CD	$C\bar{D}$
$\bar{A}\,\bar{B}$	1	1	1	1
$\bar{A}B$	1	0	0	1
AB	1	0	0	1
$A\bar{B}$	1	1	1	1

The minimizing expression in the sum of product from is given by,

$$Y = \bar{C}\,\bar{D} + C\bar{D} + \bar{A}\,\bar{B} + A\bar{B}$$

$$= \bar{D} + \bar{B}$$

$$= \overline{BD} \text{ (using De-Morgan's theorem)}$$

Therefore, the circuit diagram for the truth table is given by Fig. 19.6 and all the combination of the truth table is verified with the circuit diagram.

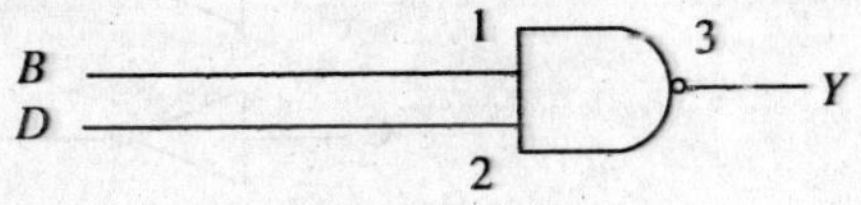

Fig. 19.6

Problem 7 Construct the basic circuit design corresponding to the truth Table 7.

Truth Table 7

A	B	C	Y
0	0	0	1
0	0	1	0
0	1	0	0
0	1	1	0
1	1	0	1
1	1	1	0
1	0	0	1
1	0	1	1

Solution : The boolean expression of the above truth table 7 is

$$Y = \bar{A}\,\bar{B}\,\bar{C} + AB\bar{C} + A\bar{B}\,\bar{C} + A\bar{B}C$$

Karnaugh map :

	$\bar{C}$	C
$\bar{A}\,\bar{B}$	1	0
$\bar{A}B$	0	0
AB	1	0
$A\bar{B}$	1	1

The minimized expression in sum of product from is

$$Y = A\bar{B} + \bar{B}\,\bar{C}$$

$$= \overline{\overline{A\bar{B}}.\overline{\bar{B}\,\bar{C}}} \text{ (using De-Morgam's theorem)}$$

Therefore, circuit diagram for the truth table is given by Fig. 19.7 and the truth table is verified.

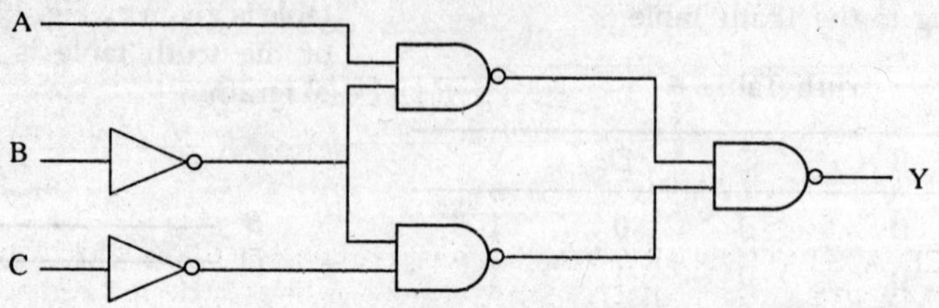

Fig. 19.7

Discussion

(i) Before doing the experiment all the ICs were checked properly.

(ii) The lengths of the connecting wires were kept as small as possible.

(iii) We must be careful about loose connections.

(iv) We used 5 V regulated power supply.

EXPERIMENT 20

TO PRODUCE A STATIONARY WAVE PATTERN WITH MICROWAVE AND TO DRAW A CRYSTAL CURRENT *VS* DISTANCE CURVE AND TO MEASURE THE FREQUENCY OF GUNN DIODE; ALSO TO FIND THE CUT-OFF WAVELENGTH RESONANT FREQUENCY AND GUIDE WAVELENGTH

Construction

A practical Gunn diode is usually made of the GaAs doped with Si, Te or selinium. The basic construction is shown in Fig. 20.1 The n+ substrate is heavily doped and active n-layer is less heavily doped.

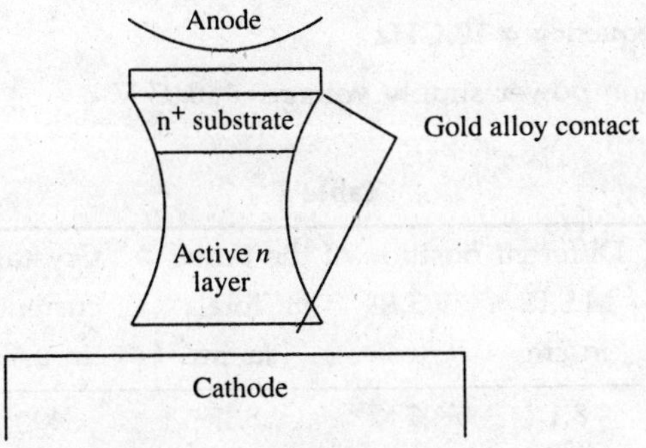

Fig. 20.1

The commercial diode requires ~ 9 V D.C. bias and the operating current is about 90 mA. These diodes are generally used for producing microwaves with frequency from 4 GHz to 100 GHz and the output efficiency is within 2.5 – 5%.

The resonant cavity with proper design and a D.C. source are required for its application as a micro wave oscillator (Fig. 20.2). The cavity is in the form of a wave guide and the effective length of the cavity is varied by sliding short plunger.

Theory

Gunn diode generates microwave when voltage is applied to the diode. The voltage applied across the diode is 8 volt. The microwave generated in the tube goes through the wave guide and reflected by the reflector at the tunnal. Stationary waves are produced and are intercepted by the antenna attached to the crystal detector. The electric field existing in the stationary wave puts a voltage across the detector and the detected (or diode) current is given by

$$I = KV^p$$

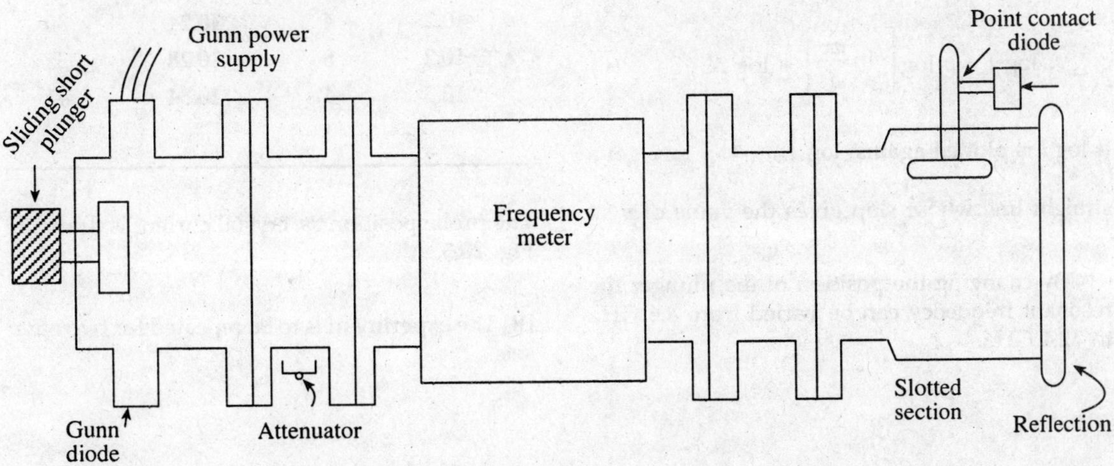

Fig. 20.2

where, V = Voltage across the diode

K = Constant

I = Average rectified current

P = Index parameter depending on the individual crystal and upon the magnitude of the voltage over the wide range of voltage; it is supposed to be constant and its value is 2.

Now, the electric field of the stationary wave in the wave guide varies with distance x in accordance with the relation

$$E = E_o\left[\sin\frac{\pi x}{d}\right]$$

where $d = \frac{\lambda}{2}$ is the distance between two consecutive null points.

Since the crystal voltage is proportional to the electric field at the travelling,

$$V = K_1 \sin\frac{\pi x}{d} \quad \text{(iii)}$$

K_1 = constant

Now from (i) and (iii), we get

$I = KV^P$

$$\text{or, } I = A\left[\sin\frac{\pi x}{d}\right]^p \quad \text{(iv)}$$

$$\therefore \log I = p \log\left[\sin\frac{\pi x}{d}\right] + \log A \quad \text{(v)}$$

If log I is plotted against log $\left[\sin\frac{\pi x}{d}\right]$, we get a straight line whose slop gives the value of p.

By changing the position of the plunger the resonant frequency can be varied from 8.6 GHz to 12.4 GHz.

To verify the standard formula

We have

$$\lambda_g = \frac{\lambda}{\sqrt{1-\left(\frac{\lambda}{\lambda_c}\right)^2}} \quad \text{(i)}$$

We measure λ_g from the slotted section. λ_c is obtained by measuring the wave guide cavity (λ_c = 2a) while λ is obtained by measuring with the frequency meter.

Experimental Data

[A] Data for drawing crystal current vs. position of the probe graph.

Fixed beam current = 22 mA

Fixed reflector voltage = 250 V

Micrometer reading = 13.27 mm

Frequency = 10 GHz

Gunn power supply voltage = 10.03 V

Table I

Different position of the probe			Crystal current in μA
M.S.R. in cm	V.S.R.	Total in cm	
8.1	7	8.17	40
8.2	4	8.24	45
8.3	0	8.30	50
⋮	⋮	⋮	⋮
10.2	4	10.24	30
10.2	8	10.28	35
10.3	4	10.34	41
⋮	⋮	⋮	⋮

The probe position *vs.* crystal current is drawn in Fig. 20.3.

[B] The experiment is to be repeated for two other sets.

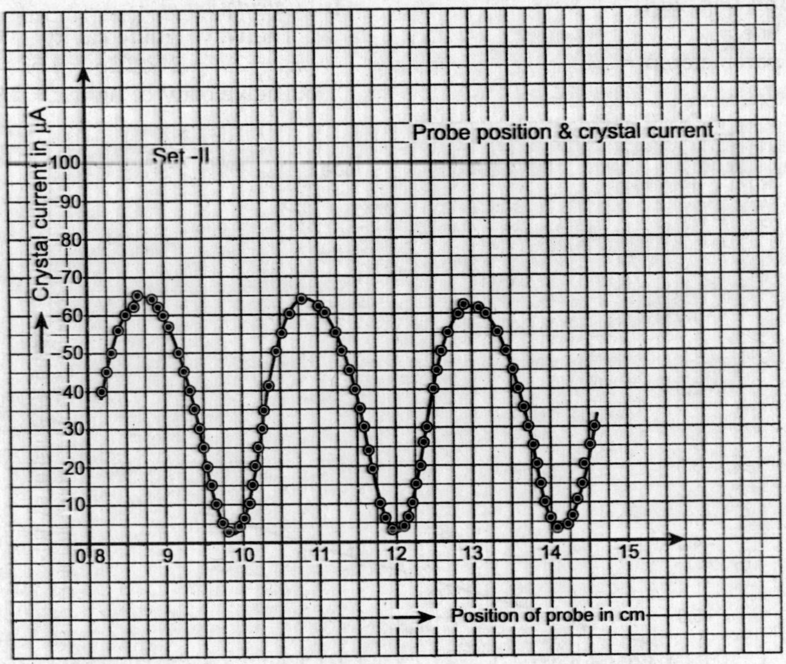

Fig. 20.3

[C] To determine the frequency of the klystron

Table II

No. of Obs.	Current in crystal before resonance (in μA)	Current in crystal at resonance (in μA)	Corresponding frequency in metre (in MHz)	Mean frequency (in MHz)
1	8.8	8.4	9890	
2	11.2	10.7	...	9891.25
3	13.4	13.0	...	
4	15.6	15.0	9890	

[D] Data for drawing $\log_e I$ *vs.* $\log_e \left\{ \sin \dfrac{\pi x}{d} \right\}$

Table III

Crystal current I (in μA)	$\log_e I$	Probe position (in cm) [from graph]	Null point (in cm) [from graph]	Distance(x) from null point (in cm)	'd' value between two null points (in cm)	value of $\left(\dfrac{\pi x}{d}\right)$	log sin $\left(\dfrac{\pi x}{d}\right)$
10	2.30	10.2		0.24	12.12	...	...
20	2.99	10.28		0.32	–9.96	...	...
30	3.40	10.34	9.96	0.38	= 2.16	...	...
⋮	⋮	⋮		⋮		⋮	⋮
80	4.38	10.8		0.84		...	...

From the graph (Fig. 20.4)
Slope of the line = ...
∴ the index parameter, p = ...

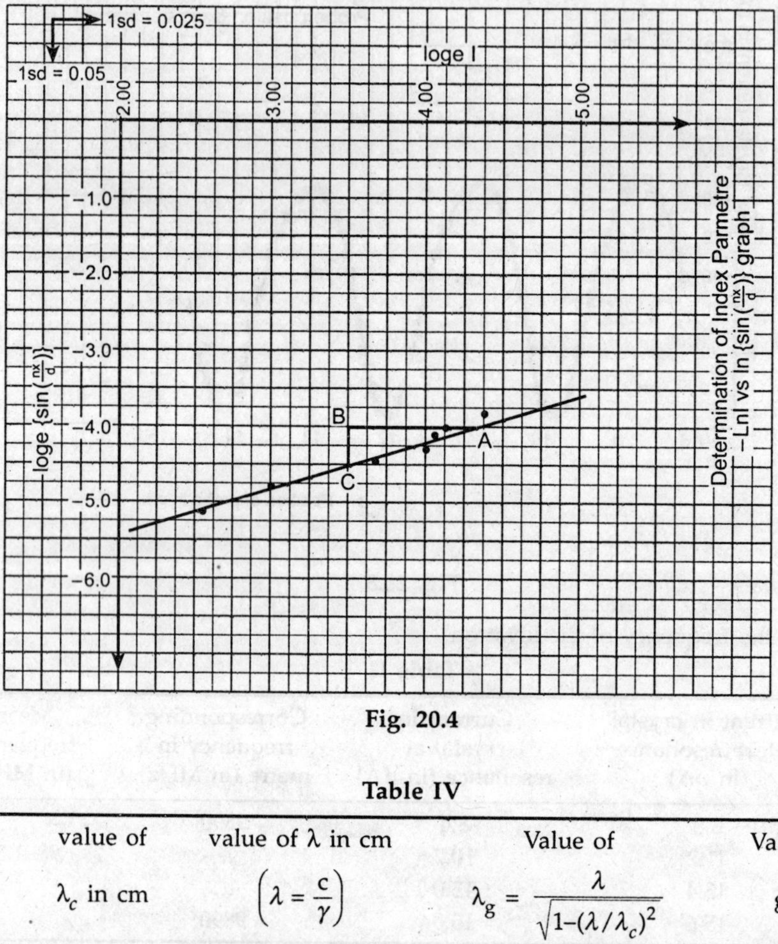

Fig. 20.4

[E] *Calculation of* λ_g :

The wave guide cavity, a = 2.26 cm

$\therefore \lambda_c = 2\,a = 2 \times 2.26 = 4.52$

The velocity of light c = 2.99×10^{10} cm/sec.

Table IV

Set No.	value of λ_c in cm	value of λ in cm $\left(\lambda = \frac{c}{\gamma}\right)$	Value of $\lambda_g = \frac{\lambda}{\sqrt{1-(\lambda/\lambda_c)^2}}$	Value of λ from graph in cm
Set I	4.52	$\frac{2.99 \times 10^{10}}{9.8 \times 10^9} = 3.051$	$\frac{3.051}{\sqrt{1-\left(\frac{3.51}{4.52}\right)^2}} = 4.135$	(12.69 – 8.32) = 4.37
Set II	4.52	$\frac{2.99 \times 10^{10}}{10 \times 10^9} = 2.99$	$\frac{2.99}{\sqrt{1-\left(\frac{2.99}{4.52}\right)^2}} = 3.986$	(12.60 – 8.30) = 4.30
Set III	4.52	$\frac{2.99 \times 10^{10}}{9.6 \times 10^9} = 3.114$	$\frac{3.114}{\sqrt{1-\left(\frac{3.114}{4.52}\right)^2}} = 4.296$	(13.15 – 8.62) = 4.53

Discussion

(i) The attenuation is used to make the current in between 80-90 μA in peak position.

(ii) The theoretical and experimental values are found to be different to some extent.

EXPERIMENT 21

TO STUDY THE CHARACTERISTICS OF A SOLAR CELL UNDER DARKNESS AND ILLUMINATION CONDITION

Theory :

A solar cell is basically a P-N junction diode with large surface area. Due to photo-voltaic effects, it directly converts the solar radiation to electrical energy. In dark condition it works as an ordinary diode. But when one illuminates the surface of a solar cell the minority carriers are increased and diffused across the junction. So a potential is generated across the junction and also an e.m.f. in the external circuit is replaced. The e.m.f. of the current varies with intensity of the incident radiation and also with the variation of external load. The equivalent circuit of solar cell is given in Fig. 21.1.

Short circuit current : The current through the circuit when zero load resistance is inserted called the short circuit current (I_{sc}).

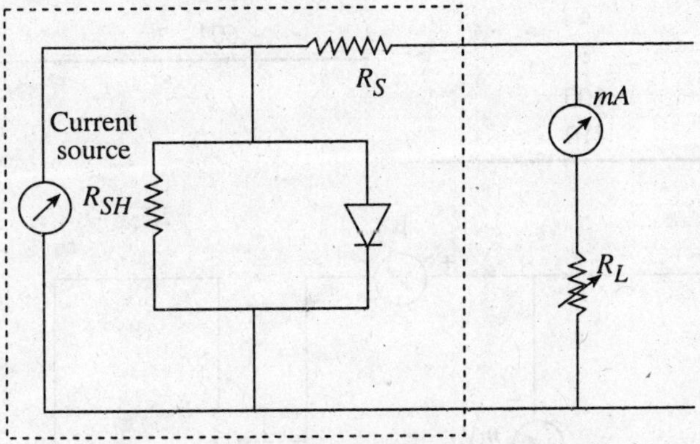

Fig. 21.1 : Equivalent circuit of solar cell

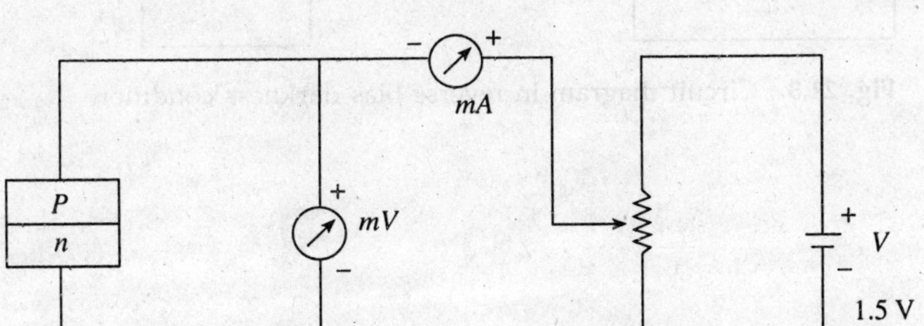

Fig. 21.2 : Circuit diagram in forward bias under darkness condition

Open circuit voltage : The voltage across the solar cell when load is infinitely large is called the open circuit voltage (V_{oc}).

Experimental Data

The circuit diagram in forward bias under darkness condition is shown in Fig. 21.2.

[A] Characteristics of solar cell in forward bias under darkness condition.

Table I

Voltage accross the solar cell in mV	current through the solar cell in mA
0	0
225	1
275	2
334	4
⋮	⋮
563	100
567	110

The circuit diagram in reverse bias darkness condition is shown in Fig. 21.3.

[B] Characteristics of solar cell in reverse bias under darkness condition :

Table II

Voltage across the solar cell in mV	Current through the solar cell in μA
0	0
432	56
445	60
495	70
⋮	⋮
594	96
604	100

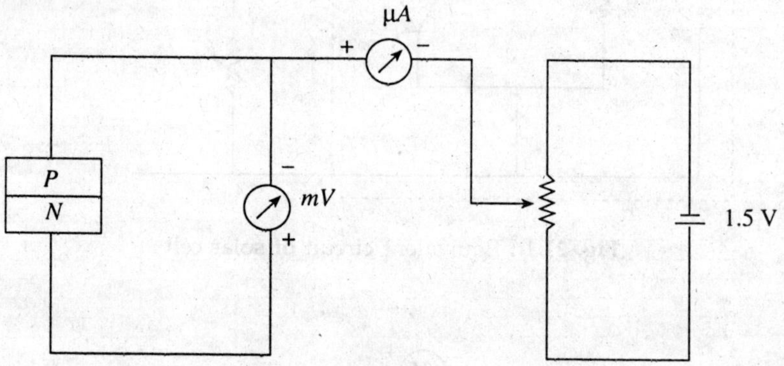

Fig. 21.3 : Circuit diagram in reverse bias darkness condition

The *V-I* characteristics of a solar cell under forward bias and reverse bias conditions can now be drawn. Fig. 21.4 illustrates V-I characteristics under forward bias condition.

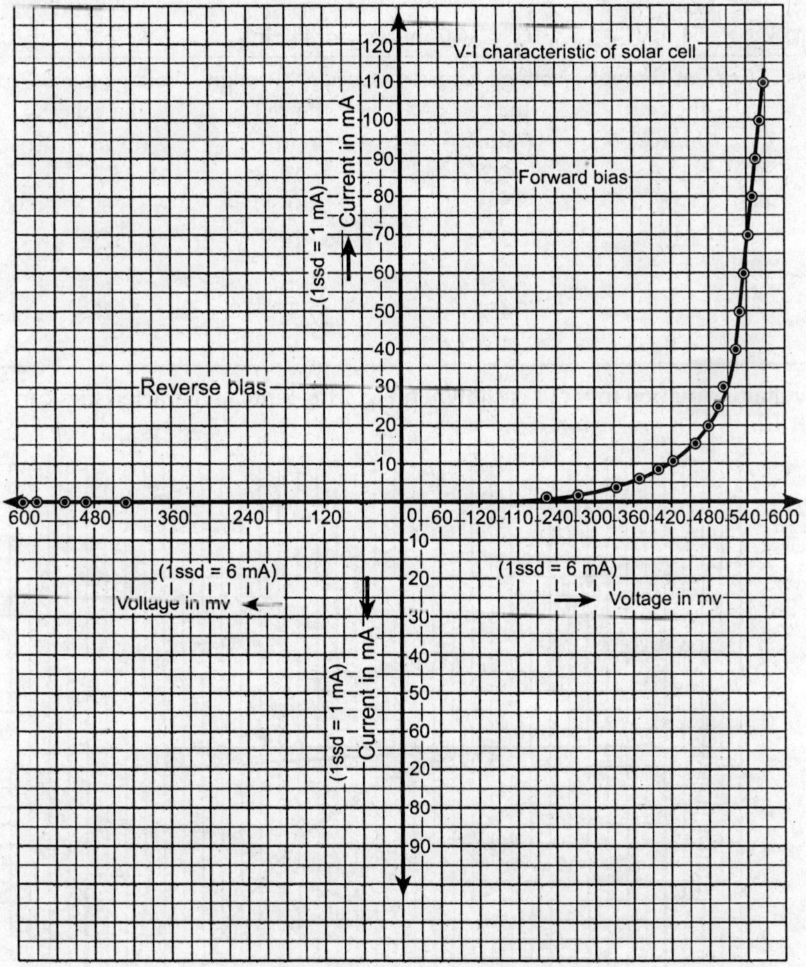

Fig. 21.4

The characteristics of a solar cell under different illumination conditions can be obtained from Fig. 21.5.

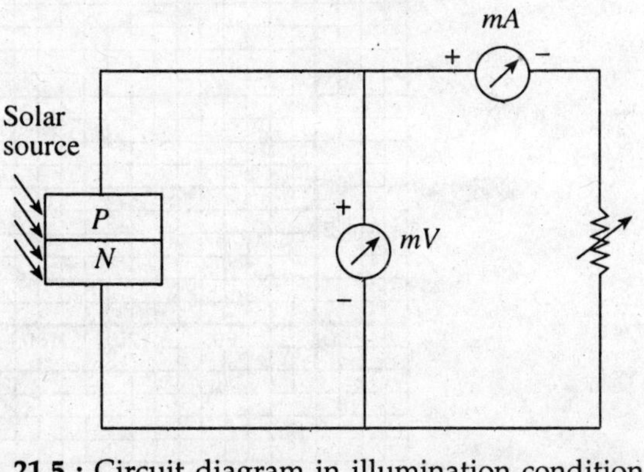

Fig. 21.5 : Circuit diagram in illumination condition

[C] Characteristics under different illumination condition & measurement of I_{sc} and V_{oc}

Table III

Set I		Set II		Set III	
Illumination voltage V_{illu} = 12 V		Illumination voltage V_{illu} = 12 V		Illumination voltage V_{illu} = 12 V	
I_{sc} = 14.5 mA; V_{oc} = 454 mV		I_{sc} = 16.0 mA; V_{oc} = 458 mV		Isc = 72.5 mA; Voc = 534 mV	
Voltage across the cell (in mV)	Current through the cell (in mA)	Voltage across the cell (in mV)	Current through the cell (in mA)	Voltage across the cell (in mV)	Current through the cell (in mA)
16	14.5	18	16	81	72.5
186	14	235	15	259	70
256	13	275	14	372	64
⋮	⋮	⋮	⋮	⋮	⋮
448	1	452	1	527	2
454	0	458	0	534	0

The current-voltage variation for Set I is shown in Fig. 21.6. Similar characteristics can be obtained from Set II and Set III.

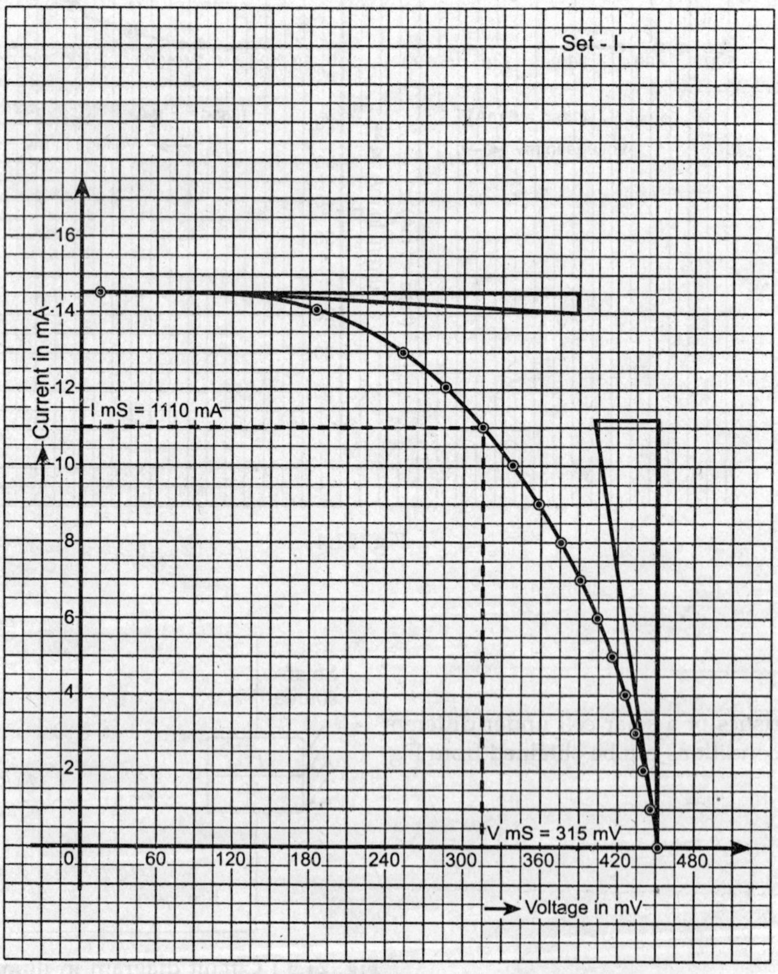

Fig. 21.6

[D] To determine the input power of the solar cell

A standard solar cell is illuminated by solar radiation at the time and short circuit current I_{sc} is noted for standard solar cell and test cell.

For standard cell, I_{sc} = 84 mA

For test cell, I_{sc} = 72 mA.

We know for standard solar cell 120 mA current (Isc) corresponds to 100 mW/cm² intensity.

Now area of our cell = 2 × 2 = 4 cm².

So input power of the test cell for a given I_{sc} is

$$P_{in} = \frac{100 \times 10^{-3}}{120 \times 10^{-3}} \times \frac{84}{72} \times I_{sc} \times 4 \text{ mW}$$

$$= 3.88 \times I_{sc} \text{ mW.}$$

[E] To determine the maximum power output of the solar cell for different sets and efficiency

We have $P_{max} = I_{max} \times V_{max}$

Also fill factor, $A = \dfrac{P_{max}}{I_{sc} \times V_{oc}}$

$\therefore$ Efficiency $\eta = \dfrac{P_{max}}{P_{in}} \times 100\%$

Now, from the graph we determine the rectangle of maximum area and we take the values of I_{max} and V_{max} for different sets.

Table IV

Set No.	I_{sc} (mA)	V_{oc} (mV)	I_{max} from graph (mA)	V_{max} from graph (V)	P_{max} (mW)	Fill factor $A = \frac{P_{max}}{I_{sc} \times V_{oc}}$	Pin $= 3.88 \times I_{sc}$ (mw)	Efficiency $\eta = \frac{P_{max}}{P_{in}} \times 100\%$
I	14.5	0.454	11.0	0.315	3.465	0.524	56.26	6.15
II	...	...	...	...	...	...	...	...
III	...	...	...	...	...	...	...	...

[F] Evalution of R_s : (from graph)

Table V

Set No.	Value of ΔV in mv	Value of ΔI in mA	$R_s = \frac{\Delta V}{\Delta I}$ in Ω
I	51.0	11.2	4.55
II	...	...	...
III	...	...	...

[G] Evalution of R_{sh} (from Graph) :

Table VI

Set No.	Value of ΔV in mv	Value of ΔI in mA	$R_{sh} = \frac{\Delta V}{\Delta I}$ in Ω
I	390	0.5	780
II	...	...	...
III	...	...	...

Discussion

(i) For the reverse bias characteristics under dark condition the variable resistance is connected across the batterry to get the maximum resolution in voltage difference. Actually here we get the current for small voltage.

(ii) Under illumination condition voltage will not be zero at I_{sc}. So we have extrapolated from I_{sc} to get the current at zero voltage.

(iii) The term 'fill factor' acually means the part of the area covered by the curve by the rectangle area $V_{oc} \times I_{sc}$.

SECTION C

ADVANCED QUESTIONS FOR VIVA VOCE

UNIT 1
DIODES AND TRANSISTORS : AMPLIFIERS

1. What are point contact diodes, PIN diode, varactors and hot carrier diode?

Ans. :

(a) *Point contact diodes* are intended for RF applications due to their extremely small internal capacitance, considerably less than that of a junction diode designed for same applications.

(b) *PIN diode* is made by diffusing heavily doped P^+ and N^+ regions into an intrinsically pure silicon. Characteristics of PIN diodes are obtained by the semiconductor nature of the chip especially that of *I*-layer (a lightly doped N-region). PIN diodes can be used as RF switches, attenuators and many types of phase shifting devices.

(c) *Varactors* can be used as voltage variable capacitors. They depend on change in capacitance that occurs across their depletion layers when the reverse bias voltage is altered. These are designed to provide various capacitance ranges from a few pF to 100 pF. In different varactor types, the minimum and maximum capacitances may change, the maximum to minimum capacitance ratio is typically 2.5 to 3 in commercial types.

(d) *Hot carrier diode* is a Metal-to-Semiconductor majority carrier conducting device with a single rectifying junction. The carriers are high mobility electrons in an *N*-type semiconductor material. These are useful in particular as mixers and detectors at VHF and higher frequency range. It is both mechanically and electrically superior when compared to a point contact diode.

2. 'Transistors are current operated devices while tubes are voltage operated devices'. Justify the statement.

Ans. : The amplification factor (μ) of a tube is the ability of the grid voltage and the plate voltage for producing an equal change in the output current. The voltage in the output circuit is directly related to the input voltage and for this reason tubes are considered as voltage-operated devices. The characteristics of a transistor are such that output current is a function of input current. That is why, they are called current operated devices.

3. What is a transistor? Explain the transistor action for the *N-P-N* and *P-N-P* types.

Ans. : A transistor is a semiconductor device that is formed when two *P-N* junction diodes are tied together. If a thin *P*-type material is sandwiched between two thick *N*-type materials we get an *N-P-N* transistor. When a thin *N*-type material is sandwiched between two thick *P*-type materials, we get a *P-N-P* transistor.

If the emitter base junction is forward-biased, normal diode action takes place. Electrons from *N*-type emitter and holes from *P*-type base cross the junction and tend to recombine. Base region being extremely thin, only a small number of holes are available for recombination. Electrons that do not recombine may be powerfully attracted towards *N*-type collector, if the second junction is reverse biased (with *N*-type collector positive). This constitutes a current.

The action in a *P-N-P* transistor is similar to that in the *N-P-N* transistor with the difference that it is the holes that act as charge carriers within the semiconductor material.n

4. Derive an expression for the power gain to be provided by transistor Q of Fig. 1

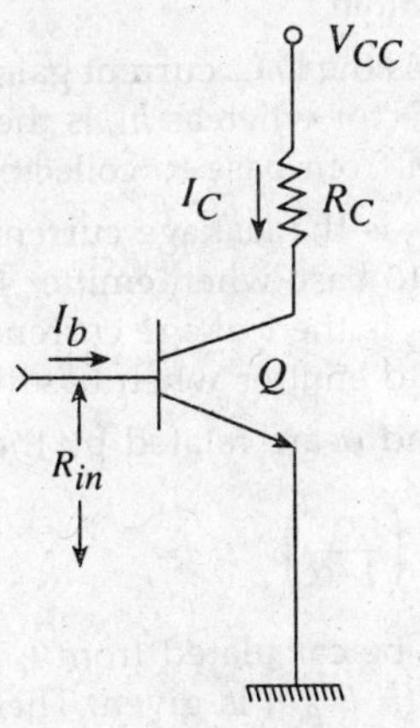

Fig. 1

Ans. : Input power = Input voltage × Input current.

Input voltage = $I_b R_{in}$

Input power = $I_b R_{in} \times I_b = I_b^2 R_{in}$

Again, output power = Output voltage × Output current

$= I_C R_C \times I_C$

$= \beta I_b R_C \times \beta I_b$

$= \beta^2 I_b^2 R_C$

So, Power gain = $\dfrac{\text{Output power}}{\text{Input power}}$

$$= \left[\frac{\beta^2 I_b^2 R_C}{I_b^2 R_{in}}\right] = \beta^2 \frac{R_C}{R_{in}}$$

5. How do you distinguish between—

(i) β and α ; (ii) h_{FE} and h_{fe}; (iii) I_{CBO} and I_{CEO}?

How do you calculate beta(β), if alpha (α) is given and I_{CEO} is supplied?

Ans. :

(i) Beta (β) is the current gain of the transistor from base to collector. It may be defined as the current gain of the transistors when used in common-emitter (*CE*) configuraton. Alpha (α) is a general term for current gain of the transistor when used in common-base (*CB*) configuration. It is the current gain from emitter to collector terminals as in a common-base configuration, input current flows in the emitter terminal and output current flows in the collector terminal.

(ii) h_{FE} is the D.C. current gain from base to collector whereas h_{fe} is the A.C. current gain from base to collector.

(iii) I_{CBO} is the leakage current from collector to base when emitter is open while I_{CEO} is the leakage current from collector to emitter when base is open.

β and α are related by the expression,

$$\beta = \left(\frac{\alpha}{1-\alpha}\right)$$

I_{CEO} can be calculated from I_{CBO} if h_{FE} at a base current (= I_{CBO}) is given. Then

$$I_{CEO} = (h_{FE} \times I_{CBO}).$$

6. What is meant by f_T of a transistor? What is its significance? Why is it called unity gain crossover frequency?

Ans. : The gain of transistor falls at higher frequencies. The gain falls at a rate of 6 dB per octave. The frequency at which gain falls to unity (or 0 dB) is called unity gain crossover frequency and is denoted by f_T (Fig. 2). The gain bandwidth product at this frequency is f_T. This gain bandwidth product remains same throughout the 6 dB/octave slope as decrease in gain is accompanied by an increase in bandwidth. So f_T is the gain bandwidth product and is indicative of high frequency response of the transistor.

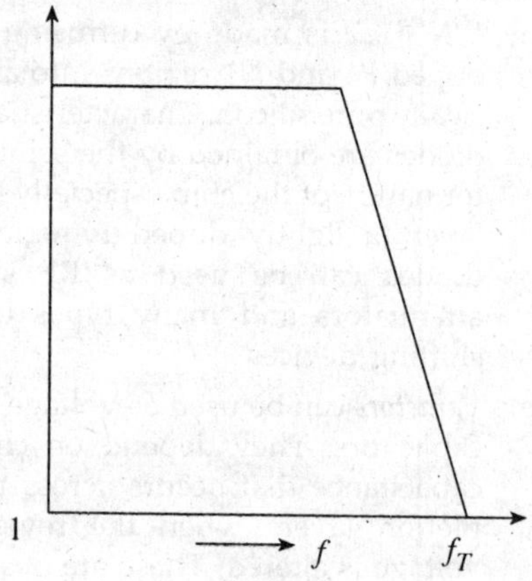

Fig. 2

7. The input impedance effects the power gain provided by the transistor. How is it then that input impedance parameter, does not figure in the transistor manuals?

Ans. : The input impedance parameter does figure in the usual list of parameters. However, it is not given as such as it comprises of a number of components. It is specified as a maximum base to emitter voltage (V_{be}) under specified input current conditions.

8. What is the significance of thermal resistance of a power transistor? How can you define (θ_{F-C}) and (θ_{C-A})?

Ans. : The thermal resistance indicates the power handling capability of a transistor. Quantitatively, it is given by the ratio of rise in transistor junction temperature with respect to some reference to the power dissipated. It is measured in °C/watt.

There is another parameter known as maximum junction temperature. It is a transistor rating to tell at what temperature of the transistor junction, the device will retain its properties. An estimate of transistor junction temperature is made when the thermal resistance parameter is also known in addition to the quantum of power dissipated. One of the thermal resistances encountered with power transistors is (θ_{J-C}) called thermal resistance from junction to case. If θ_J and θ_C are the junction and case termperature in °C respectively for a power dissipated equal to P watts, then

$$\theta_{J-C} = \left(\frac{\theta_J - \theta_C}{P}\right)$$

This expression is used to get the value of (θ_J) from known values of (θ_{J-C}), (θ_C) and (P).

Another thermal resistance is (θ_{C-A}) which is case-to-free air thermal resistance. Mathematically,

$$\theta_{C-A} = \left(\frac{\theta_C - \theta_A}{P}\right)$$; where θ_A = ambient temperature.

So, if θ_{C-A} and θ_{J-C} are known, one can find the junction temperature from the temperature and power dissipated.

9. Interpret the following transistor ratings and specifications : V_{CEO}, V_{CBO}, V_{EB}, $I_{C(max)}$, $P_{D(max)}$, t_{ON}, t_{OFF}, C_{ob}.

Ans. : V_{CEO} is the maximum collector to emitter voltage with base open. It is the collector-emitter breakdown voltage and is an important specification particularly in switching transistors such as those used for designing of switching mode power supplies. This rating can be specified in different ways. V_{CER} (collector-emitter breakdown voltage with a specified resistance usually 100 to 200 ohms connected between base and emitter) and V_{CES} (collector-emitter breakdown voltage with base shorted to emitter) are given in addition to V_{CEO} for switching transistors particularly designed for use in switching supplies. V_{CEO} is the lowest of all and if design is such that V_{CEO} rating is not exceeded, the possibility of collector-emitter junction breaking down gets eliminated.

V_{CBO} is the collector to base breakdown voltage with emiitter open. It determines the worst case output breakdown voltage when the transistor is connected in common-base configuration. V_{EB} is the maximum base-emitter voltage of the transistor.

$I_{C(max)}$ is the peak collector current rating of the transistor, either continuous or pulsed.

$P_{D(max)}$ is the maximum power dissipation capability of the transistor. It is specified at a given ambient temperature. It should be derated at higher temperatures as per the derating curve supplied by the manufacturer. Also, in the absence of the heat-sink, the dissipation capability will be much less. The actual dissipation capability can be determined from the known values of maximum operating junction temperature, ambient temperature and the thermal resistances involved.

The turn-on time (t_{ON}) is the sum of delay time (t_d) and rise time (t_r). Delay time is the time taken by the collector current to change from its OFF-state value to 10 per cent of its final ON-state value after the turn-on pulse is applied at the base. The rise time (t_r) is the time taken for the collecttor current to rise from 10 per cent to 90 percent of its final ON-state value.

The turn-off time (t_{OFF}) is the sum of storage time (t_s) which is the time for the collector current to change from its final ON-state value to 90 per cent of its final ON-state value and the fall time (t_f) which is the time for the collector current to change from 90 per cent of its final ON-state value to 10 per cent of its final ON-state value.

C_{ob} is the output capacitance of the transistor when connected in common-base configuration. It is particularly useful in switching and high frequency applications.

10. Is it theoretically possible to realise the maximum dissipation capability of a transistor? If so, under what conditions can we do so? What are the practical limitations?

Ans. : It can be possible theoretically only if the ambient temperature is the one at which the maximum dissipation capability has been specified.

Secondly, there should be no temperature differential between the transistor case and the ambient, i.e., the thermal resistance between transistor case and free air is zero which in turn necessitates that the transistor is mounted on an infinite heat-sink. This is practically impossible.

We consider the case of 2N3055 with junction temperature (max) rating of 200 °C and maximum dissipation capability of 115 watts (at 25 °C ambient) and a thermal resistance from junction to case of 1.5 °C/watt. Power of 115 watts can be realised only if case temperature is at ambient temperature as—

$$1.5 = \left(\frac{\theta_J - \theta_A}{115}\right) \text{ gives } \theta_C = 25\ °C$$

θ_J = junction temperature

θ_C = case temperature

Again, θ_{C-A} = thermal resistance from case to air

$$= \left(\frac{\theta_C - \theta_A}{115}\right)$$

[θ_A = Ambient temperature]

Putting the value of θ_C and θ_A in the above expression, $\theta_{C-A} = 0$.

11. What are the ill effects of finite values of input and output resistances on the voltage gain of an amplifier?

Ans. : The input resistance of an ideal voltage amplifier should be infinite. If it is not so, the amplifier input loads on to the source of input signal. If the source has a finite internal resistance, the signal that actually reaches the amplifier input is less than what is available at the source. As a result the voltage gain reduces. However, if the source is an ideal one and has a zero internal resistance, then the amplifier's input resistance makes no difference to the overall voltage gain.

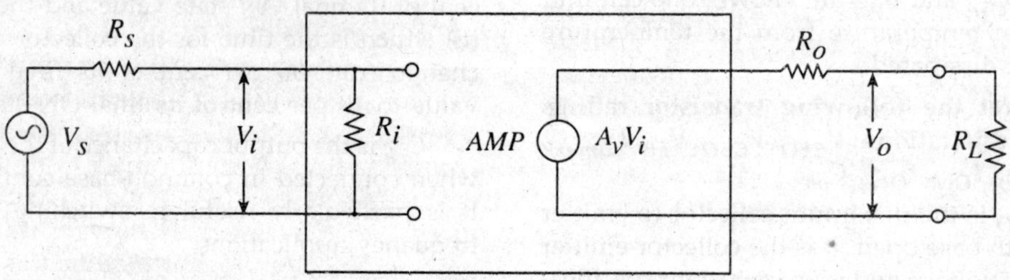

Fig. 3

The output resistance of an ideal voltage amplifier should be zero. If it is not so, the load resistance would load on to the output, i.e., the voltage appearing across the load would be less than the open circuited output voltage of the amplifier. The figure depicts the effects of input and output resistances on the voltage gain of the amplifier.

12. Point out the Dynamic Range and Noise Figure specifications of amplifiers.

Ans. : The dynamic range of an amplifier is defined as the ratio of its maximum used output signal without exceeding the maximum acceptable specified distortion limit (usually 10%) to the minimum usuable output signal having a certain specified minimum signal to noise ratio (usually 20 dB). It is expressed in decibels. Dynamic range of 50 dB is considered reasonably good, while an amplifier with more than 70 dB dynamic range is considered to be exceptionally good.

The noise figure of an amplifier is defined as the ratio of signal to noise ratio at the input to the signal to noise ratio at the output. That is,

$$N_F = \frac{(S_i/N_i)}{(S_o/N_o)}$$

It is expressed in decibels and is calculated from

$$N_F = 10 \log \frac{(S_i/N_i)}{(S_o/N_o)}$$

An ideal amplifier (a noiseless amplifier) has a noise figure of zero decibel.

13. What do you mean by thermal stability of the operating point? What are its consequences? Define the overall stability factor for an amplifier.

Ans. : When a particularly chosen operating point in an amplifier circuit shifts on its own due to change in temperature, the phenomenon is called *thermal stability.* Shifting of operating point leads to distortions in the output waveform.

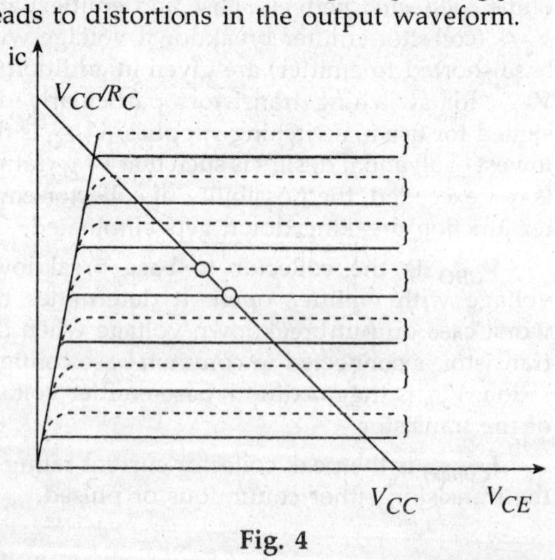

Fig. 4

We consider the case of common-emitter amplifier to illustrate the phenomenon of shifting of operating point. Refer to Fig. 4 for the output characteristics, the D.C. load line and the location of operating point.

The collector current (I_C) for the CE-amplifier configuration is given by

$I_C = \beta I_B + I_{CEO}$

where β = D.C. current gain

I_B = base current

I_{CEO} = reverse collector-emitter current with base open

Again, $I_{CEO} = (\beta + 1)I_{CBO}$,

where I_{CBO} = reserve collector base current with emitter open.

Hence, we get $I_C = \beta I_B + (\beta + 1)\ I_{CBO}$,

where I_C is dependent on β, I_B and I_{CBO}. All the three parameters are observed to be strongly temperature dependent, particularly the I_{CBO}. In fact, I_{CBO} doubles for every 10 °C rise in temperature for a silicon transistor. The effect becomes more severe as increase in I_{CBO} is getting multiplied by a large magnitude in $(\beta + 1)$ to affect the collector current. Since (β) also increases with temperature, the overall effect sometimes becomes cumulative and leads to thermal runaway if amplifier has not been designed to counter this thermal instability. The result is that the output characteristics shift upwards (as shown by dotted lines) when the temperature increases and the operating point which was chosen to be initially in the centre may find itself somewhere near the saturation region of the characteristics.

The stability factor explains change in collector current due to changes in β, I_B (or V_{BE}) and I_{CBO} with temperature or even due to change of transistor. The overall stability factor is the combined effect of the stability factors resulting from all the three causes, It may be expressed as

$$\Delta I_C = S\ \Delta I_{CBO} + S'\Delta\ V_{BE} + S''\ \Delta\beta,$$

where ΔI_C = total change in collector current

$$S = \frac{\Delta I_C}{\Delta I_{CBO}},\ \beta \text{ and } V_{BE} \text{ held constant}$$

$$S' = \frac{\Delta I_C}{\Delta V_{BE}},\ I_{CBO} \text{ and } \beta \text{ held constant}$$

$$S'' = \frac{\Delta I_C}{\Delta \beta},\ I_{CBO} \text{ and } V_{BE} \text{ held constant.}$$

Smaller values for S, S′ and S″ imply a smaller change in collector current due to variation in respective parameters. Hence, smaller the magnitude of stability factor, better the stability.

14. What should be the design approach so that the amplifier does not have the thermal instability or the consequent thermal runaway problem?

Ans. : If the amplifier circuit is so modified that the increase in $(\beta + 1)\ I_{CBO}$ term is compensated by a decrease in $(\beta\ I_B)$ term, the collector can be maintained at a constant value irrespective of the changes of β, I_B and I_{CBO}. A reduction in the value of $(\beta\ I_B)$ product can be brought out only by reducing I_B. The circuit is modified in a manner that any tendency for the collector current to increase reduces I_B which in turn does not allow the collector current to increase. There are different type of bias stability schemes employed with common-emitter amplifier configuration.

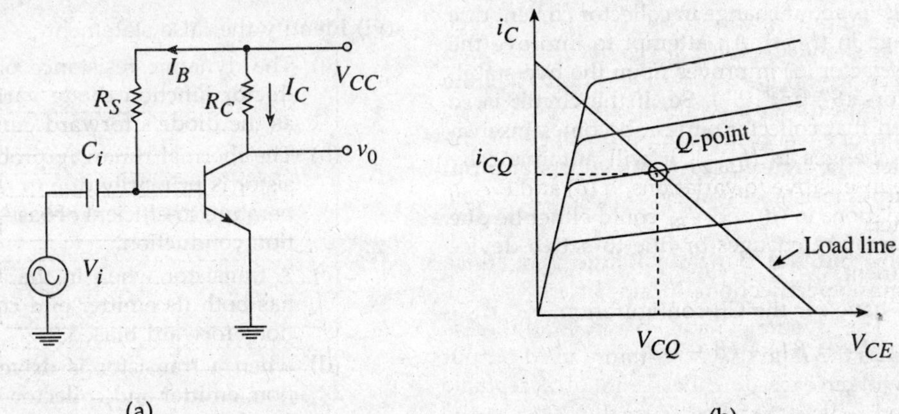

Fig. 5

15. What is a load line with reference to amplifier? Draw its load line. What is Q-point?

Ans. : The load line with reference to an amplifier is the straight line drawn and superimposed on its output characteristics that enables to choose an appropriate D.C. operating point, also called the *Q*-point. It helps in calculating output currents and output voltages for given input signal swing.

From the circuit of Fig. 5(a).

$$V_{CC} = I_C R_C + V_{CE}$$

$$\text{or,} \quad I_C = \left(-\frac{V_{CE}}{R_C} + \frac{V_{CC}}{R_C}\right).$$

This is the equation of a straight line of the type $y = mx + b$ with y, x, m and b representing I_C, V_{CE}, $(-1/R_C)$, and V_{CC}/R_C. This straight line is the load line having slope of $(-1/R_C)$ and a y-axis intercept of (V_{CC}/R_C).

The point of intersection of load line and the quiscent value of base current is called the *Q*-point [Fig. 5(b)]. The choice of *Q*-point depends on the characteristics desired. For example, to operate the amplifier in class-A, the *Q*-point is chosen somewhere around middle of output characteristics. Having chosen a proper operating point, the value of base resistance R_B to get the desired base current can be calculated from

$$R_B = \left(\frac{V_{CC} - V_{BE}}{I_B}\right)$$

16. Clarify the operating point (*Q*-point) stabilization schemes commonly employed with CE-amplifier circuits.

Ans. : The factor that leads to operating point instability is considered as the stability factor (*S*) that tells us about change in collector current due to change in (I_{CBO}). An attempt to improve the stability factor (*S*) improves upon the two stability factors (*S*′) and (*S*″). So, if the circuit is so modified that collector current becomes insensitive to changes in (I_{CBO}), it will automatically become insensitive to variations in (β) and (V_{BE}). The variations in (β) and V_{BE} could either be due to temperature changes or due to active device replacement.

We have for the CE-configuration,

$$I_C = \beta I_B + (\beta + 1)\, I_{CBO}$$

$$\text{or,} \quad \frac{dI_C}{dI_{CBO}} = S = (\beta + 1).$$

Thus stability factor becomes very high for the basic configuration. The collector current change is $(\beta + 1)$ times the change in I_{CBO}.

17. Multiple Choice Questions

(i) The semiconductor rectifier specification that indicates subcycle current rating of the rectifier is

(a) Surge Current (I_{FSM}).

(b) Peak Inverse Voltage (PIV).

(c) $I^2\,t$ (Ampere squared seconds).

(d) Transition capacitance.

(ii) The junction diode's capacitance that is made use of in varactors is known as

(a) Transition capacitance.

(b) Diffusion capacitance.

(c) Space charge capacitance.

(d) Inter-electrode capacitance.

(iii) The static resistance of a junction diode

(a) is its opposition to the flow of D.C. currents.

(b) is its opposition to the flow of A.C. currents.

(c) is much smaller than its dynamic resistance.

(d) determines the diode's resistance when it is used as a rectifier.

(iv) Cut-in voltage $(V\gamma)$ of a silicon diode is

(a) 0.1 V (b) 0.2 V (c) 0.3 V (d) 0.6 V

(v) In a semiconductor diode, P-side is grounded and *N*-side is applied a potential of –5 V through a resistance of 100 ohms. The diode shall

(a) conduct fully. (b) not conduct.

(c) conduct partially. (d) None of these.

(vi) Identify the false statement :

(a) The dynamic resistance of a semiconductor junction diode varies inversely as the diode's forward current.

(b) The thermal runaway problem in transistor is primarily due to negative temperature co-efficient of base-emitter junction conduction.

(c) A transistor when in the active region has both its emitter and collector junctions forward biased

(d) when a transistor is driven to saturation, emitter and collector junctions are forward biased.

(vii) A power transistor has the following specifications : P_D = 100 watts, maximum junction temperature = 200 °C, case to ambient thermal resistance of 10 °C/W and junction to case thermal resistance of 2 °C/W. This transistor can dissipate a maximum power of ...... watts in the absence of any heat-sink at an ambient temperature of 50 °C.

(a) 100 (b) 50 (c) 12.5 (d) 22.5.

(viii) Transistor's α and β are related by

(a) $\beta = \left(\frac{\alpha}{1-\alpha}\right)$ (b) $\alpha = \left(\frac{\beta}{1+\beta}\right)$

(c) $\beta = \left(\frac{\alpha}{1+\alpha}\right)$ (d) $\alpha = \left(\frac{1+\beta}{\beta}\right)$

(ix) When current through a zener diode increases by a factor of 2, voltage across its terminals

(a) is halved.

(b) is doubled too.

(c) is practically unchanged.

(d) None of these.

(x) The dynamic resistance of a zener diode

(a) increases with increase in current throught it.

(b) decreases with increase in current through it.

(c) is independent of current passing through it.

(d) None of these.

(xi) Typical value of impurity concentration in a tunnel diode is

(a) 1 part in 10^8 parts. (b) 1 part in 10^6 parts.

(c) 1 part in 10^3 parts. (d) None of these.

(xii) In Zener and Avlanche breakdown diodes, current flow is due to flow of

(a) minority carriers. (b) majority carriers.

(c) minority as well as majority carriers.

(d) minority carriers (in Zener diodes) and majority carriers (in Avlanche diodes).

(xiii) Zener diodes have

(a) a positive temperature co-efficient for their breakdown voltages.

(b) a negative temperature co-efficient for their breakdown voltages.

(c) a breakdown voltage that is independent of temperature.

(d) None of these.

(xiv) The Backward diode

(a) is a current controllable device.

(b) is a tunnel diode operated in negative resistance region of its *V-I* characteristics.

(c) is another name for a tunnel rectifier.

(d) None of these.

(xv) A tunnel diode

(a) is a current controllable device.

(b) exhibits characteristics that are represented by a single valued function of current.

(c) is a voltage controllable device.

(d) has a cut-in voltage of 0.3 V.

ANSWERS

(i) (c), (ii) (a) and (c), (iii) (a), (iv) (d), (v) (a), (vi) (c), (vii) (c), (viii) (a), (ix) (c), (x) (b), (xi) (c), (xii) (a), (xiii) (a), (xiv) (c), (xv) (c).

UNIT 2
FIELD EFFECT TRANSISTORS

1. Draw the circuit symbols for an *N*-channel FET and a *P*-channel FET. Also draw their characteristics.

Ans. : The circuit symbols for *N*-channel and *P*-channel FETs are shown in Fig. 1. When there is

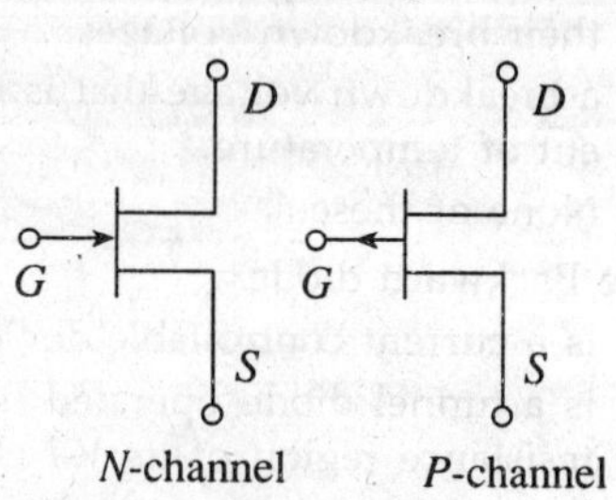

Fig. 1

no voltage at the gate terminal (i.e., for $V_{GS} = 0$), then the current increases as the drain to source potential is increased. The current saturates after some time for a particular value of V_{ds}. This increase in current with increase in V_{ds} is expected as a greater V_{ds} which means a stronger force for the current carriers emitted by the source.

When a certain gate potential is present so as to reverse bias the gate source junction, it restricts the flow of carriers by the electric field

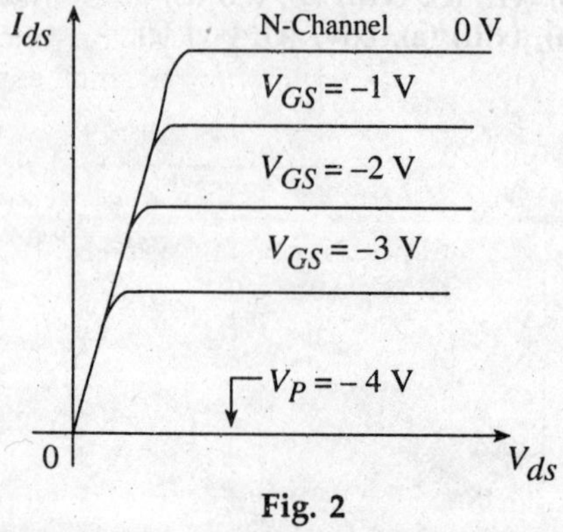

Fig. 2

produced due to *P-N* junctions being reverse biased. So the drain current saturates at a lower value. More the reverse bias, lesser is the saturation value for drain current. The behaviour is indicated in the characteristic curves shown in Fig. 2.

2. What is Pinch-off voltage?

Ans. : It is that value of gate to source potential (V_{GS}) for which the drain current is zero. Then the electric field becomes strong enough to block the flow of electrons completely or pinch off the channel. The pinch-off voltage V_P is indicated in Fig. 2 to be – 4 V.

3. Draw the drain current (I_{ds}) versus drain voltage (V_{ds}) characteristics of a *P*-channel junction FET and show the pinck-off phenomenon.

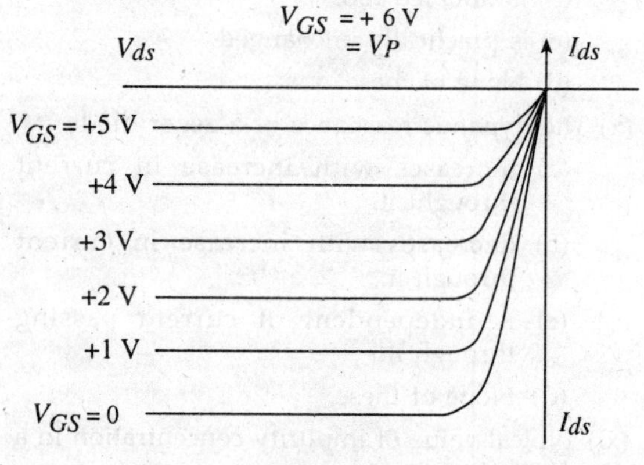

Fig. 3

Ans. : The (I_{ds}) versus (V_{ds}) characteristics are shown in Fig. 3. (I_{ds}) has been considered negative as the current in this case is constituted by holes and not electrons as is the case in an *N*-channel FET.

4. Give one important reason for which *N*-channel FETs are preferred over *P*-channel FETs.

Ans. : The mobility of electrons which constitute the current in *N*-channel FET is more than that of holes (which constitute the current in *P*-channel FETs). That is why *N*-channel FETs are preferred over P-channel FETs.

Mobility is the case with which a charge carrier can be made to move. If electrons have more mobility, this signifies that it is easier to make them mobile as compared to holes.

For the same applied drain to source voltage, number of electrons reaching the drain in *N*-channel FET will be more than the number of holes in *P*-channel FET reaching the drain.

5. **We can see from the drain characteristics of the FET that the drain current (I_{ds}) tends to level off or saturate beyond a certain value of drain to source voltage (V_{ds}) even if the gate to source voltage is kept at a fixed value including V_{GS} = 0. Expain this phenomenon.**

Ans. : The FET channel is considered to be composed of a resistance between gate and source terminals and another resistance between gate and drain terminals. This is assumed for the sake of explanation. The applied drain voltage (V_{ds}) presents a positive voltage on the *N*-side of the gate source junction in an *N*-channel FET and a negative voltage on the *P*-side of the gate source junction in a *P*-channel FET. In both cases, it tends to reverse bias the junction. The amount of this reverse bias increases with increase in (V_{ds}) even for a fixed (V_{GS}). A stage appears where an increase in (I_{ds}) resulting from an increase in (V_{ds}) is nullified by a greater channel constriction due to increase in reverse bias. This is where the saturation sets in. The channel constriction becomes greater as we move towards the drain as the amount of reverse bias which is the product of drain current and the channel resistances at that point is also increasing.

6. **How do the junction FETs have a very high input impedance?**

Ans. : Junction FETs operate with a reverse biased gate source junction. So, the input impedance is the impedance offered by a reverse biased *PN* junction. For a sufficient reverse bias, the current drawn by the gate (equal to reverse biased leakage current) may be a few nanoamperes, which yields an input impedance in the region of 100 M.

7. **Indicate the channel ohmic region and the channel pinch-off region on the drain characteristics of a junction FET. List the typical applications when you need to operate your FET in the channel ohmic region and channel pinch-off-region.**

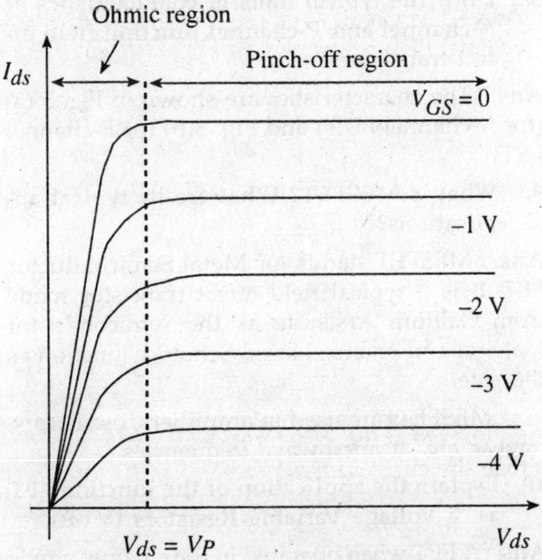

Fig. 4

Ans. : The channel ohmic region is the region where the drain current varies linearly with the drain voltage as shown in Fig. 4. The slope of this linear region depends on the gate to source voltage with the slope for V_{GS} = 0 being widely different from what is at V_{GS} near the pinch-off. The FET ON-resistance may be as small as a few

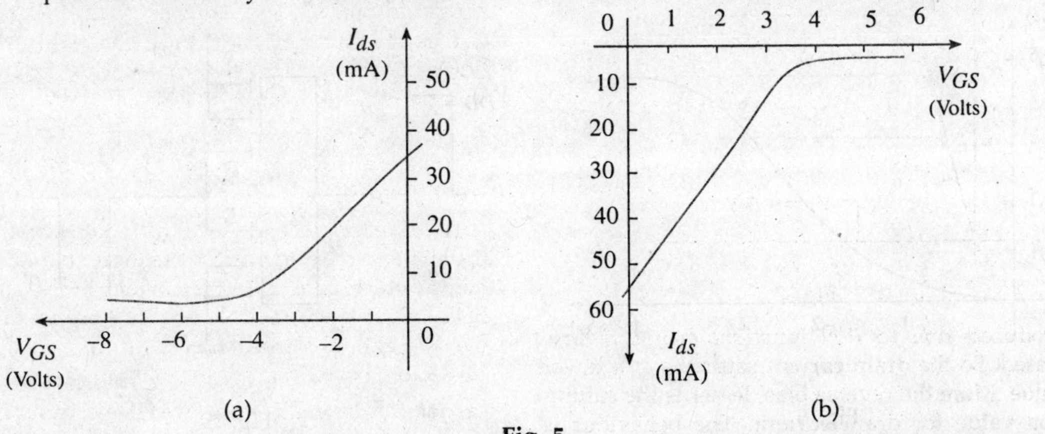

Fig. 5

hundred ohms for $V_{GS} = 0$ and the FET OFF-resistance may be in hundreds of meg-ohms.

So for small values of (V_{ds}) FET can be used as a voltage variable resistor where the FET ON-resistance can be made to vary with a varying gate to source voltage. The applications based on the use of FET in the channel ohmic region are the voltage controlled attenuator, automatic gain control circuit, FET as a switch in chopper circuits, FET voltmeters, etc.

8. Draw the typical transfer characteristics of *N*-channel and *P*-channel junction field effect transistors.

Ans. : The characteristics are shown in Fig. 5 (a) (for *N*-channel FETs) and Fig. 5(b) for P-channel FETs).

9. What is MESFET? What are its typical applications?

Ans. : MESFET stands for Metal Semiconductor FET. It is a typical field effect transistor made from Gallium Arsenide as the semiconductor material and reverse biased Schottky junction as the gate.

MESFETs are used in amplifiers, oscillators, mixers, etc. at microwave frequencies.

10. Explain the application of the junction FET as a Voltage Variable Resistors (VVR).

Ans. : A FET when operated in the channel ohmic region for smaller values of (V_{ds}) it can be used as a variable resistor with the drain source resistance (r_d) varying almost linearly with gate source reverse bias voltage [Fig. 6(a)]. The circuit diagram is given in [Fig. 6(b)]. The output voltage (V_O) can be varied from near zero to the maximum value as the gate source reverse bias is varied from zero to near pinch-off.

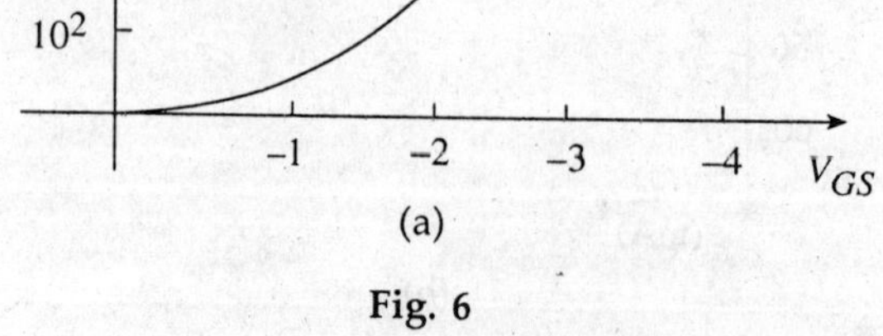

(a)

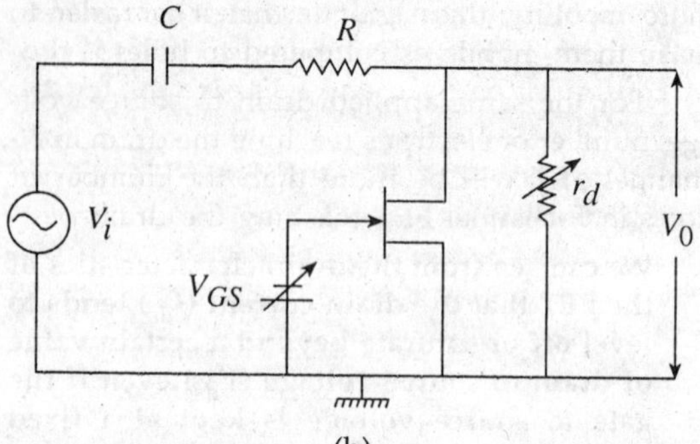

(b)

Fig. 6

11. The internal circuitries of CMOS digital integrated circuits make abundant use of *N*-channel and *P*-channel enhancement MOSFETs. The CMOS inverter is one of the most important building blocks of all these ICs. Draw the basic configuration of a CMOS inverter and briefly explain its functional principle.

Ans. : Fig. 7 shows the basic configuration. When the input to the circuit is a logic LOW, a GND at the gate of *N*-channel MOSFET forces it to the non-conducting state while the *P*-channel MOSFET goes into conduction. So, (V_{DD}) appears at the output through the conducting *P*-channel device. The output is logic HIGH for a logic LOW input.

When the input is a logic HIGH, a (V_{DD}) at the gate of *N*-channel MOSFET turns it ON while *P*-channel device goes to non-conducting state. The output goes to GND through the conducting *N*-channel MOSFET. The output is thus logic LOW for a logic HIGH input.

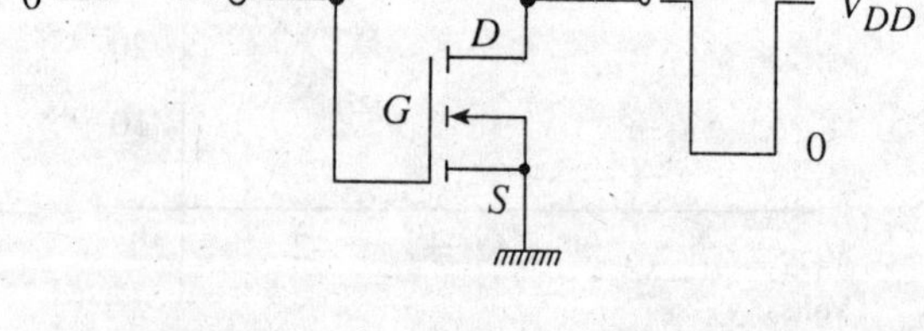

Fig. 7

12. Current regulating diodes often comprise a junction FET inside the device. What is the configuration used and what are the terminals brought out?

Ans. : A junction field effect transistor (JFET) with gate shorted to the source constitutes a current regulating diode where the drain current varies by an extremely small amount for large variation in the drain voltage [Fig. 8]. Junction FET is preferred over a MOSFET as the former has a much flatter characteristics to provide better current regulation.

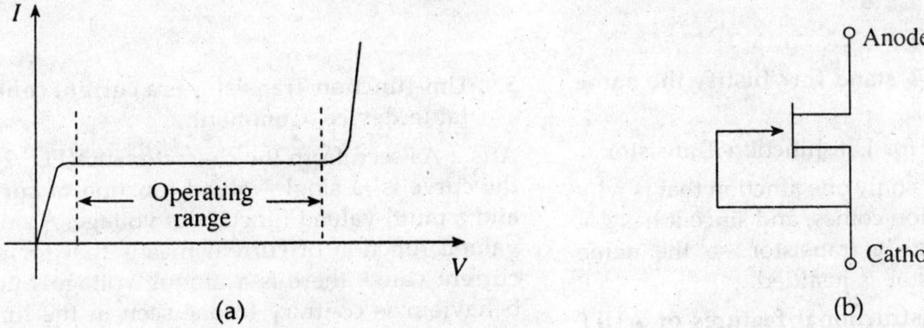

Fig. 8

13. What is VMOS?

Ans. : VMOS stands for vertical MOS. It is a type of MOSFET structure capable of higher current ratings than those possible with the conventional MOSFETs. These are basically power field effect transistors. Typical VMOS devices have a maximum drain current rating of a few tens of amperes and a drain to source voltage rating in the range of 200 to 300 volts.

UNIT 3
UJT AND THYRISTORS

1. What does UJT stand for? Justify the name UJT.

Ans. : UJT stands for Uni-Junction Transistor.

This device has only one junction that is why the name unijunction comes, and since it has got three terminals like a transistor, so the name unijunction transistor is justified.

2. Show the constructional features of a UJT and draw its circuit symbol.

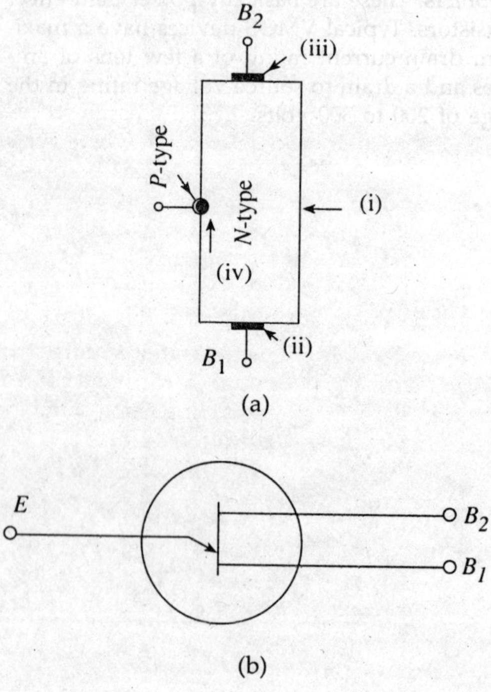

Fig. 1

Ans. : The constructional features of a UJT are shown in Fig. 1(a). Indices are :

(i) *N*-type silicon bar (ii) Ohmic contact, B_1 (iii) Ohmic contact, B_2 (iv) P-type material.

Fig. 1(b) shows the circuit symbol.

3. Uni-Junction Transistor is a current controllable device. Comment.

Ans. : As seen from the characteristics[Fig. 2(b)], the curve is a single-valued function of current and a multi-valued function of voltage. A single-valued function of current means that for each current value, there is a unique voltage. Such a behaviour is contrary to one seen in the tunnel diode characteristics wich could be expressed by a single valued function of voltage. This makes tunnel diode a voltage controllable device. Due to single valued function of current nature of its *V-I* characteristics, a UJT is called a current controllable device.

4. Draw the input diode characteristics of UJT. Explain the occurrence of negative resistance

Ans. : Fig. 2(a) shows the circuit for determining the input characteristics. The characteristics are shown in Fig. 2(b). It is required to forward bias a *P-N* junction diode by a voltage $V\gamma$ (= 0.5 V), if it is to exhibit a very low resistance path. Thus the voltage across the *P-N* junction diode has to be 0.5 V with *P*-side positive to make the conduction possible.

The voltage appearing on *N*-side of the diode is equal to ηV_{BB}, where $\eta = \frac{R_2}{R_1 + R_2}$, R_1 being the resistance of the semiconductor bar between *E* and B_1, and R_2 the resistance between *E* and B_2. This input diode shall conduct only when V_E exceeds 5.5 V ($V_E = \eta V_{BB} + V\gamma$). The emitter voltage V_E at which diode starts conducting is termed as V_P.

From the characteristics, as the voltage is increased, the current remains very small in magnitude till a voltage V_P is attained at which diode is forward biased. After that current rapidly increases. As the current increases, the voltage decreases giving a negative resistance region. Such a behaviour can be explained considering the fact that due to increase in forward current, the resistance of E_{B1} region falls rapidly resulting in the reduction in the voltage. Finally, when I_E becomes too large, it may be considered to be much greater than I_{B2}. Under these conditions the curve approaches asymptotically the curve for $I_{B2} = 0$ giving rise to a valley point. The curve for $I_{B2} = 0$ is the same as that for ordinary *P-N* junction diode.

The region before peak point (V_P, I_P) is called cut-off region as the input diode remains reverse biased in this region. The region after the valley point (V_V, I_V) is known as saturation region as there is not much increase in the emitter voltage for large change in emitter current.

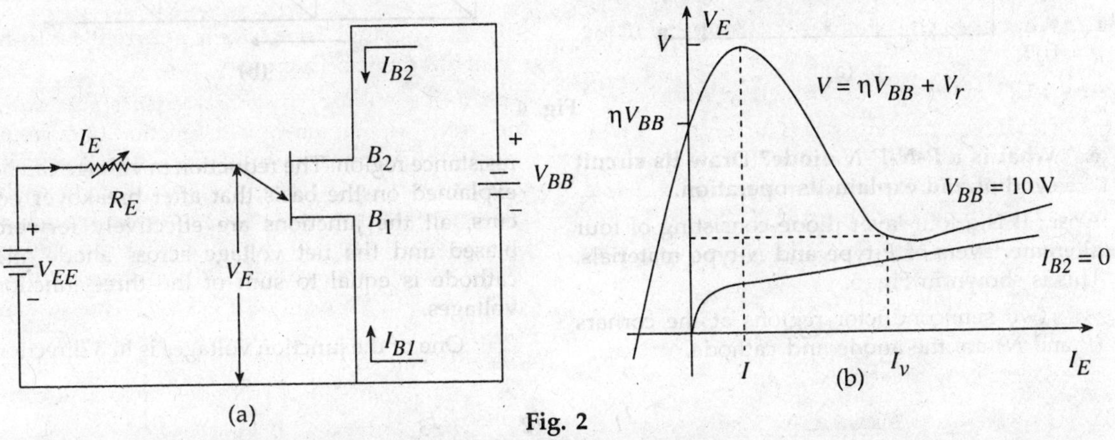

Fig. 2

5. Draw UJT relaxation oscillator circuit and explain its operation.

Ans. : Fig. 3(a) shows the basic UJT relaxation oscillator circuit. The sawtooth waveform that appears across the capacitor is shown in Fig. 3(b). Initially, UJT is in the cut-off region, i.e., the input diode is reverse biased. The capacitor starts charging from V_{CC} through *R*. When the voltage V_C across the capacitor becomes sufficiently high to forward bias the input diode, the capacitor starts discharging through the low resistance between the emitter-base B_1 region. This discharge process continues until it reaches a point where input diode is again reverse biased. At this point, the capacitor again starts charging. The process of charging through *R*, comparatively a higher resistance, and discharging through low forward resistance of input diode continues and gives rise to a sawtooth signal as shown in Fig. 3(b).

If the capacitor in the relaxation oscillator of Fig. 3(a) is charged through a constant current source as shown in Fig. 4(a), the result will be a perfect sawtooth waveform [Fig. 4(b)].

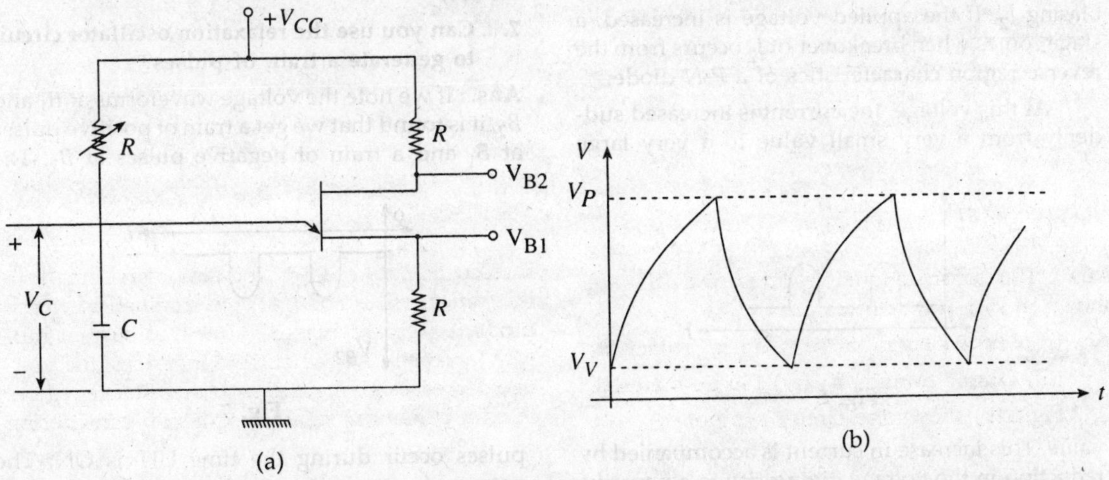

Fig. 3

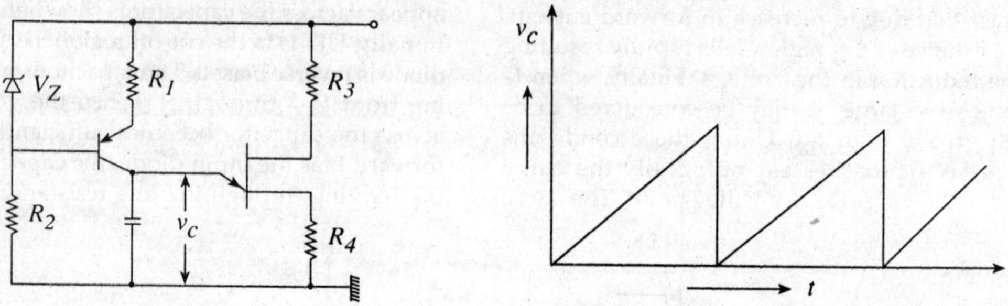

(a) (b)

Fig. 4

6. What is a *P-N-P-N* diode? Draw its circuit symbol and explain its operation.

Ans. : It is a four-layer diode consisting of four alternate layers of *P*-type and *N*-type materials. This is shown in Fig. 5.

Two semiconductor regions at the corners (*P* and *N*) are the anode and cathode.

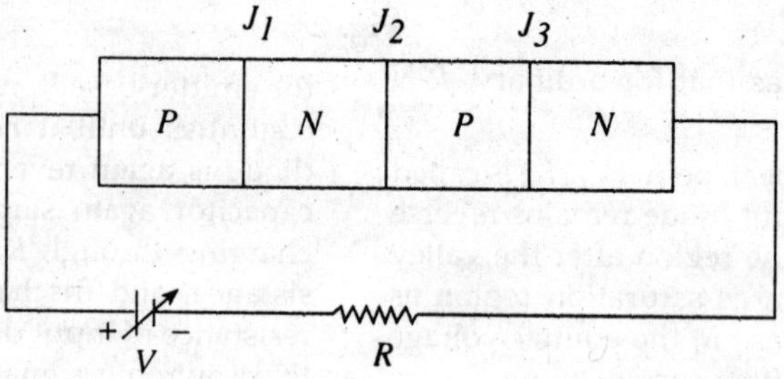

Fig. 5

If the anode (*P*-type) is positive with respect to the cathode (*N*-type), then junction J_1 and J_3 are forward biased. This applied voltage then effectively appears across junction J_2 thus reverse biasing J_2. If the applied voltage is increased, a stage comes when breakover of J_2 occurs from the reverse region characteristics of a *P-N* diode.

At this voltage, the current is increased suddenly from a very small value to a very large value. This increase in current is accompanied by reduction in the voltage giving rise to a negative resistance region. The reduction in voltage can be explained on the basis that after breakover occurs, all the junctions are effectively forward biased and the net voltage across anode and cathode is equal to sum of the three junction voltages.

One of the junction voltages is in a direction opposite to that of other two junction voltages. As a result the net voltage is equal to that across one of the junction diodes (= 0.6 V for silicon diodes).

7. Can you use the relaxation oscillator circuit to generate a train of pulses?

Ans. : If we note the voltage waveforms at B_1 and B_2, it is found that we get a train of positive pulses at B_1 and a train of negative pulses at B_2. The pulses occur during the time UJT is ON. The nature of waveforms are shown in Figs. 6 & 7.

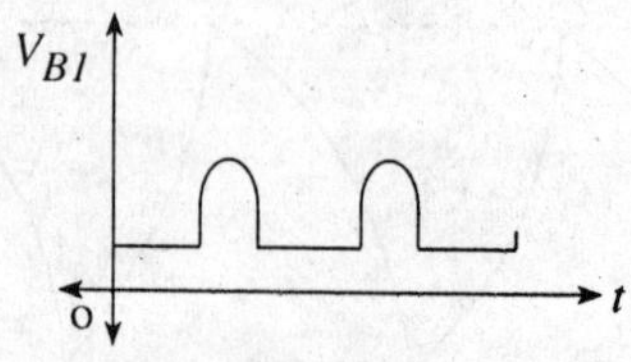

Fig. 6

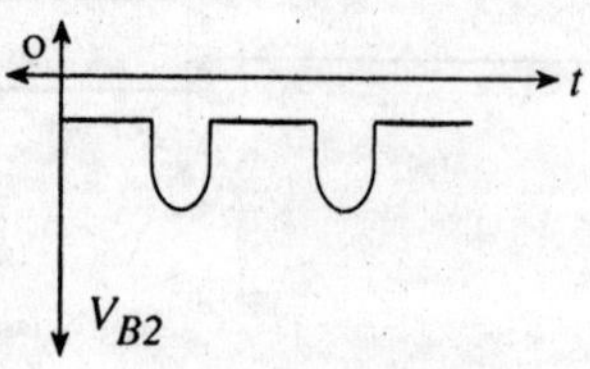

Fig. 7

8. **Which semiconductor material is used for constructing *P-N-P-N* diodes?**

Ans. : Silicon

9. **From the expression I = $\dfrac{I_{CO}}{1-(\alpha_1+\alpha_2)}$ it appears that when ($\alpha_1 + \alpha_2$) exceeds unity, current I should reverse, but that does not happen actually. Why so?**

Ans. : The moment ($\alpha_1 + \alpha_2$) becomes unity or approaches unity, current becomes too large. This brings all of three junction diodes into saturation region, i.e., Q_1 and Q_2 go to saturation due to their collector junctions now forward biased. This saturation then diminishes the values of α_1 and α_2 or limits their magnitudes so that ($\alpha_1 + \alpha_2$) never exceeds unity. The transistors Q_1 and Q_2 enter saturation only upto the extent that ($\alpha_1 + \alpha_2$) remains unity.

10. **Is it impossible to make four-layer diodes with Germanium as the semiconductor material?**

Ans. : For making *P-N-P-N* diodes, we always use silicon. Germanium is not used as it does not have a stable Off-state. With Germanium as the semiconductor material, the magnitudes of α may become large enough to give ($\alpha_1 + \alpha_2$) equal to unity for a very small applied voltage. The device has thus a tendency to switch to its 'ON' state from 'OFF' state very soon, giving an unstable 'OFF' state.

11. **Draw the characteristics of a *P-N-P-N* diode and mark the various regions. Give the typical values of resistances offered by the device when operating in various regions.**

Ans. : The *V-I* characteristics of a *P-N-P-N* diode are shown in Fig. 8. Resistance when operating in cut-off region, becomes hundreds of meg- ohms.

Resistance in 'ON' state is a few ohms.

12. **Justify the name Silicon Controlled Switch.**

Ans. : The *P-N-P-N* device conducts heavily when the breakover voltage is attained. The breakover voltage further depends on ($\alpha_1+\alpha_2$) becoming unity. By inducing current at these gate terminals, ($\alpha_1+\alpha_2$) may be made to approach unity at a reduced applied voltage. The breakover voltage may be in fact reduced to an extent depending upon strengths of currents induced. Thus, we have a sort of control over breakover voltage. As the *P-N-P-N* diodes are always made from silicon and these devices perform the switching operation, hence the name Silicon Controlled Switch comes.

13. **What is SCS? How does it resemble a four-layer diode?**

Ans. : SCS stands for silicon controlled switch. The construction of a silicon-controlled switch is

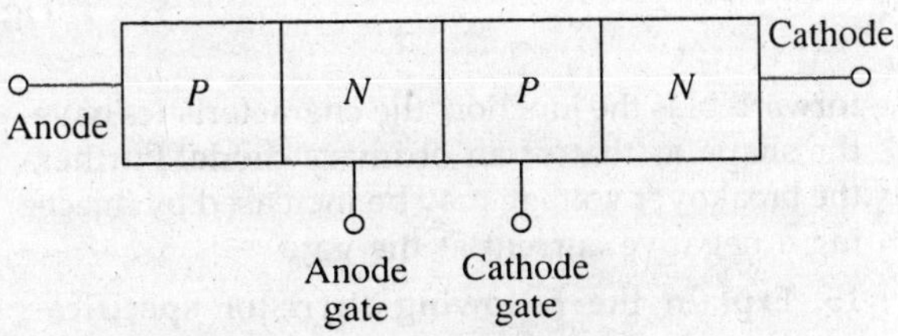

Fig. 9

the same as that of a four-layer diode with the only difference that the connections to the inner layers that are not accessible in a 4-layer diode are

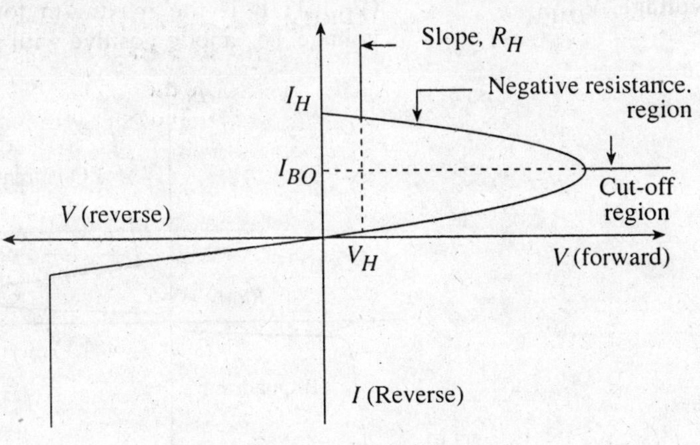

Fig. 8

available in the case of an SCS. Fig. 9 shows the junction diagram of an SCS.

N-type base near the anode is the anode gate and the *P*-type base near the cathode is the cathode gate. In some devices, both the connections are available but in most cases, only one connection is available, i.e., the cathode gate.

14. Draw the circuit symbol and characteristics of a silicon controlled switch.

Ans. : The characteristics show that when gate current becomes very large and in a direction to forward bias the junction, the characteristics have the shape as that of an ordinary diode. Further, the breakover voltage may be increased by injecting a negative current at the gate.

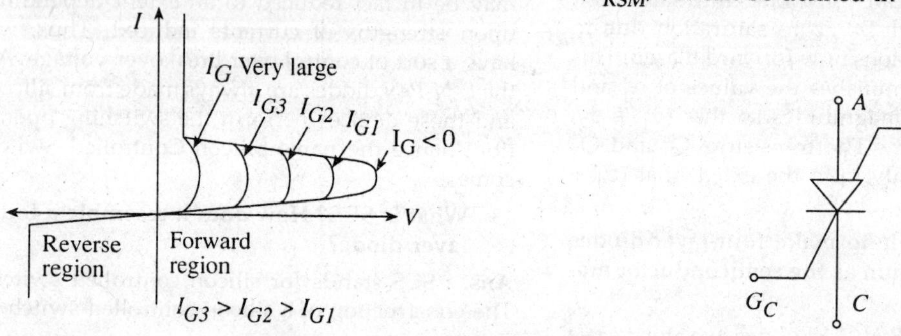

Fig. 10

15. Explain the following thyristor specifications :

(i) Repetitive peak reverse voltage, V_{RRM}

(ii) Non-repetitive peak reverse voltage, V_{RSM}

(iii) Repetitive peak off state voltage, V_{DRM}

(iv) Peak off-state voltage, V_{DM}.

Ans. : (i) **Repetitive Peak Reverse Voltage (V_{RRM}) :** It is defined for reverse blocking mode type of thyristors (e.g. SCRs) for operation in the third quadrant (Fig. 11). It is the maximum repetitive reverse voltage (i.e. anode negative with respect to the cathode) that can be applied to the device safely and still keeping it in the blocked state. The current that flows through the device in this state is called the Repetitive Peak Reverse Current (I_{RRM}).

(ii) **Non-Repetitive Peak Reverse Voltage (V_{RSM}) :** It can be defined in a similar way. This rating is little higher than V_{RRM}. If the application is such that the reverse voltage does not appear across the device repetitively, which may happen because the device is being operated in the forward off-state and if it is to be protected against any occasional reverse transients, then (V_{RSM}) and not the (V_{RRM}) is the rating we should look for. For practical purposes, if we have not allowed any reverse voltage to exceed (V_{RRM}), we are very safe.

(iii) **Repetitive Peak Off-state Voltage (V_{DRM}) :** It is the maximum forward off-state voltage, i.e., anode positive with respect to cath-

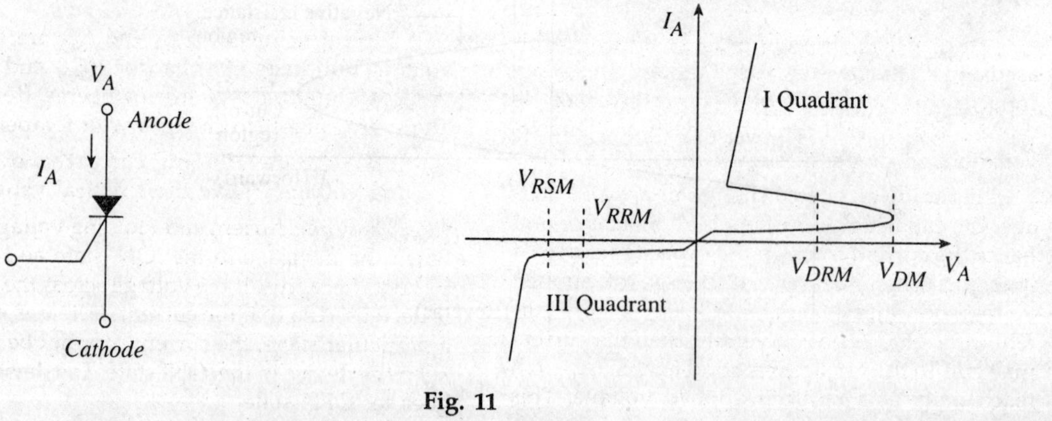

Fig. 11

ode) that can be allowed to appear across the device when it is being operated in the first quadrant, i.e., in the forward blocking mode. This voltage is usually much less than the actual breakover voltage of the device which is basically a characteristic of the construction of the thyristor. The maximum value of current that can flow in the forward blocking state is represented as (I_{DRM}).

(iv) **Peak Off-state Voltage (V_{DM})** : It is the maximum off-state voltage. A forward voltage greater than this breaks the junction and switches the device to the ON-state even in the absence of any gate signal. It does not depend on the voltage grades of different thyristors in the same family. For instance, SCRs TY 505 F, TY 1005 F, TY 2005 F, TY 3005 F, TY 4005 and TY 5005 F respectively have (V_{DRM}) rating of 50, 100, 200, 300, 400 and 500 volts whereas breakover voltage (V_{BO}) for all these device belonging to the same class may be larger than the highest available (V_{DRM}) rating in this class.

16. What is rate-effect? How does it affect the *P-N-P-N* diode operation?

Ans. : The effect of rate of change of applied votage on the breakover voltage is known as the Rate-Effect.

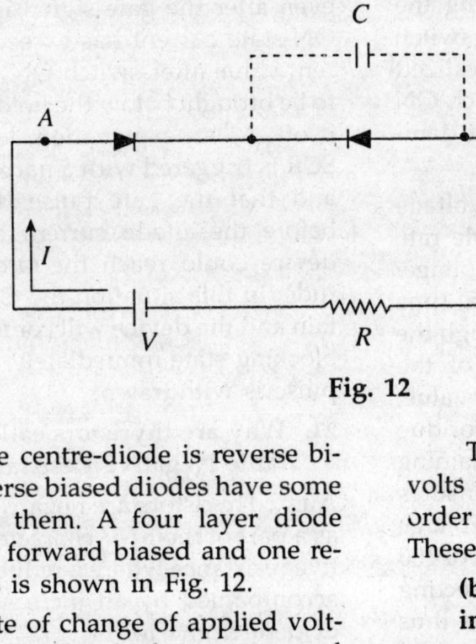

Fig. 12

In off-State, the centre-diode is reverse biased and all the reverse biased diodes have some capacitances across them. A four layer diode represented by two forward biased and one reverse biased diodes is shown in Fig. 12.

For a lower rate of change of applied voltage, the capacitance C offers high reactance and hence the current through the capacitance can be ignored. When the rate of change of applied voltage becomes exceedingly large so that current through C is not negligible, then the current may add to the total current and enable ($\alpha_1 + \alpha_2$) reach unity at a reduced applied voltage. This higher rate of change of applied voltage reduces the breakover voltage. This phenomenon known as the rate-effect. Thus rate effect may couse premature firing of a four-layer device.

17. (a) Define breakover voltage (V_{BO}) and breakover current (I_{BO}). Give their typical values.

Ans. : The forward applied voltage at which the centre junction breaks over to provide a sudden

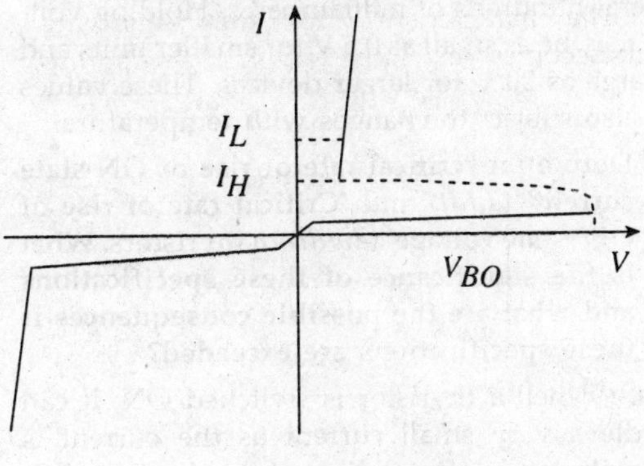

Fig. 13

rise of current is known as the *Breakover Voltage*. The current corresponding to breakovr voltage (V_{BO}) is the *Breakover Current.*

Typical values of V_{BO} and I_{BO} are tens of volts to hundreds of volts (for V_{BO}) and of the order of hundreds of microamperes (for I_{BO}). These are subject to changes with temperature.

(b) What are 'Holding current' and 'Holding voltage'? Give their typical values.

Ans. : Holding current and Holding voltage arise when the device is in the 'ON' state and has to be kept there only. If the voltage across the device is decreased so that the current decreases, then at a particular stage, the current may not be able to keep the device in the 'ON' state. The device then goes to 'OFF' state.

Holding current is the minimum current necessary to keep the device in the 'ON' state once triggered. It is indicated in the characteristics shown in Fig. 13.

Voltage corresponding to holding current is called the *'Holding voltage'*.

The values of these parameters vary from one type to another. Typical value of holding current may lie between several milliamperes to several hundreds of milliamperes. Holding voltage may be as small as 0.5 V for smaller units and as large as 20 V for larger devices. These values are also subject to changes with temperature.

18. Define the 'critical rate of rise of ON-state current' (*di/dt*) and 'Critical rate of rise of OFF-state voltage' (*dv/dt*) in thyristors. What is the significance of these specifications and what are the possible consequences if these specifications are exceeded?

Ans. : When a thyristor is switched ON, it can handle a very small current as the current is concentrated in a small area of the device pellet. Gradually, the load current carrying capability increases to its specified rating. Critical rate of rise of ON-state current (*di/dt*) is the maximum rate of change of ON-state current, the device can handle safely. If this rating exceeds during the device switch ON, may be due to a faster switch ON, there could be a development of localized hot spots in the pellet at the time of switch ON and this could result in the device getting damaged. Typical (*di/dt*) rating is 100 A/μs.

The critical rate of rise of OFF-state voltage (*dv/dt*) determines the maximum allowable rate of change of applied forward OFF-state voltage. When the applied forward voltage is a time varying one, a current (*C dv/dt*) flows through the reverse biased junction capacitance (C) of the thyristor. This current if exceeds a certain value, it can cause premature firing of the thyristor due to decrease in its breakover voltage. Maintaining (*dv/dt*) within (*dv/dt*) rating of the device ensures that there is no change in the breakover voltage (V_{BO}) of the device. Typical (*dv/dt*) rating is of 100 V/μs. However, it can be enhanced by connecting a low resistance betwen gate and cathode thus providing a low resistance shunt path for the charging current. Majority of high (*dv/dt*) rating thyristors use a 'Shorted Emitter Construction' which is nothing but an internally created low value resistance across gate cathode junction.

19. What does the Amperes Squared Seconds (I^2t) rating of a thyristor signify?

Ans. : The Amperes Squared Seconds (I^2t) rating of a thyristor speaks about the surge current handling capability of the device for sub-cycle time periods when the device is being used as a rectifier. In general, (I^2t) rating is specified or it can be calculated from the maximum peak surge current that the device can sustain for a given time period which is one full cycle of a applied-50 Hz/60 Hz waveform. (I^2t) rating of the thyristors is used to calculate the surge current capability when the thyristor conducts for only a part of the full-cycle time period.

20. Distinguish between Holding Current (I_H) and Latching Current (I_L) in thyristors.

Ans. : Holding Current (I_H) is the minimum ON-state current necessary to kept the thyristor in the low impedance state once it has been triggered to the ON-state (Fig. 13). To switch off the thyristor the anode current must be brought below the holding current value.

Latching current (I_L) is slightly larger than the holding current. ON-state current equal to or more than the latching current ensures that a switched ON device remains in the ON-state even after the gate signal is removed. Once the ON-state current has exceeded the latching current value after switch ON, the current then has to be brought below the holding current to switch it off. As an explanation, let us assume that an SCR is triggered with a narrow gate trigger pulse and that the gate pulse has been withdrawn before the anode current of the switched ON device could reach the latching current amplitude. In this situation the ON-state will not sustain and the device will come back to the forward blocking state immediately after the gate trigger pulse is withdrawn.

21. Why are thyristors called Current Controllable Negative Resistance Devices?

Ans. : Thyristors are negative resistances devices as a part of their *I-V* characteristics exhibits negative resistance where a reduction in the voltage is accompanied by an increase in current. This is evident from Fig. 14.

Thyristors are current controlled devices because their *I-V* characteristics can be expressed by a Single Valued Function of Current, i.e., for every current values, there is one and only one corresponding voltage value (Fig. 14). As an example for currents equal to 30 mA and 40 mA, the corresponding voltages are 0.5 V and 0.7 V respectively. On the contrary, these characteristics are representative of a Multivalued Function of

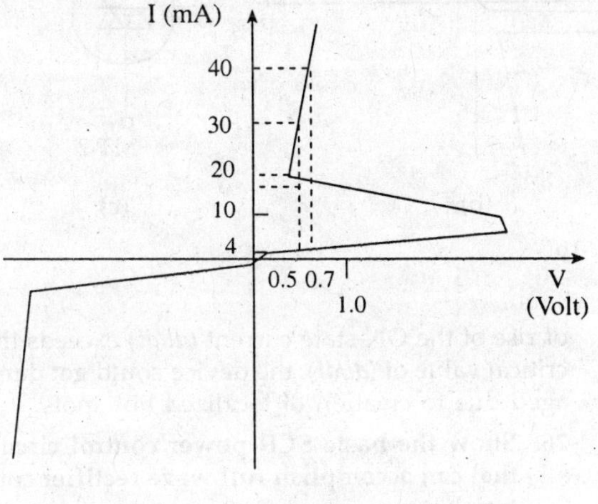

Fig. 14

Voltage, i.e., for every voltage value, there could be one or more than one corresponding current values. For instance, for a voltage of 0.7 volt, the current could be 4 mA or 20 mA or 40 mA. However, 20 mA point is in the unstable region and cannot sustain. It is because of 'single valued function of current' and 'multivalued function of voltage' nature of the *I-V* characteristics that thyristors are current controlled devices.

22. When do we need to connect thyristors in series or in parallel? What are the basic considerations to be looked after while doing so?

Ans. : Thyristors are connected in series to enhance the voltage ratings of the individual devices. For example, two SCRs each having a peak repetitive OFF-state voltage rating of 1000 volts could be connected in series to get an overall (V_{DRM}) of 2000 volts provided that steps are taken to ensure that the applied voltage is divided equally between the individual devices. The devices to be connected in series should preferably have closely matched OFF-state characteristics like forward breakover voltage, OFF-state resistance and so on.

Thyristors are connected in parallel to enhance the current capability of the individual devices. Care should be taken to ensure that the total On-state current is divided equally amongst individual devices connected in parallel. To achieve this, the devices should have closely matched forward conduction characteristics like the ON-state dynamic impedance, ON-state voltage and so on. Figs. 15(a) and (b) respectively reveal series and parallel connection.

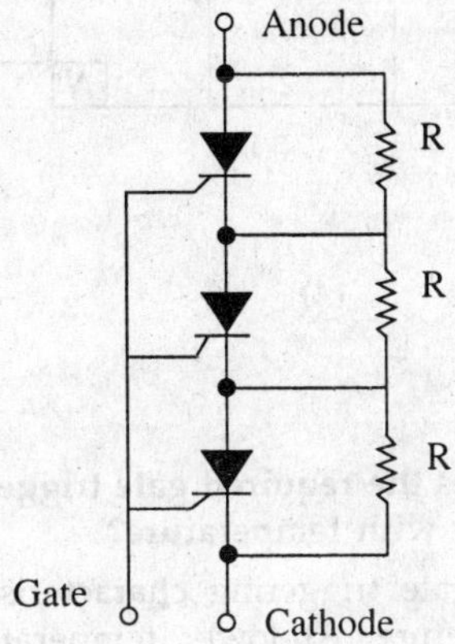

(a)

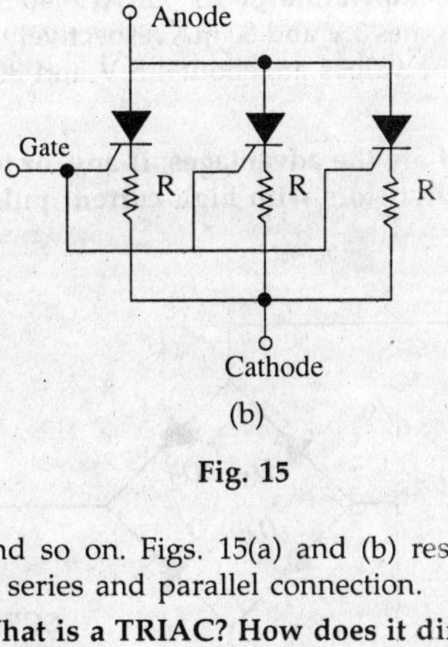

(b)

Fig. 15

23. What is a TRIAC? How does it differ from an SCR?

Ans. : A TRIAC is a bi-directional thyristor which can be considered to be equivalent to two P-N-P-N devices connected back to back [Fig. 16 a]. The voltage-current characteristics of a triac are similar to those of an SCR with the difference that a triac exhibits identical characteristics both in the forward (first quadrant) and reverse (third quadrant) directions [Fig. 16 (b)]. Like an SCR, a triac too is a three-terminal device and the terminals are designated as Main Terminal-1 (MT-1), Main Terminal-2 (MT-2) and Gate [Fig. 16(c)].

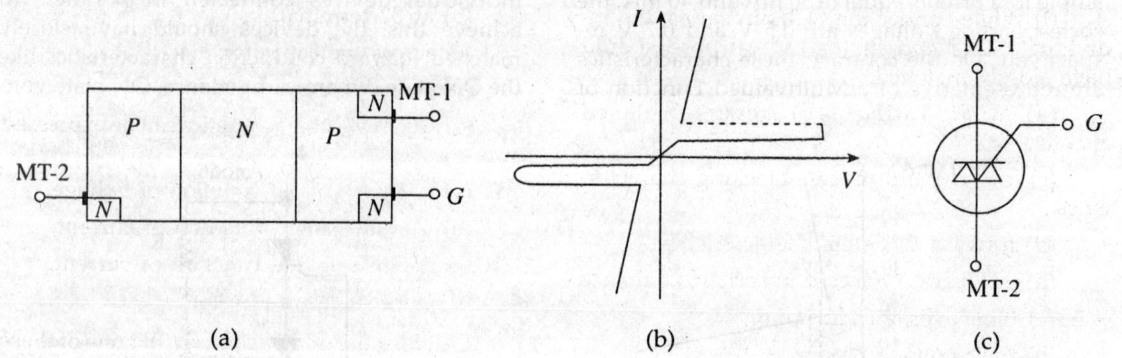

Fig. 16

24. How does the required gate triggering current vary with temperature?

Ans. : The gate triggering characteristics vary with temperature. At lower temperatures, we require higher gate trigger voltage and current. For SCR 2N3668, the maximum values of V_{GT} and I_{GT} are 2 V and 40 mA respectively at a junction temperature of 25 °C. At – 40 °C, the same becomes 3 V and 80 mA respectively and at 100 °C, the values are about 1.5 V and 30 mA.

25. What are the advantages, if any, of triggering thyristors with high current pulses?

Ans. : Higher gate trigger currents yield faster turn on. But it has also come problems. If the rate of rise of the ON-state current (*di/dt*) exceeds the critical value of (*di/dt*), the device could get damaged due to creation of localized hot spots.

26. Show the basic SCR power control circuit that can accomplish full wave rectifier control.

Ans. : The basic circuit is shown in Fig. 17(a). Typical input and output waveforms are shown in Fig. 17(b). Conduction angle is 90°.

Diodes D_2 and D_4 conduct during positive half-cycles of input whereas diodes D_1 and D_3 conduct during negative half cycles of input. The power control is provided by SCR.

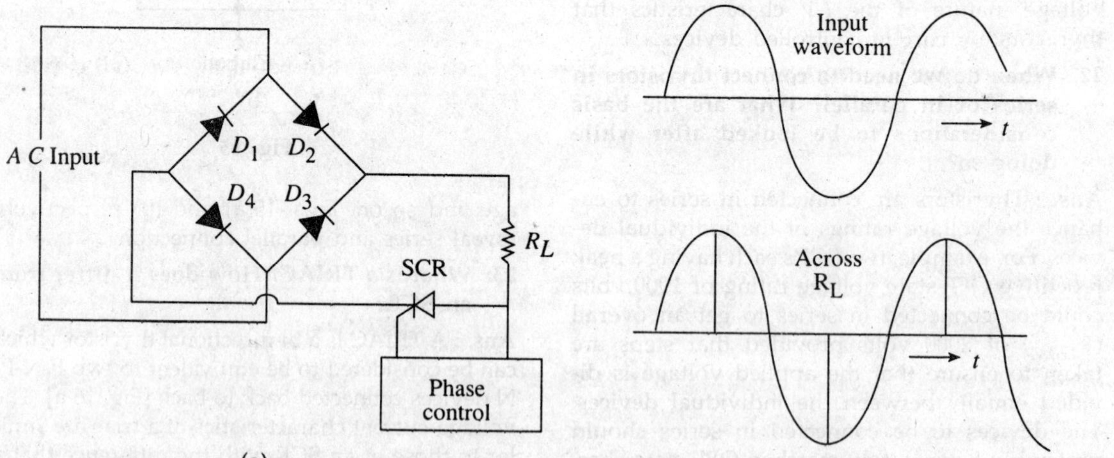

Fig. 17

27. Multiple Choice Questions

(i) While connecting SCRs in series, the individual SCRs chosen for the purpose should have closely matched.

(a) *di/dt* rating. (b) reverse biased characteriscs.

(c) forward and reverse blocking characteristics.

(d) forward ON-state characteristics.

(ii) SCRs are connected in series to enhance

(a) their overall *dv/dt* rating.

(b) their voltage ratings.

(c) their current handling capability.

(d) None of these.

(iii) SCRs are connected in parallel to enhance

(a) votage handling capability.

(b) current handling capability

(c) *di/dt* rating.

(d) None of these.

(iv) While connecting SCRs in parallel, the SCRs to be connected in parallel should have closely matched

(a) forward ON-state characteristics.

(b) forward and reverse OFF -state characteristics.

(c) forward OFF-state and ON-state characteristics.

(d) SCRs cannot be connected in parallel.

(v) A UJT relaxation oscillator circuit produces a sawtooth like waveform

(a) at B1 terminal.

(b) at B2 terminal.

(c) across the capacitor.

(d) None of these.

(vi) UJT's *V-I* characteristics can be expressed by

(a) a single-valued function of voltage.

(b) a multi-valued function of current.

(c) a single-valued function of current.

(d) none of these.

(vii) Pick the odd one out (**Hint** : one of these is not a thyristor)

(a) 2 N2646; (b) OE 104; (c) TY 1005 F; (d) OE 106.

(viii) Pick the odd one out (**Hint** : Think of their *V-I* characteristics)

(a) Tunnel diode; (b) UJT; (c) SCR; (d) Traic.

(ix) In half-wave SCR power control circuit, if the firing angle is 30°, then for one complete cycle of operation, the load gets power for

(a) 60°; (b) 150°; (c) 330°; (d) 30°.

(x) When we want to know the sub-cycle surge current capability of a thyristor, we should look for

(a) *di/dt* rating. (b) surge current rating, I_{FSM}. (c) I^2t rating. (d) None of these.

ANSWERS

(i) c; (ii) b; (iii) b; (iv) a; (v) c; (vi) c; (vii) a **(viii) a; (ix) b; (x) c.**

UNIT 4
OPTOELECTRONIC DEVICES

1. What is an LED? Explain its operational principle. On what factors does the wavelength of the emitted photons depend and what are the practical materials used for fabricating commercially available LEDs?

Fig. 1

Ans. : A semiconductor *P-N* junction diode designed to emit light when forward biased is known as a light emitting diode (LED). Fig. 1 shows its symbol. When a *P-N* junction is forward biased, the electrons in the *N*-type material and the holes in the *P*-type material travel towards the junction. Some of these electrons and holes would tend to recombine with each other and in the process radiate energy in the form of light. The radiant energy can also be in the form of heat and the process is named as phonon emission. The wavelength of the emitted photons is a function of the band-gap energy in the semiconductor material. It is given by the empirical formula λ (nm) = $1240/\Delta E$ where ΔE is the energy gap in eV. The expression thus relates the expected wavelength of photon emission to energy differential experienced by an electron undergoing recombination. The maximum possible energy of the emitted photons is obtained by the band-gap energy of the solid in which the *P-N* junction is formed. Table I shows a few of the available materials, the associated band-gap energy and the emission wavelength to be useful for practical applications. Some have too low a conversion efficiency to be useful. Presently, the only commercially available LED devices are manufactured using Gallium Arsenide (GaAs), Gallium Phosphide (GaP) and the ternary compound Gallium Arsenide Phosphide (Ga As P). These LEDs are available in numeral and alphabetic characters also measuring in height from 1/8 inch to more than an inch or so.

Table 1

Material	*Band-gap Energy* eV	*Emission* λ(nm)
Ge	0.66	1880
Si	1.09	1140
GaAs	1.43	910
GaP	2.24	560
$GaAs_{60}P_{40}$	1.91	650
AlSb	1.60	775
InSb	0.18	6900
SiC	2.2–3.0	563–413

2. What are the characteristic curves of interest in a light emitting diode? Sketch these curves for a typical LED.

Ans. : The characteristics in an LED are the

(a) Volt-Ampere characteristics

(b) Spectral distribution

(c) Light output versus diode forward current

(d) Directional characteristics

For a typical LED these are shown in Fig. 2.

In operation, LED is forward biased like a junction diode. Fig. 2(a) shows an example of Volt-Ampere characteristics of an LED. The threshold voltage is 1.5 V. Another curve of interest is the spectral distribution of light emanating from the LED. Such a curve is shown in Fig. 2(b). The curve reveals that the maximum intensity is obtained at 700 nm in case of the LED for which this curve has been drawn. The dependence of the brightness of the emitted light on the forward current is depicted in Fig. 2(c). The curve of Fig. 2(d) exhibits the distribution of light for different viewing angles.

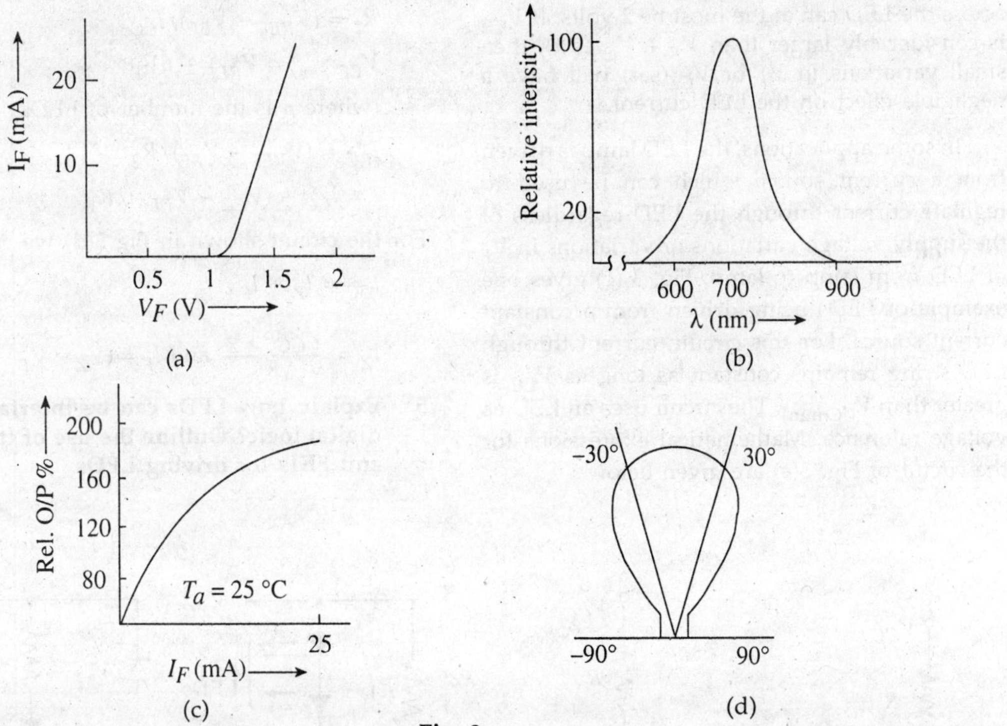

Fig. 2

3. Define the typical LED parameters.

Ans. : The LED parameters of interest are the

(a) Forward voltage (V_F)

(b) Candle Power (CP)

(c) Radiant power output (P_o)

(d) Peak spectral emission (λ_{peak})

(e) Spectral bandwidth (BW).

Forward voltage (V_F) is the D.C. voltage across the LED when it is ON. Candle Power (CP) is a measure of the luminous intensity or the brightness of the emitted light. Radiant power output (P_o) is the light power output of the LED. Peak spectral emission (λ_{peak}) is the wavelength of the brightest colour emitted by the LED. Spectral bandwidth is an indication as to how concentrated is the brightest emitted colour around its nominal wavelength.

4. How do you drive an LED from a regulated D.C. voltage? What are the shortcomings of driving LEDs from a constant voltage source. Explain the use of constant current source for driving LEDS.

Ans. : If an LED is driven from a regulated D.C. power supply, a resistor can be used to limit the current flowing through the LED. In Fig. 3(a), the current that would flow through LED can be written as —

$$I = \left(\frac{V_{CC} - V_F}{R}\right), \text{ where } V_F \text{ is typically 1.5 V.}$$

To bias the LED at, say, 10 mA and operating voltage 10 volts, we require a resistance of 850 ohms. In order to use configuration like the one given in Fig. 3(b), the LED current can be expressed as

$$I_F = \frac{V_{CC} - V_F - V_{CE}(sat)}{R},$$

where $V_{CE\ (sat)}$ is the saturation collector to emitter voltage of the driver transistor. Then we can write the following expression :

$$I_{F(max)} = [V_{CC(max)} - V_{F(min)} - V_{CE(sat)(min)}]/R_{(min)}$$

and

$$I_{F(min)} = V_{CC(min)} - V_{F(max)} - [V_{CE(sat)}]_{max}/R_{(max)}$$

To ensure a minimum of 10 mA of current through the LED when forward biased and connected in the configuration of Fig. 3(b) when the supply can vary from 10 volts to 15 volts and V_{CE}(sat) could be anywhere between 0.2 volt and 0.6 volt, then the current limiting resistance in no case could exceed 740 ohms if the forward voltage drop across the LED can at the most be 2 volts. If V_{CC} is considerably larger than $V_F + V_{CE}$(sat), then small variations in V_F or V_{CE}(sat) will have a negligible effect on the LED current.

In some applications, the LED lamp is driven from a current source which can be used to regulate current through the LED regardless of the supply voltage variations or variations in V_F of LED from lamp to lamp. Fig. 3 (c) gives one example of LEDs being driven from a constant current source. For this circuit, current through LED string remains constant as long as V_{CC} is greater than $V_{CC(min)}$. The circuit uses an LED as voltage reference. Mathematical expressions for the circuit of Fig. 3(c) are given below :

V_{REF} = 1.5 V (for a GaAsP LED)

$$R_1 = \frac{(V_{CC} - V_{REF})}{1\text{mA}}$$

(if LED 1 is biased at 1 mA)

$$R_2 = (V_{REF} - V_{BE})/I_{CC}$$

$$V_{CC(min)} = V_{REF} + nV_F$$

(where n is the number of LEDs)

$$I_{CC} = (V_{REF} - V_{BE})/R_2$$

$$I = I_{CC} + (V_{CC} - V_{REF})/R_1.$$

For the circuit shown in Fig 3(d) we can write,

$$I_{CC} = I_Z + I_F.$$

$$I_Z = \frac{V_{CC} - V_Z}{R} \text{ and } I_F = (V_Z - V_F - V_{EB})/R'.$$

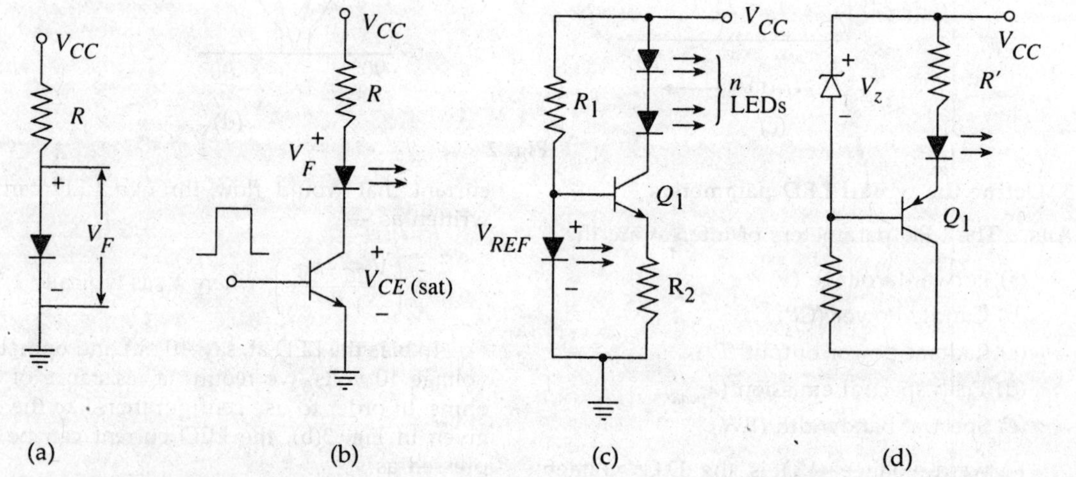

Fig. 3

5. Explain how LEDs can be interfaced with digital logic? Outline the use of transistors and FETs for driving LEDs.

Ans. : As LED lamps operate at low voltage and currents, they can directly interface to most logic families. Figs. 4(a) to (d) depict some of the ways that an LED lamp can be used to interface to digital logic. Figs. 4(a) and (b) depict LED interface with logic gates having open collector out-

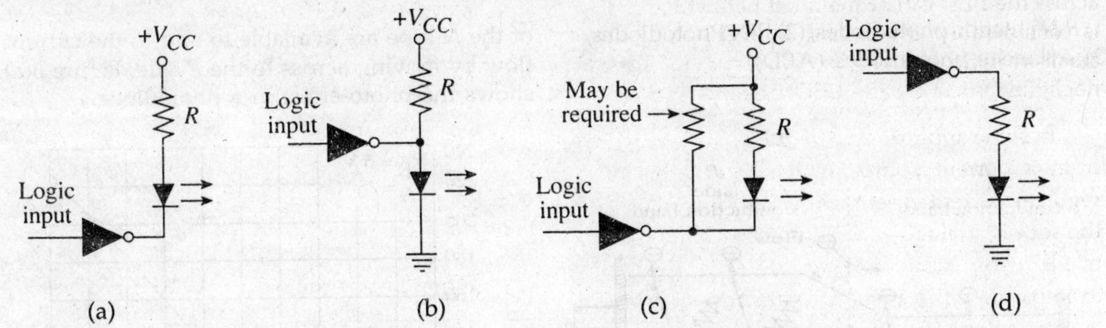

Fig. 4

puts whereas Figs. 4(c) and (d) show the interface circuits with active pull up totem pole gates. TTL

some of the commonly used drive schemes implemented with *N-P-N* and *P-N-P* transistors.

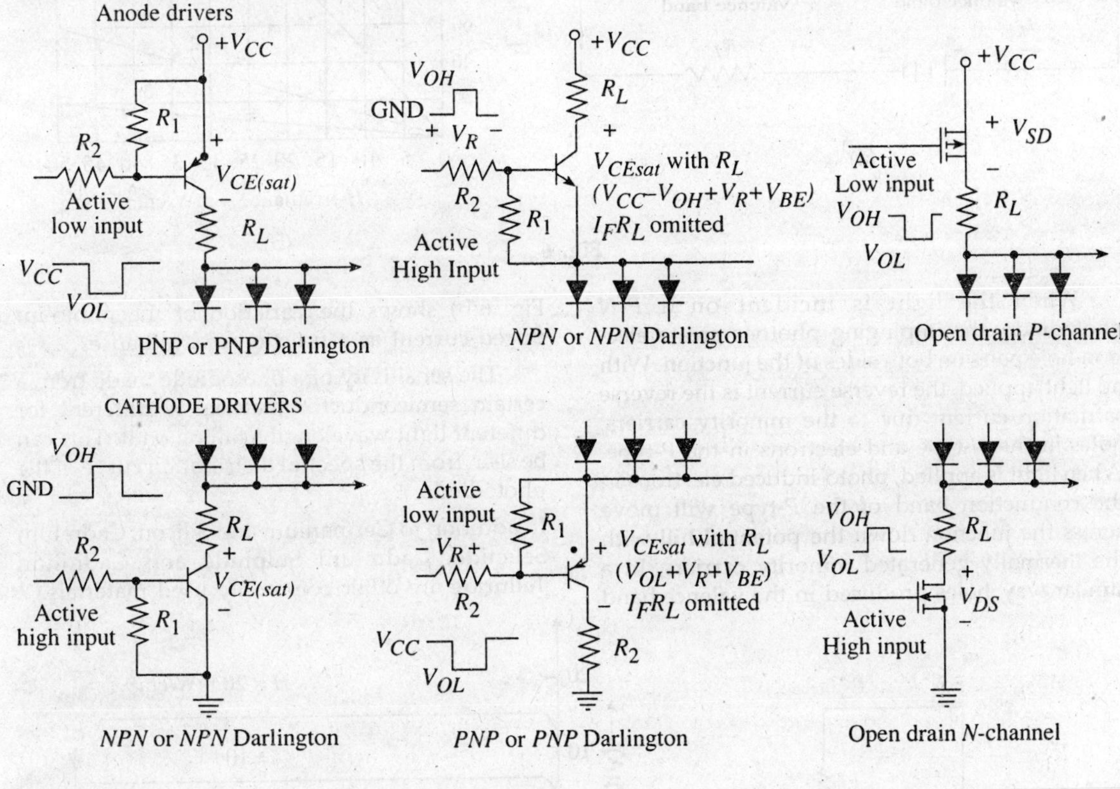

Fig. 5

logic families are guaranteed to sink a minimum of current which can drive most LED lamps without any additional buffering. If additional current is necessary, both TTL as well as CMOS buffers are available. Low cost transistors can also be used when larger currents are required than that a logic gate can supply directly. Fig. 5 shows

6. What is a photodiode? Briefly explain the *P-N* junction photodiodes, the PIN photodiodes and the Avlanche photodiodes.

Ans. : The photodiode is a *P-N* junction diode through which a significant current flows when light falls on it. There are three main types into which these photodiodes can be classified.

(1) *P-N* junction photodiodes, (2) PIN Photodiodes, (3) Avlanche photodiodes (APD).

of the *N*-type are available to add to the current flow by moving across to the *P*-side. Figure 6(a) shows the photo-effect in a photodiode.

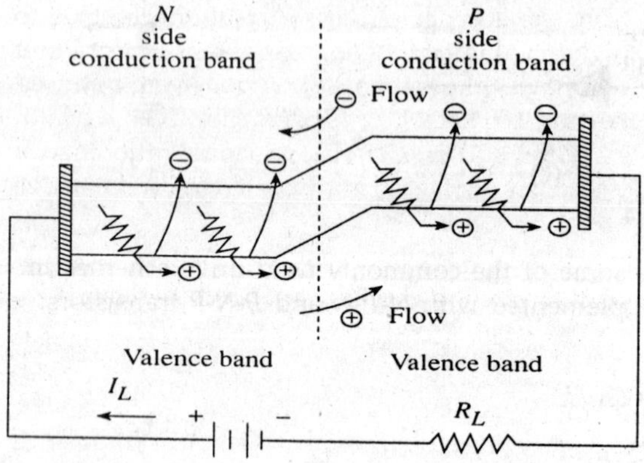

(a)

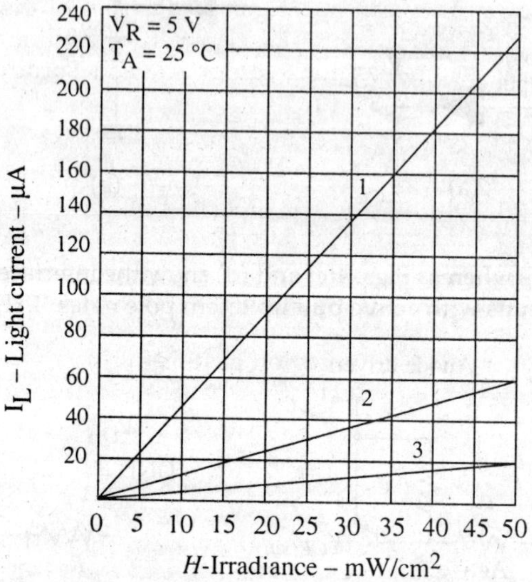

(b)

Fig. 6

When the light is incident on a *P-N* photodiode, the impinging photons create electron-hole pairs on both sides of the junction. With no light applied, the reverse current is the reverse saturation current due to the minority carriers, holes in the *N*-type and electrons in the *P*-type. When light is applied, photo-induced electrons in the conduction band of the *P*-type will move across the junction down the potential hill with the thermally generated minority carriers. In a similar way, holes produced in the valence band

Fig. 6(b) shows the variation of the photo-induced current as a function of irradiance.

The sensitivity of a photodiode made from a certain semiconductor material is different for different light wavelengths falling on it. This can be seen from the spectral distribution curve of the photodiode.

In addition to Germanium and Silicon, Cadmium Selenide, Cadmium Sulphide and Cadmium Telluride are other commonly used materials.

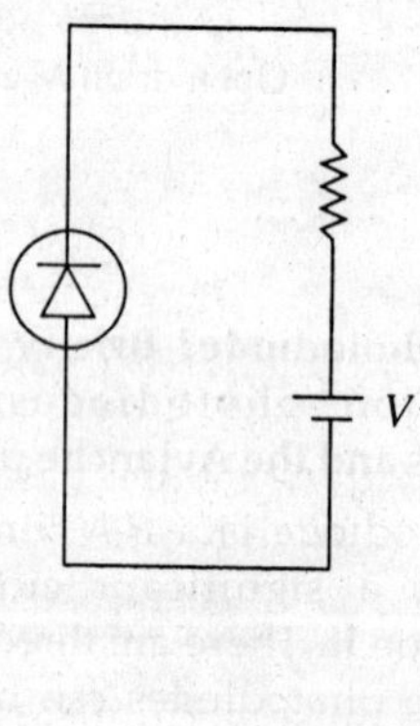

Fig. 7

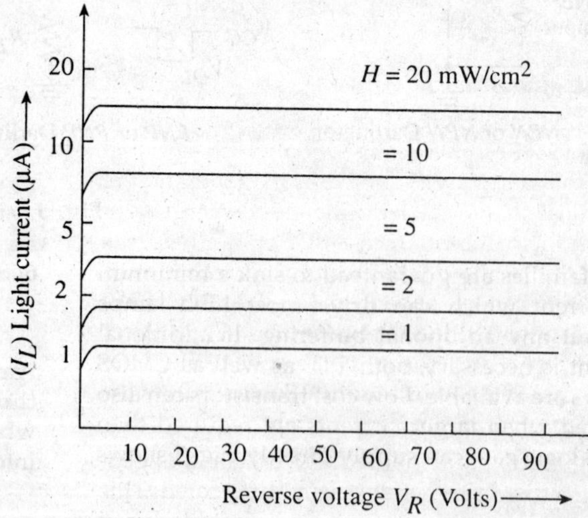

Fig. 8

In operation, the photodiode is reverse-biased (Fig. 7). The active diameter of these devices is typically in the range of 2 to 15 mm^2, and are mounted in a standard TO-5 package with a window. They can operate at frequencies of the order of 1 MHz to use in character recognition, optical communications equipment and encoders.

PIN photodiodes have a faster response. A PIN photodiode consists of a layer of intrinsic silicon between heavily doped *P* and *N*-type silicon semiconductor materials. This has the effect of reducing the transit or the diffusion time of photo-induced electron-hole pairs. That is, carriers created by photons incident on the middle of the depletion region have less distance to travel than those generated on one side or the other side of the depletion layer. The relatively thick layer of high resistivity intrinsic silicon ensures the absorption of most of the photons. Fig.8 shows the electro-optical characteristics of typical PIN photodiode. The device is a constant current generator at a given light level.

Avalanche photodiode is a photodiode operated in the reverse breakdown or avalanche breakdown region. By doing this, the current sensitivity may be increased 30 to 100 times. In the avlanche photodiode, the construction provides a very uniform junction that exhibits the avlanche effect at voltage typically between 30 V and 300 V. The electron-hole pairs that are generated by incident photons are accelerated by the high electric field to knock out more electrons in the valence band to make them available for conduction.

7. Distinguish between Photoconductive and Photovoltaic modes of operation of photodiodes. What are their preferred application areas?

Ans. : Photoconductive photodiode operates on the principle of photoconduction which says that electrical conductivity increases by generation of additional charge carriers when high energy light falls on the material. In case of a photodiode, the sensitivity of the photo-induced current to incident light can be enhanced considerably if the diode is operated in the reverse biased condition. The current that flows in the absence of any light in this case is the reverse saturation current of the diode, known as the dark current.

In photovoltaic photodiode an e.m.f. is generated across the junction when illuminated. This photovoltaic e.m.f. produces the photo-induced current when the external circuit is closed. As a photovoltaic diode needs no external reverse bias, it is free from any dark current. Photovoltaic photodiodes are preferred in low noise, low frequency applications. Low noise mainly results from the absence of dark current effects. Due to higher value of junction capacitance and shunt resistance, photovoltaic photododes are not used beyond 100 kHz. Below 1 kHz, however, it exhibits a far superior signal, to noise ratio than a photoconductive photodiode having same active area.

Photoconductive photodiodes are used in high frequency applications. Their frequency response easily extends beyond a few GHz. This is made possible due to reduced junction capacitance resulting from application of reverse bias.

8. How does a phototransistor differ from a photodiode? What are its advantages? What is a photodarlington?

Ans. : A phototransistor differs from a photodiode in the sense that the photo induced current of the photodiode gets internally multiplied by transistor action resulting in a larger sensitivity. If (I_λ) is photo-induced current, then the collector in a phototransistor would be $I_\lambda(1+h_{fe})$. Some

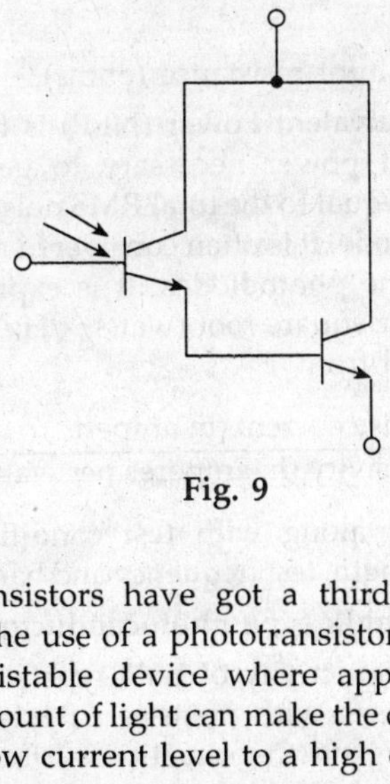

Fig. 9

phototransistors have got a third lead which enables the use of a phototransistor as a switching or bistable device where application of a small amount of light can make the device switch from a low current level to a high current level.

A photodarlington (Fig 9) is darlington configuration with the input transistor as light sensitive. The sensitivity of a photo-darlington is an order of magnitude higher than phototransistors.

9. Compare the noise performance of photoconductive and photovoltaic photodiodes. What is Noise Equivalent Power (NEP)?

Ans. : Noise in a photoconductive photodiode is a combination of shot noise mainly due to the dark current and the Johnson or thermal noise due to series resistance, load resistance and channel resistance if any. The shot noise in photoconductive photodiode is written as,

I_s (noise current) = $(2eI_d\Delta f)^{1/2}$,

where e = electronic charge = 1.6×10^{-19} coulomb

I_d = dark current (amperes)

Δf = noise equivalent bandwidth (hertz)

The thermal noise contribution (I_t) is expressed by

$$I_t = \left(\frac{4KT\Delta f}{R}\right)^{1/2},$$

where K = Boltzmann's constant

$= 1.38 \times 10^{-23}$ J/K

T = temperature (K)

Δf = noise equivalent bandwidth (hertz)

R = resistance (series, load, channel)

Noise in a photovoltaic photodiode is mainly the thermal noise due to device's shunt resistance. It can be written as

$$\left(\frac{4KT\Delta f}{R_{sh}}\right)^{1/2},$$

where R_{sh} = shunt resistance (ohms).

Noise Equivalent Power (NEP) is the minimum incident power necessary to generate a photocurrent equal to the total RMS noise current of the photodiode. It is often considered as figure-of-merit for the photodiodes. It is expressed in watts per hertz square root (watts/$\sqrt{\text{Hz}}$) and can be determined from

$$\frac{\text{Noise current (in amperes)}}{\text{Responsivity (in amperes per watt)}}\cdot$$

It is specified along with test conditions like source wavelength, test frequency and bandwidth.

10. What are bulk type photoconductors?

Ans. : The conductivity of bulk-type photoconductors increases with increase in intensity of light impinging on the sensitive surface of the device. The conductivity increases due to release of electrons caused by impinging photon energy.

11. What is a light activated SCR or a photo SCR?

Ans. : It is an SCR that can be switched into conduction by light falling on gate cathode junction. Like a conventional SCR, once switched ON, the light loses control on the anode current.

12. What are photoemissive type of photo-detectors?

Ans. : Photoemissive detectors are based on photoemissive effect. Due to this effect, there occurs electron emission from the surface of a material when the impinging photon energy exceeds the work function of the material. In a typical photoemissive detector, electrons are ejected from the surface of a photocathode placed in vacuum and then collected by a charged anode. The important devices in this category include vaccum photodiodes, photomultiplier tubes (PMT) and microchannel multipliers.

13. What are Thermal Detectors? What are different types of Thermal Detectors?

Ans. : Thermal detectors absorb radiation and operate on the resulting temperature rise. There are three major categories of thermal detectors. They are :

1. Thermocouple (or Thermopile) type
2. Bolometric type
3. Pyroelectric detectors.

14. Outline major characteristic parameters of photosensors.

Ans. : Major characteristic parameters of photosensors include the following :

1. Responsivity
2. Noise Equivalent Power (NEP)
3. Quantum Efficiency
4. Sensitivity usually expressed as either Detectivity (D) or Dee-Star (D*)
5. Detector noise comprising of Johnson noise, Shot noise and Flicker noise
6. Response time
7. Active area
8. Spectral Response.

15. What is an optocoupler or an optosolator? What are its different constituents?

Ans. : An optocoupler is a photon emitting device whose flux is coupled through an optically transparent insulation to some sort of photodetector. The photon emitting device may be an incandescent or neon lamp, or an LED. The transparent insulation may be air, glass, plastic or fibre-optic. The photodetector may be a photoconductor, a photodiode, phototransistor or an integrated combination of photodiode/amplifier.

UNIT 5
WAVESHAPING CIRCUITS

1. How does the basic *RC* circuit respond to a STEP or a PULSE input of given time duration? Under what conditions can this circuit be called an integrator circuit

Ans. : STEP INPUT

The basic *RC* circuit is drawn in Fig. 1. For a step input of Fig. 1(b), the output voltage (v_o), which is also voltage across *C*, rises exponentially towards the final value of *V* with a time constant (*RC*). The output voltage (V_o) is expressed as

$$v_o = V(1 - e^{-t/RC}).$$

The output (V_o) during the low time of the pulse is then written as

$$v_o = V_{(t_p)}\left[e^{-(t-t_p/RC}\right]$$

If the circuit shown in Fig. 1(a) is an integrator circuit, the output (v_o) should be integral of the input (v_i), i.e., $v_o = k\int v_i\, dt$, where *k* is a constant. In the given *RC* circuit, if the product (*RC*) is much larger than the time period of the

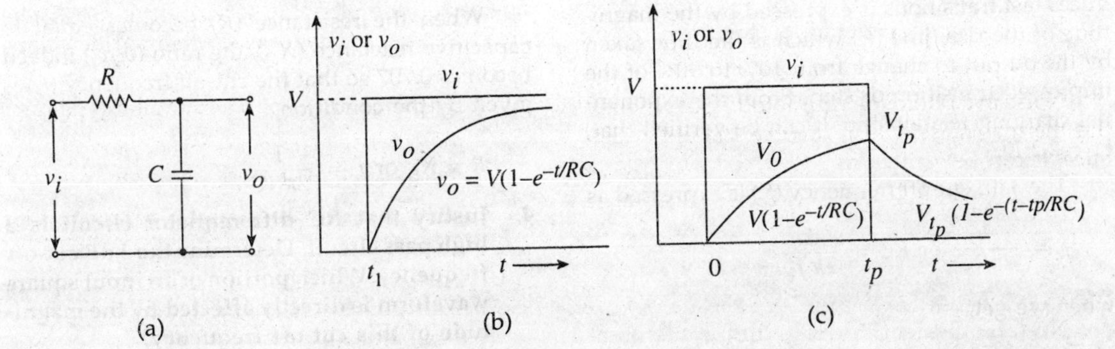

Fig. 1

This expression is valid only when the capacitor is initially fully discharged. If the capacitor is initially charged to a voltage (V_o) less than (*V*), then the exponential charging equation becomes

$$v_o = \left[V-(V-V_o)e^{-t/RC}\right].$$

If this input step occurs at time $t = t_1$, then

$$v_o = V\left(1-e^{-(t-t_1)/RC}\right)$$

PULSE INPUT

The output(v_o) during the high time of the pulse is expressed as

$$v_o = V\,(1-e^{-t/RC}).$$

when $t = tp$, we have

$$v_o\,(\text{at } t= tp) = V(1-e^{-tp/RC}) = V_{(t_p)}$$

applied input, the capacitor voltage would change by only a very small amount as the input goes through a complete cycle. Then it is correct to assume that whole of (v_i) appears across (*R*) only. As a result current (*i*) can be written as

$$i = (V_i/R)$$

$$v_o = \frac{1}{C}\int idt\ \frac{1}{C}\int\left(\frac{V_i}{R}\right)dt = \frac{1}{RC}\int v_i dt$$

Therefore, (v_o) is integral of (v_i) provided that $RC >> T$. When $RC \geq 15T$, integration is almost ideal.

2. When a pulsed waveform is applied to an *RC* low pass circuit of the type shown in Fig. 2, we are often interested in knowing the quality with which the leading and the trailing edges or the flat top portions will be passed. Such a situation arises particularly

in the analysis of high frequency amplifiers where the amplifier input has a finite capacitance and this capacitance along with

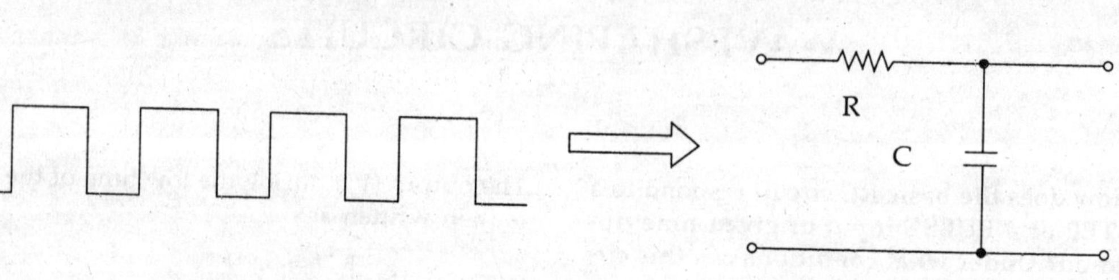

Fig. 2

source resistance constitutes the RC network. How is the 3 dB cut-off frequency of such a network related to the capability of the network to pass fast transitions? On what factors will the faithful reproduction of flat top portion depend?

Ans. : The quality with which the network reproduces fast transitions is expressed by the magnitude of the rise time (t_r) which is the time taken by the output to change from 10% to 90% of the impressed transition or step. From the exponential charging relationship, it can be verified that, $t_r = 2.2\ RC$.

The 3 dB cut-off frequency (f_c) is expressed as

$$f_c = \frac{1}{2\pi RC} \quad \text{or} \quad RC = \frac{1}{2\pi f_c}$$

when we get,

$$t_r = \frac{2.2}{2\pi f_c} = \left(\frac{0.35}{f_c}\right).$$

This expression suggests that higher the 3 dB cut-off, smaller is the rise time. So, for faithful reproduction of fast transitions, f_c should be as high as possible. If the sharp transitions are reproduced, flat portions would automatically be passed undistorted. As long as 3 dB cut-off frequency is equal to or larger than the frequency of the input square waveform or pulse train, flat portion will be passed with not too much of distortion. In order to allow the transitions to pass, the cut-off frequency should be much larger than the input frequency.

3. **How does the RC circuit of Fig. 1(a) behave as a low pass circuit? How is the 3 dB cut-off frequency of this circuit related to (R) and (C)?**

Ans. : Qualitatively, since the output is taken across the capacitor and the reactance of a capacitor is inversely proportional to the frequency, the output voltage will fall with increase in frequency. That is how an *RC* network of the type shown in Fig. 1(a) behaves as a low pass circuit. The 3 dB cut-off is the frequency at which output amplitude is 0.707 times or 3 dB below the nominal maximum amplitude. The nominal maximum output amplitude is same as the input amplitude so that ratio (v_o/v_i) equals 0.707 at 3 dB cut-off frequency.

When the resistance (R) becomes equal to capacitive reactance (X_C), the ratio (v_o/v_i) indeed becomes 0.707 so that the cut-off frequency (f_c) is given by the condition,

$$R = X_C \text{ or } f_c = \frac{1}{2\pi RC}.$$

4. **Justify that *RC* differentiator circuit is a high pass circuit. Determine the 3 dB cut-off frequency. Which portion of an input square waveform is directly affected by the magnitude of this cut-off frequency?**

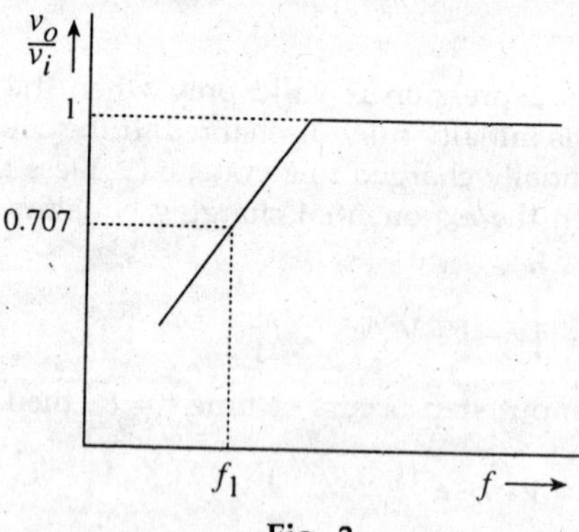

Fig. 3

Ans. : Qualitatively, as the reactance of a capacitor is inversely proportional to the frequency, the output voltage (v_o) must fall with decrease in frequency of the input waveform and hence the high pass characteristics. The frequency where the ratio (v_o/v_i) falls to 0.707 is known as the lower 3 dB cut-off (Fig. 3). We can write,

$$\frac{v_o}{v_i} = \frac{R}{\sqrt{R^2 + X_c^2}} = 0.707$$

for $R = X_C$ or $f_1 = \dfrac{1}{2\pi RC}$,

where f_1 is lower 3 dB cut-of frequency.

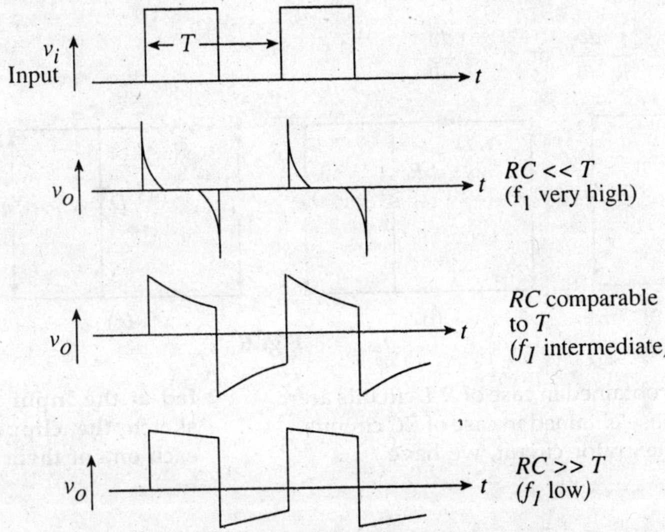

Fig. 4

The lower 3 dB cut-off frequency affects the low frequency response due to the high pass nature of the circuit. Lower the lower 3 dB cut-off frequency, less severe is its effect on the flatter portions of the waveform. In a differentiator circuit where (f_1) is very high, the flat portion of square wave input may be completely missing.

The effect of different lower 3 dB cut-off frequencies in a high pass RC circuit on a square waveform input are shown in Fig. 4.

5. Prove that the R-L circuit of Fig. 5(a) behaves as an integrator circuit provided the time constant (L/R) is larger than the time period of the input waveform.

Ans. : If the time constant (*L*/*R*) is much larger than the time period of the input waveform, it is justified to assume that the current (*i*) is exclusively determined by (*L*) as the input goes through one complete cycle.

i.e., $i = \frac{I}{L}\int v_i dt$.

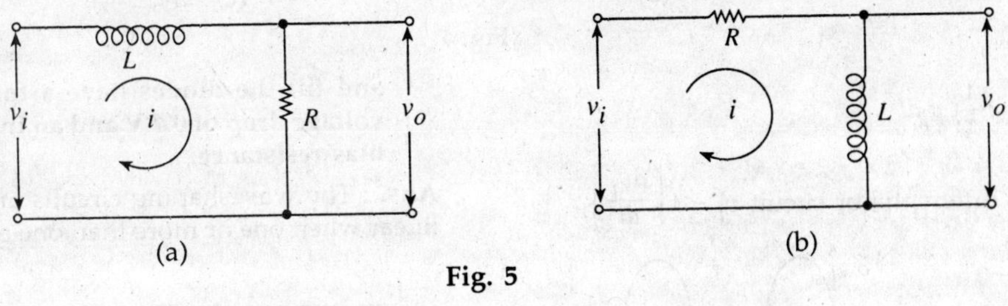

(a) (b)

Fig. 5

$$v_o = i.\ R = \frac{R}{L}\int v_i dt \text{ or } v_o \propto \int v_i dt$$

6. Prove that the R-L circuit of Fig. 5(b) is a differentiator provided the time constant (L/R) is much smaller than the time period of the input waveform.

Ans. : If the time constant (L/R) is much smaller than the input waveform time period, the current (i) very quickly attains the steady state and the transient portion takes negligible time. Then it can be assumed that the current (i) is exclusively determined by (R).

i.e., $i \propto \dfrac{v_i}{R}$

$$v_o \propto L\frac{di}{dt} = \frac{L}{R}\frac{dv_i}{dt} \text{ or, } v_o \propto \frac{dv_i}{dt}.$$

where, τ = time constant = RC (for R–C circuit) = $\dfrac{L}{R}$ (for R–L circuit).

7. Fig. 6 shows some diode clipper circuits. Why are these called non-linear waveshaping circuits? If these circuits are

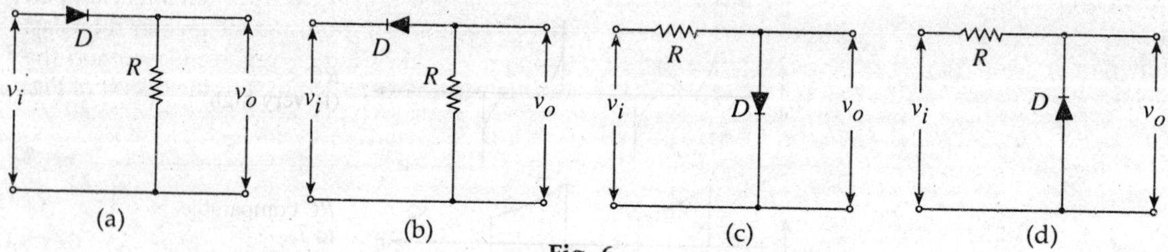

Fig. 6

The results obtained in case of R-L circuits are analogous to those obtained in case of RC circuits. For a simple integrator circuit, we have

fed at the input with a sinusoidal signal, sketch the clipped output waveform for each one of them when (i) diodes are ideal

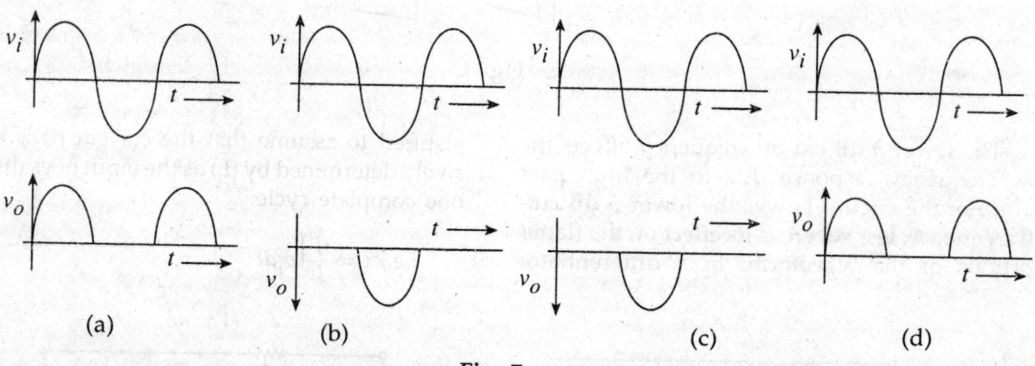

Fig. 7

$$v_o = \frac{1}{\tau}\int v_i dt$$

and for differentiator circuit $v_i = \tau\left(\dfrac{dv_i}{dt}\right)$,

and (ii) the diodes have a forward biased voltage drop of 0.7 V and an infinite reverse bias resistance.

Ans. : The waveshaping circuits are called non-linear when one or more than one elements have

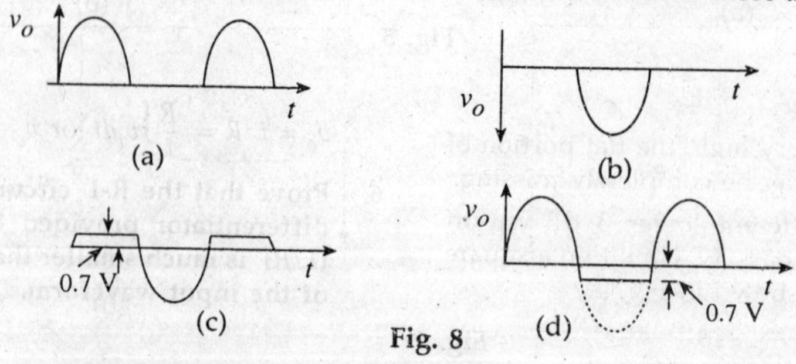

Fig. 8

non-linear current-voltage characteristics. The non-linear element in the circuits of Fig. 6 is the diode.

(i) The input and output waveforms for the clipping circuit of Fig. 6 (a) are shown in Fig. 7(a).

Similarly, the output waveform for the circuit of Fig. 6(b) is shown in Fig. 7 (b). The clipping circuit of Fig. 6(c) is functionally identical to the one shown in Fig. 6(b). The positive half-cycles are clamped to zero and the negative half-cycles are passed as the diode would then be reverse biased. The circuit of Fig. 6(d) is identical to that shown in Fig. 6(a). In this case, the negative half-cycles would be clamped to zero.

8. Sketch the output waveform for the clipping circuits of Fig. 9. Assume the diodes to be ideal and the peak amplitude V_m of the input sinusoidal signal to be greater than the battery voltage (V_{BB}).

Ans. : The output waveforms are shown in Fig. 10 (a) for the circuit of Fig. 9(a) and in Fig. 10(b) for the circuit of Fig. 9(b).

In case of the circuit shown in Fig. 9(a), the diode gets forward biased when the input exceeds V_{BB} and thus the output remains clamped at V_{BB} as long as input remains greater than V_{BB}. For $v_i < V_{BB}$, the diode is reverse biased and the output is same as the input. For the circuit of Fig.

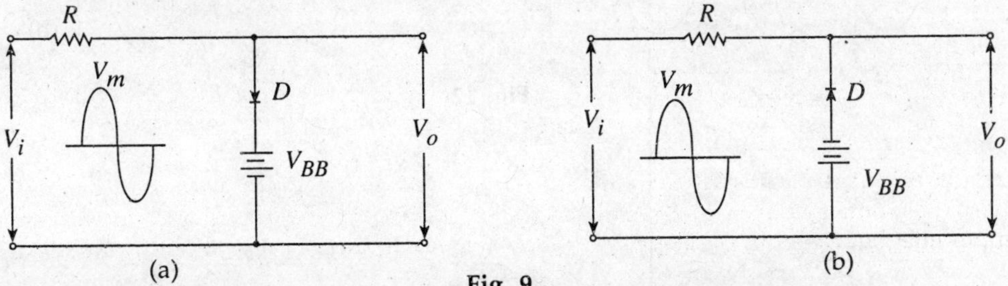

Fig. 9

(ii) The output waveforms would be similar to those shown in Fig. 7 except that the diode would have a forward biased drop of 0.7 V

9(b), the diode is reverse biased for (v_i) greater than ($-V_{BB}$) and forward biased for (v_i) more negative than (V_{BB}).

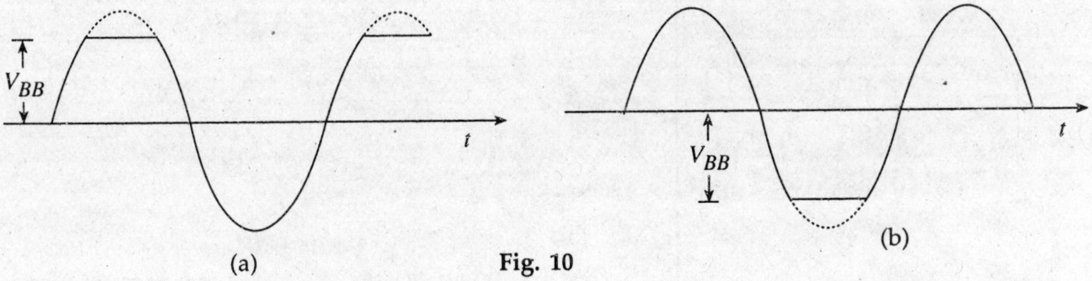

Fig. 10

instead of zero assumed in case of ideal diodes. The waveforms are shown in Fig. 8.

9. Sketch the clipped output waveforms for the circuits of Fig. 11. Assume the diodes to

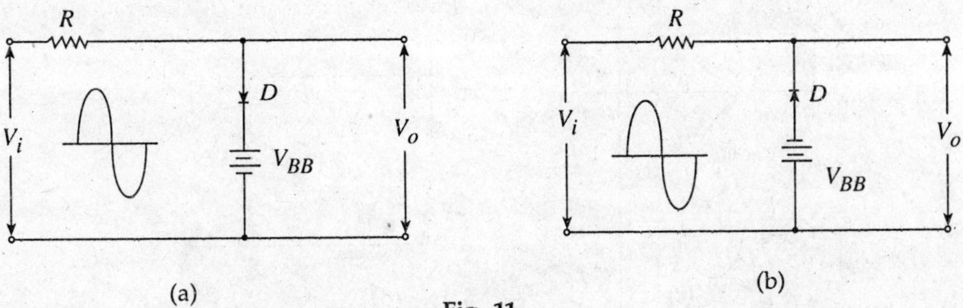

Fig. 11

be ideal and the peak amplitude of v_m of the input sinusoid to be greater than battery voltage V_{BB}.

Ans. : For the circuit shown in Fig. 11(a), the diode is forward biased for all input voltages except for the period when v_{in} is more negative than the battery voltage $-V_{BB}$ where it is reverse biased. For the circuit shown in Fig. 11(b), the diode is forward biased for v_{in} less than equal to V_{BB} and reverse biased for v_{in} greater than V_{BB}. The output waveforms for the two cases in Fig. 11(a) and (b) are shown in Figs. 12(a) and (b) respectively.

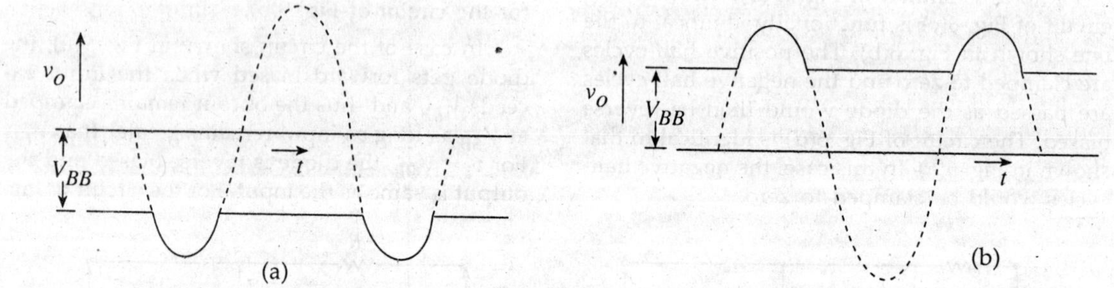

Fig. 12

UNIT 6
FLIP-FLOPS

1. Draw the NOR-gate of active HIGH input and active LOW input RS flip flops.

Ans. : Active HIGH input RS flip flop built around NOR gates in shown in Fig. 1(a) and the same with active LOW inputs is shown in Fig. 1(b).

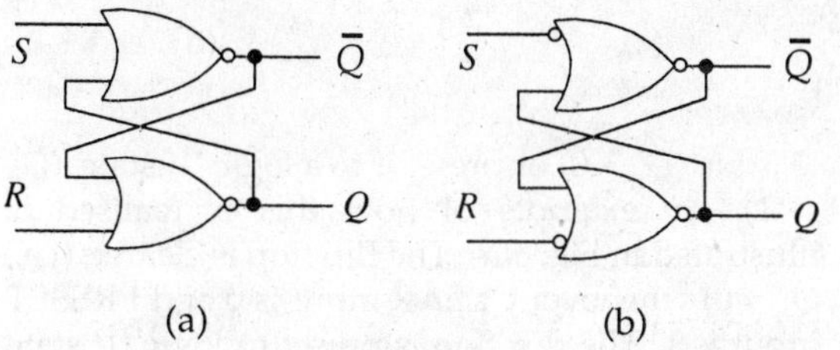

Fig. 1

2. What is the need for having a CLOCKED FLIP FLOP? Draw the logic representation, the circuit symbol and the truth table of a clocked *R-S* flip flop.

Ans. : In most of the digital application circuits like counters, registers or a combination of these, it is required that operation of all flip flops being used in the overall system be synchronised by a master clock. The flip flops then change their states as per the inputs only on the occurrence of the clock pulse. Such flip flops are called the clocked flip flops.

Figs. 2(a) and (b) show the logic representation of a clocked R-S flip flop. The truth table is the same is given in Fig. 2(c). When the clock is LOW, the AND gates are disabled and as a result the output cannot respond to the input. When the clock goes HIGH, the AND gates are enabled and the *R* and *S* inputs are passed on the inputs of and NAND gates that constitute the flip flop. The output changes state as per the logic states of the *R* and *S* inputs.

3. Differentiate between EDGE TRIGGERED flip flops and the LEVEL TRIGGERED flip flops. Draw the circuits symbols for the following flip flops.

(i) Positive edge triggered clocked R-S flip flop with active LOW inputs.

(ii) Level triggered R-S flip flop with active HIGH inputs.

(iii) Negative edge triggered clocked R-S flip flop with active HIGH inputs.

(iv) Level triggered clocked R-S flip flop with active LOW inputs.

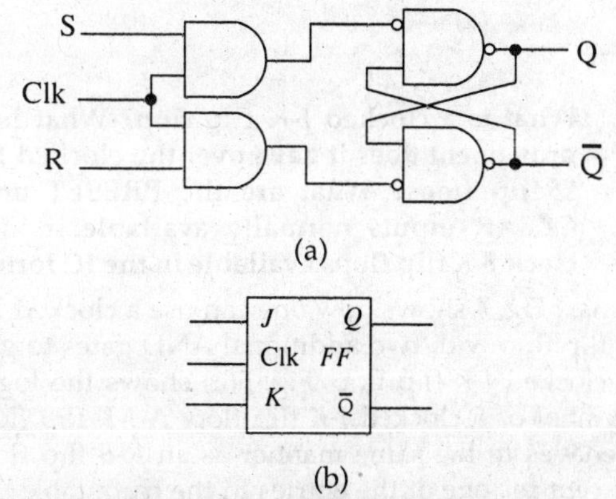

S	R	Clk	Q_{n+1}
0	0	0	Q_n
0	0	1	Q_n
0	1	0	Q_n
0	1	1	0
1	0	0	Q_n
1	1	1	1
1	1	0	Q_n
1	1	1	Not permitted

(c)

Fig. 2

Ans. : In an EDGE TRIGGERED flip flop, the output responds to the data at the inputs only on the LOW to HIGH (or HIGH to LOW) transition of the clock pulse. Any change in the input during the time the clock pulse is HIGH (or LOW) does not have any effect on the output. For all flip flops, it will be known whether the flip flop responds to the positive going transitions or the negative going transitions.

In a LEVEL TRIGGERED flip flop, the output responds to the data present at the inputs during the time when the clock pulse level is HIGH (or LOW). That is, any change at the input during the time the clock is HIGH (or LOW) is reflected at the output as per the truth table.

The circuit symbols of different flip flops from (*i*) to (*iv*) are given in Fig. 3(a) to (d).

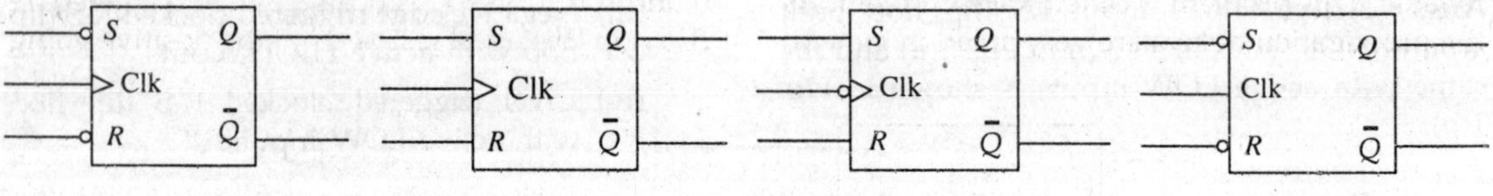

(a) (b) (c) (d)

Fig. 3

4. What is a clocked *J-K* flip flop? What improvement does it have over the clocked R-S flip flops? What are the PRESET and CLEAR inputs normally available in the clock *J-K* flip flops available in the IC form?

Ans. : Fig. 4 shows how one can use a clocked *R-S* flip flop with two additional AND gates to get a clocked *J-K* flip flop. Fig. 5(a) shows the logic symbol of a clocked *J-K* flip flop. A *J-K* flip flop behaves in the same manner as an *R-S* flip flop except for one of the entries in the truth table. In

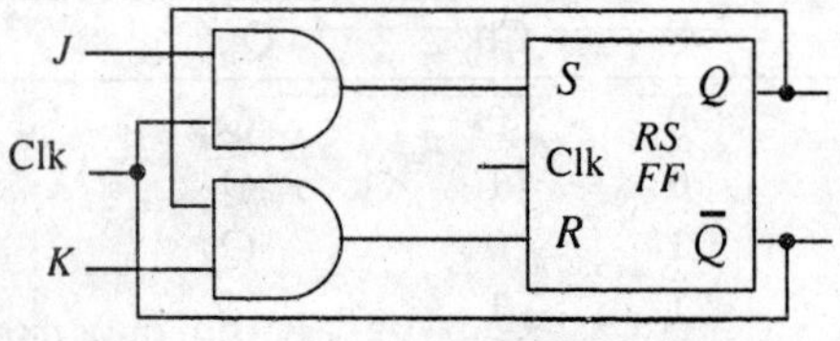

Fig. 4

case of an R-S flip flop, the input combination *S* = 1, *R* = 1 (for a flip flop with active HIGH inputs) and the input combination S = 0, R = 0 (for a flip flop with active LOW iputs) are prohibited. For a *J-K* flip flop with active HIGH inputs, the output of the flip flop toggles (i.e., it goes to the opposite logic state) for *J* = 1, *K* = 1. The output toggles for the input J = 0, K = 0 for the flip flop having active LOW inputs. So the problem of a prohibited input entry in an R-S flip flop is not there for *J-K* flip flop. Fig. 5(b) shows the truth table of a clocked J-K flip flop.

It is often required to clear a flip flop to logic '0' state ($Q_n = 0$) or preset it to a logic '1' state ($Q_n = 1$). An example of how this is realised is illustrated in Fig. 6(a). The flip flop is cleared (i.e., $Q_n = 0$) whenever CLEAR input is 0 and PRESET input is 1. The flip flop is preset to logic '1' state whenever PRESET input is 0 and the CLEAR

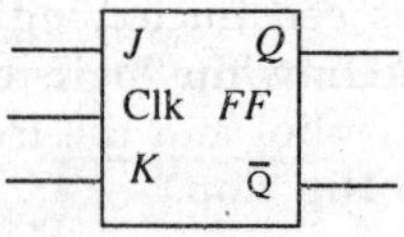

(a)

Operating Mode	*J*	*K*	*Clk*	*Q*
Set	1	0	1	1
Reset	0	1	1	0
No Change	0	0	1	Q
Toggle	1	1	1	$\bar{Q}$

(b)

Fig. 5

input is 1. In both the cases, the clock is in logic '0' state. Here the CLEAR and PRESET inputs are

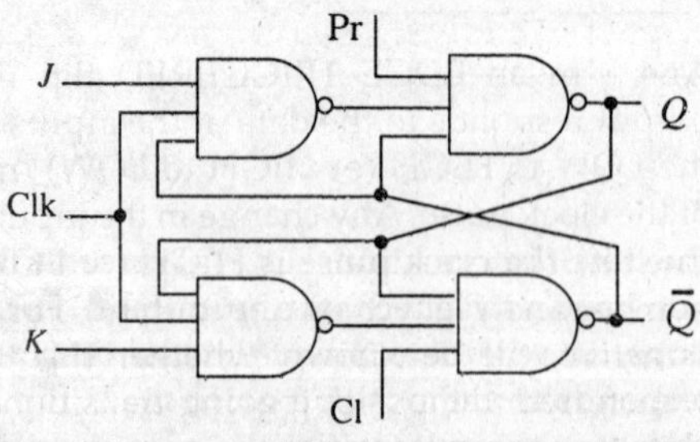

(a)

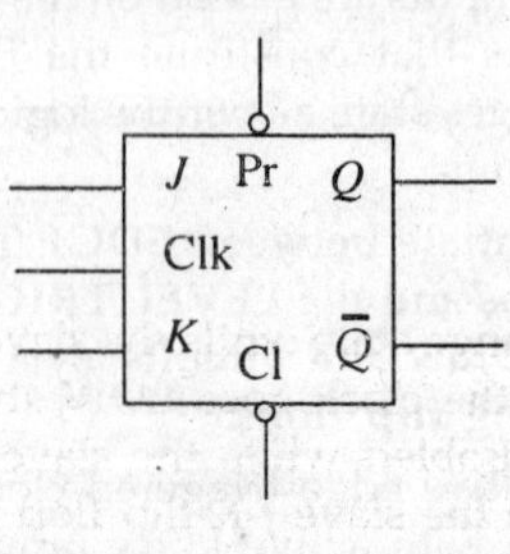

(b)

Fig. 6

active when LOW. Fig. 6 (b) shows the logic symbol for a Clearable-Presettable clocked *J-K* flip flop.

5. What is Race Problem in flip flops? How do you overcome that?

Ans. : Whenever the width of the trigger pulse is higher than the propagation time of the flip flop, the change of state at the output is not reliable. This problem is called Race Problem. As the propagation delays are very small in general there is a reasonable likelihood of race condition existing in flip flops. One technique to get over this problem in flip flows is to use a Master-Slave flip flop which is constructed with two *J-K* flip flops (Fig. 7). The first flip flop is the master flip flop and the second one is the slave. The clock to the slave flip flop is the complement of the clock to the master flip flop. When the clock pulse is present, the master flip flop is enabled while the slave flip flop is disabled. Thus the master flip flop can change state while the slave flip flop can not. When the clock goes LOW, the master flip flop gets disabled while the slave flip flopis is enabled. So the slave *J-K* flip flop changes state as per the logic states at its J and K inputs. The contents of the master flip flop are transferred to the slave flip flop, the master flip flop being disabled can acquire new inputs without affecting the output.

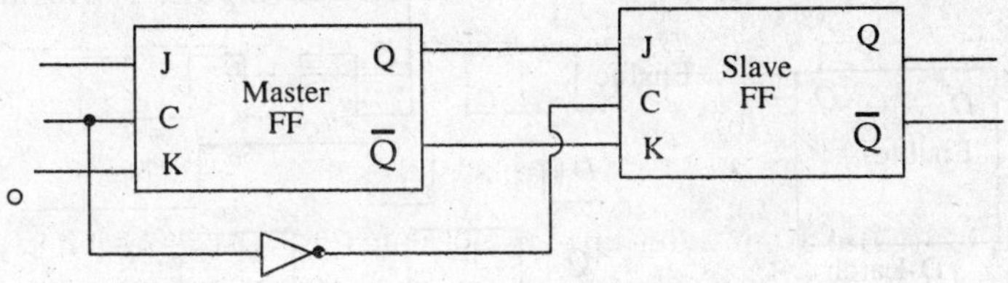

Fig. 7

6. What is a T-flip flop? Explain its operation with the help of its truth table.

Ans. : The output of *T*-flip flop changes state every time it is triggered at its *T*-input, called the toggle input. Thus the output becomes 1 if it was 0 and 0 if it was 1. A positive edge triggered *T*-flip flop [Fig. 8(a)] responds to the positive going edges of the clock waveform while a negative triggered flip flop [Fig. 8(b)) responds to the negative going edges of the clock waveform. Fig. 8(c) represents its truth table.

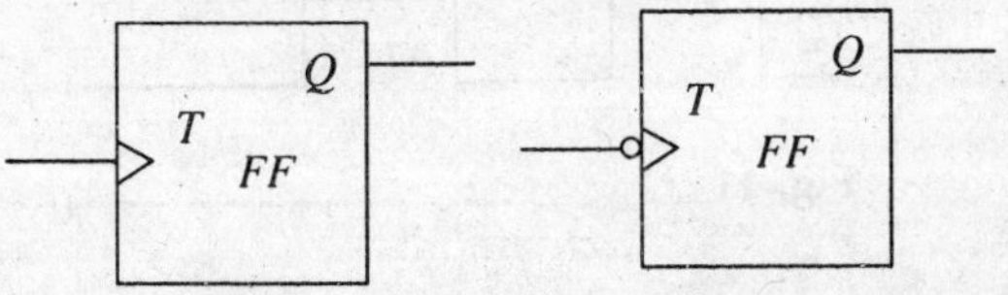

(a) (b)

T	Q_n	Q_{n+1}
⎍	0	1
⎍	1	0

(c)

Fig. 8

7. Explain the operation of a D-flip flop with the help of its truth table.

Ans. : D-flip flop is also called Delay Flip Flop. It can be used to provide temporary storage of one bit of information. Fig. 9(a) presents its circuit symbol and Fig. 9(b) the truth table. If the clock is active, the data bit present at the *D*-input is transferred to the *Q*-output. In the *D*-flip flop this data transfer will take on the negative going edge of the clock inpput. *D* inputs can acquire new

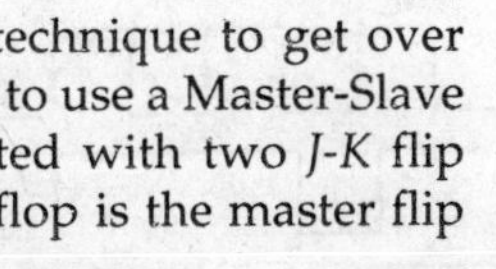

(a)

D	Clk	Q
0	↓	0
1	↓	1

(b)

Fig. 9

input data when the clock is inactive. In this way *D*-flip flop can provide a maximum delay of the clock period. Digital integrated circuits with two *D*-type flip flops or four *D*-type flops integrated witin the same IC package are commercially available. A Quad D-flip flop, for instance, can provide a temporary storage of 4-bit information.

8. What is D-type Latch? How does it differ from a D-type flip flop?

Ans. : In *D*-type latch, output *Q* follws the *D*-input so long the clock input is HIGH or LOW depending upon which clock level it responds to. If the clock goes to the inactive level, the output holds on to the logic state during the entire time period the clock is inactive.

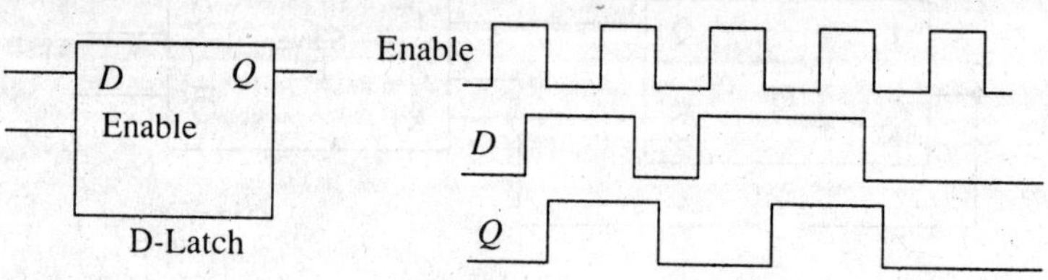

Fig. 10

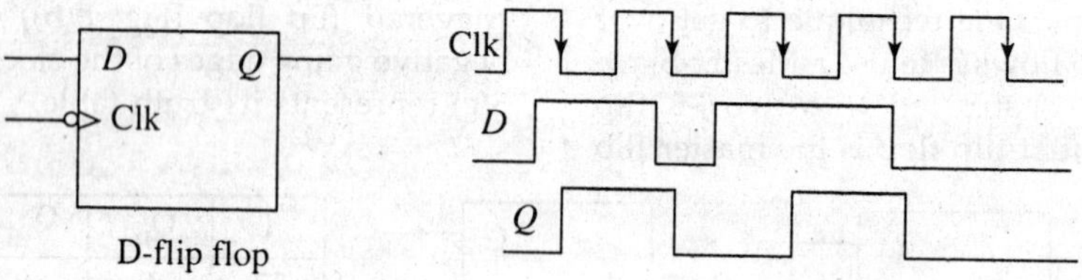

Fig. 11

In a *D*-flip flop, the data on *D*-input is transferred to the *Q*-output on the positive or negative going transition of the clock waveform depending on the flip and this logic state is held at the output till we get the next effective clock transition. Figs. 10 and 11 represent the two situations.

9. Why is a J-K flip flop called a universal flip flop? Show the logic implementation of different flip flop usisng J-K flip flop.

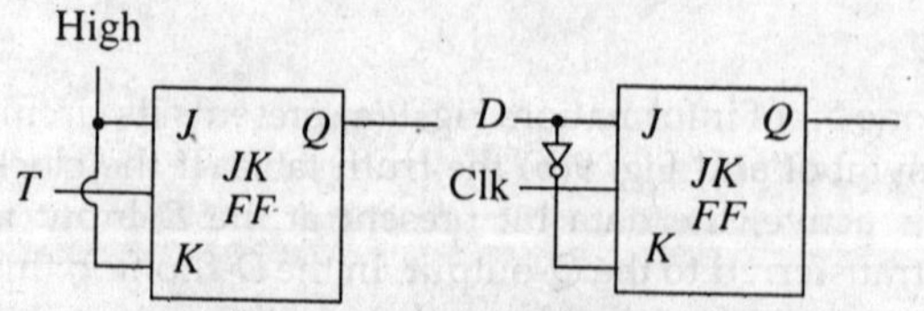

Fig. 12

Ans. : *J-K* flip flop is used to construct most of other flip flops. That is why it is called a universal flip flop. Fig. 12 shows use of J-K flip flops to construct different flip flops.

10. What do you mean by Synchronous and Asynchronous inputs with reference to flip flops?

Ans. : Synchronous inputs are those inputs whose effect on the flip flop output is synchronized with the clock input. *R, S, J, K, D* inputs are all synchronous inputs. Asynchronous inputs are those inputs which operate indepedently of the synchronous inputs and the input clock signal.

11. List important timing parameters of a flip-flop.

Ans. : (a) Set-up and Hold times

(b) Progagation delay

(c) Clock pulse HIGH and LOW times

(d) Asynchronous input active pulse width

(e) Clock transition time

(f) Maximum clock frequency.

12. Multiple Choice Questions

(i) In a presettable, clearable positive edge triggered *J-K* flip flop, the *Q* output when the flip flop is clocked for J = 1, *K* = 0, Preset = 1 and Clear = 0 will be

(a) 1; (b) 0; (c) Indeterminate; (d) complement of what it was at the time of clocking (Assume all the inputs to be active LOW inputs).

(ii) In an *R-S* flip flop with active LOW inputs, when R = S = 0, the output Q when the flip flop is clocked is

(a) 0, (b) 1, (c) Indeterminate, (d) 0, if it was 1 earlier and 1 if it was 0 earlier.

(iii) For one of the following conditions, clocked J-K flip flop can be used as DIVIDE BY 2 circuit when the pulse train to be divided is applied at clock input.

(a) *J* = 1, *K* = 1 and the flip flop should have active HIGH inputs

(b) *J* = 1, *K* = 1 and the flip flop should have active LOW inputs

(c) *J* = 0, *K* = 0 and the flip flop should have active HIGH inputs

(d) *J* = 1, *K* = 1 and the flip flop should be a negative edge triggered one.

(iv) We have two negative edge triggered *J-K* flip flops with active LOW inputs. The *J* and *K* inputs of both the flip flops have been permanently tied to logic 0. The *Q*-output of first flip flop (Q_1) feeds the clock input of the second flip flop. What will be the logic status of outputs Q_1 and Q_2 at the end of five clock cycles if the two outputs were cleared to logic 0 each before starts.

(a) $Q_1 = 1, Q_2 = 0$ (b) $Q_1 = Q_2 = 1$ (c) $Q_1 = 0, Q_2 = 1$, (d) $Q_1 = 0, Q_2 = 0$.

Ans. : (i) (b), (ii) (c), (iii) (a), (iv) (a).

13. Fill in the blanks

1. In a negative edge triggered *D*-flip flop, the data at *D*-input is transferred to the Q-output during .... transition of the clock input. (L to H, H to L)
2. "D-type latch is another name of a D-type flip-flop". This statement is .... (true, false).
3. The flip flop is the fundamental building block of .... logic circuits. (sequential, combination)
4. The clocked R-S flip flop is said to operate .... (synchronously, asynchronously)
5. The Race Problem does not exist in ...... (R-S flip flop, *J-K* flip flop, M/S J-K flip flop)

Ans. : 1. H to L, 2. False, 3. Sequential, 4. Synchronously, 5. M/s J/K flip flop.

UNIT 7
COMBINATIONAL LOGIC CIRCUITS

1. What is a Half-Adder? Starting wih the Boolean expressions for SUM and CARRY outputs, show how a half adder circuit can be implemented by using either only NAND or only NOR gates.

Ans. : Half-Adder is an arithmetic circuit block that can be used for adding only two bits. This circuit has two inputs that represent the two bits to be added and two outputs with one output giving the SUM and the other giving the CARRY. Fig. 1(a) shows the block of a Half-Adder while Fig. 1(b) gives its truth table.

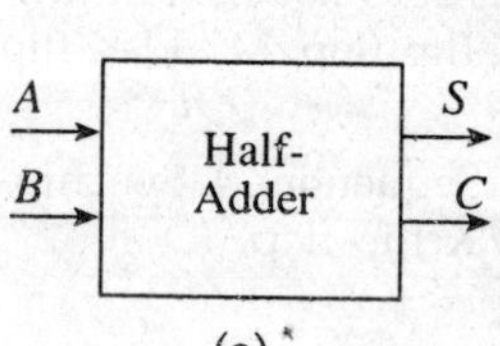

(a)

A	B	S	C
0	0	0	0
0	1	1	0
1	0	1	0
1	1	0	1

(b)

Fig. 1

The Boolean expressions for the two outputs are as follows :

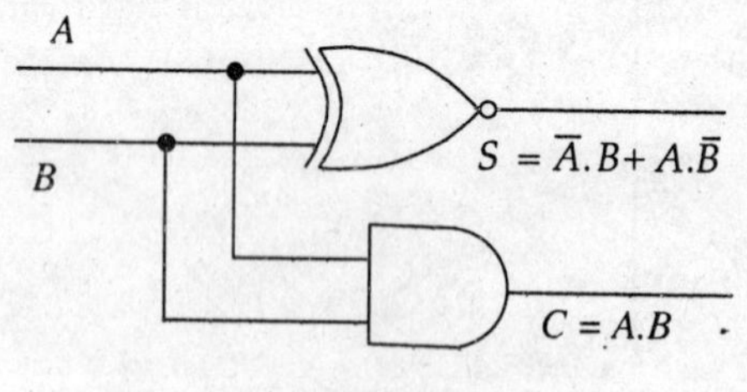

Fig. 2

$S = \bar{A}.B + A.\bar{B}$ and $C = A.B$

Fig. 2 shows the simplest form of hardware implementation of a half-adder.

Considering the two expressions we find that there is no scope for further simplification. While the first one representing the SUM output is that of an EX-OR gate, the second one representing the CARRY output is that of an AND gate. However, these two expressions can be represented in different forms using various laws and theorems of Boolean algebra.

From a consideration of the Boolean expressin for SUM output, this can be written in many ways, e.g.,

$$\bar{A}.B + A.\bar{B} = \overline{[\overline{(\bar{A}.B)}.\overline{(A.\bar{B})}]}$$

The hardware implementation of this expression is shown in Fig. 3 using only NAND gates. Further,

$$\bar{A}.B + A.\bar{B} = A.(\bar{A}+\bar{B}) + B.(\bar{A}+\bar{B})$$
$$= A.\overline{(A.B)} + B.\overline{(A.B)}$$
$$= \overline{[\overline{\{A.\overline{(A.B)}\}}.\overline{\{B.\overline{(A.B)}\}}]}$$

The hardware implementation off this transformed expression needs only four NAND gates and is shown in Fig. 4.

Also, $\bar{A}.B + A.\bar{B} = \overline{[\overline{(A.\bar{B})} + \overline{(\bar{A}.B)}]}$

The hardware implementation of this expression using only the NOR gates is shown in Fig. 5.

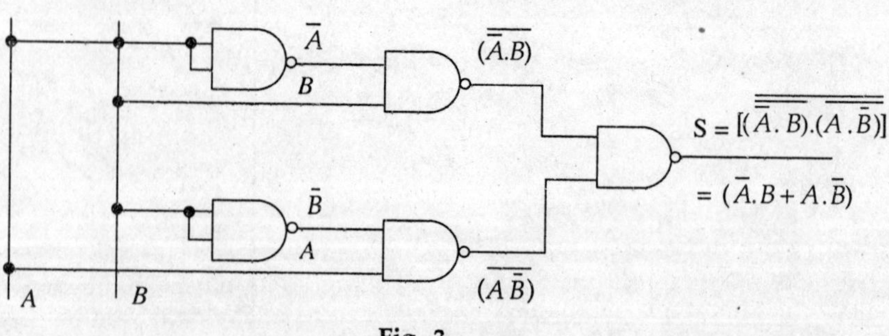

Fig. 3

The CARRY output can be implemented in a similar way by using either a single AND gate (Fig. 6(a) or two NAND gates as shown in Fig 6(b) or three NOR gates as shown in Fig. 6(c).

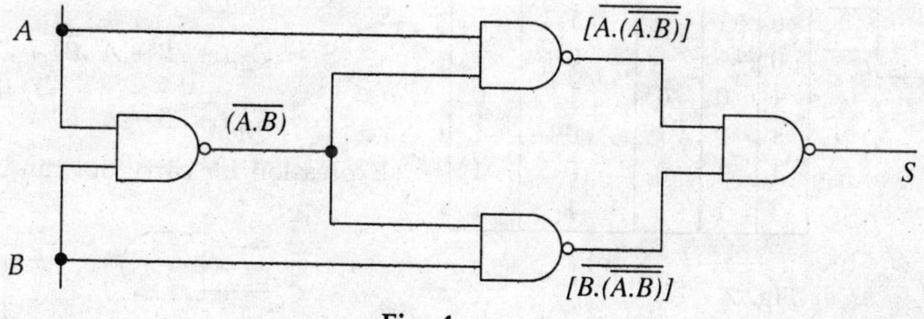

Fig. 4

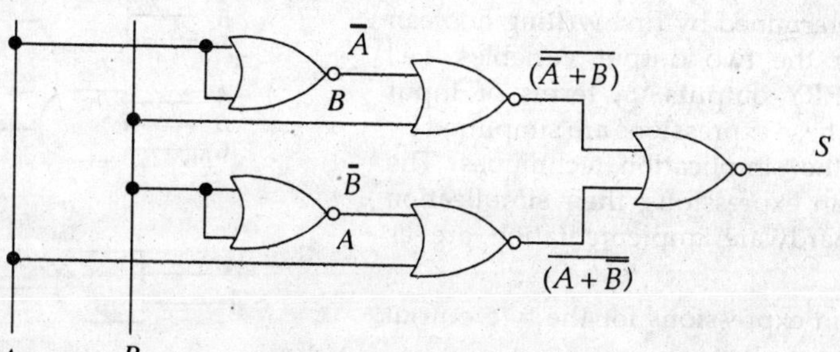

Fig. 5

2. **Explain the functioning of a full adder circuit. How does it differ from a half-adder? Give the block schematic arrangement of Q-4-bit binary adder using full-Adder building blocks.**

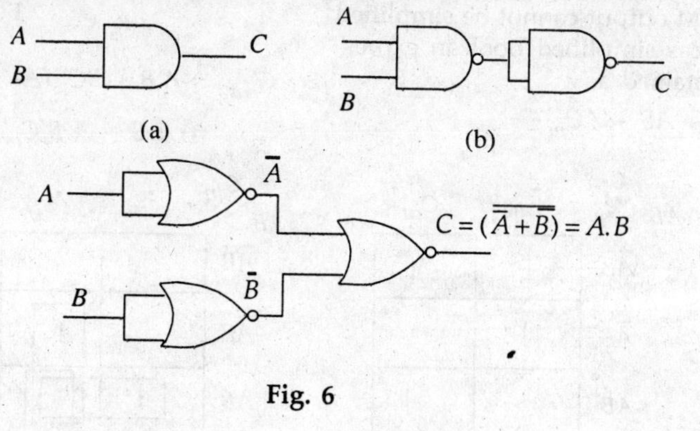

Fig. 6

Ans. : Full Adder circuit is an aithmetic circuit block which is capablee of adding three binary digits to produce a SUM and a CARRY output. This building block is very handy when it comes to adding binary numbers with large number of bits. We begin with addition of LSBs of the two numbers and record the sum under the LSB column and take the carry, if any, forward to the next higher column bits. Consequently, when add the bits of the next adjacent higher column bits, we need to add three bits if there was a carry from the previous addition. In the hardware implementation of an adder circuit capable of adding larger binary numbers, we need a Full-adder (as a Half-adder can be used only for the addition of LSBs).

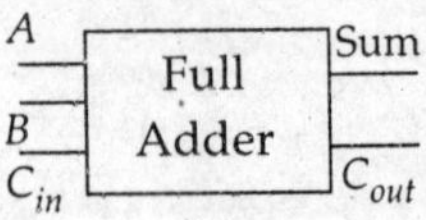

(a)

A	B	C_{in}	SUM(S)	C_{out}
0	0	0	0	0
0	0	1	1	0
0	1	0	1	0
0	1	1	0	1
1	0	0	1	0
1	0	1	0	1
1	1	0	0	1
1	1	1	1	1

(b)

Fig. 7

Fig. 7(a) shows the block representation of a full-adder circuit while its truth table is given in Fig. 7(b).

The combinational logic implementation of a full adder is determined by first writing Boolean expressions for the two output variables, i.e., SUM and CARRY outputs, in terms of input variable. Then these expressions are simplified by using any of the simplication techniques. The relevant Boolean expressions, their simplication followed by hardware implementation are as follows :

The Boolean expressions for the two output variables can be written as

$$S = \bar{A}.\bar{B}.C_{in} \;\; \bar{A}.B.\bar{C}_{in} + A.\bar{B}.\bar{C}_{in} + A.B.C_{in}$$

and

$$C_{out} = \bar{A}.B.C_{in} + A.\bar{B}.C_{in} + A.B.\bar{C}_{in} + A.B.C_{in}$$

The Karnaugh maps for the two expressions are given in Fig. 8(a) for SUM output and 8(b) for C_{out} output. The two maps indicate that the expression for SUM output cannot be simplified any further whereas simplified Boolean expression for C_{out} is obtained as

$$C_{out} = B.\,C_{in} + AB + AC_{in}$$

The hardware implementation is shown in Fig 9. the expression for the SUM output can be expressed as,

$$S = \bar{C}_{in}(\bar{A}\,.\,B + A\,.\,\bar{B}) + C_{in}(A.\,B + \bar{A}.\,\bar{B})$$

$$= \bar{C}_{in}(\bar{A}\,.\,B + A\,.\,\bar{B}) + C_{in}(\overline{\bar{A}.\,B + A.\,\bar{B}}).$$

Expression for carry out can be rewritten as :

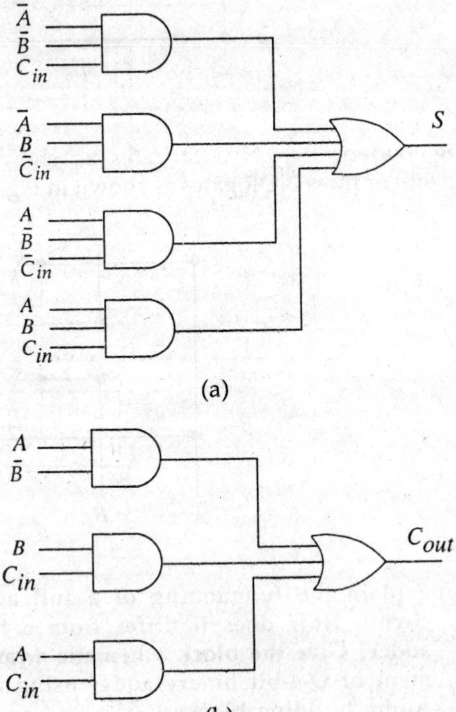

(a) (b)

Fig. 9

$$C_{out} = A.B + BC_{in}(A+\bar{A}) + AC_{in}(B+\bar{B})$$

$$= A.B + A.B.C_{in} + \bar{A}.B.C_{in} + A.\bar{B}.C_{in}$$

AB \ C_{in}	$\bar{C}_{in}$	C_{in}
$\bar{A}\bar{B}$		1
$\bar{A}B$	1	
AB		1
$A\bar{B}$	1	

(a)

AB \ C_{in}	$\bar{C}_{in}$	C_{in}
$\bar{A}\bar{B}$		
$\bar{A}B$		1
AB	1	1
$A\bar{B}$		1

(b)

Fig. 8

$$= A.B + A.B.C_{in} + \overline{A}.B.C_{in} + A.\overline{B}.C_{in}$$

$$= A.B + (1 + C_{in}) + C_{in} + (\overline{A}.B.A.\overline{B}$$

$$= A.B + C_{in}(\overline{A}.B.A.\overline{B}).$$

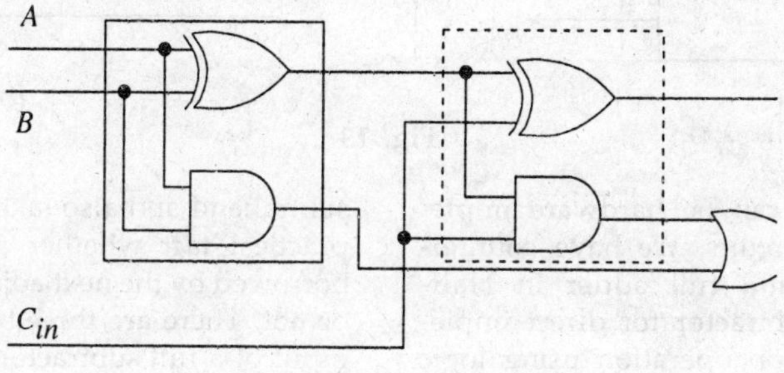

Fig. 10

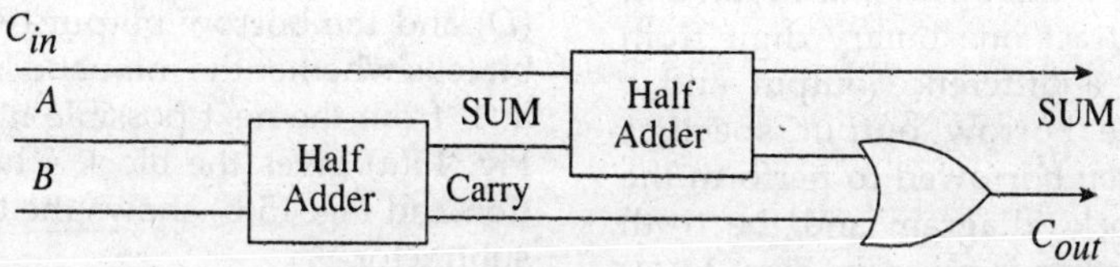

Fig. 11

In hardware to implement these re-written Boolean expression for SUM and CARRY outputs, we get a familiar logic circuit as shown in Fig. 10 wherefrom we find that it comprises of two half-adders and an OR-gate. The logic circuit of Fig. 10 is re-drawn in Fig. 11 where the portion of the circuit representing a half-adder has been replaced by a block.

The FULL-ADDER forms the basic building block for constructing binary adders. The full adder is capable of adding only 1-bit binary numbers. These adders can be added in cascade to form adders capable of adding binary numbers with larger number of bits. For example, a 4-bit binary adder would require four full-adders of the type shown in Fig. 11 to be connected in cascade. Fig. 12 shows the arrangement. In this case, ($A_3A_2A_1 = A_0$) and ($B_3B_2B_1 = B_0$) are the two binary numbers to be added with A_0 and B_0 representing LSBs of the two numbers.

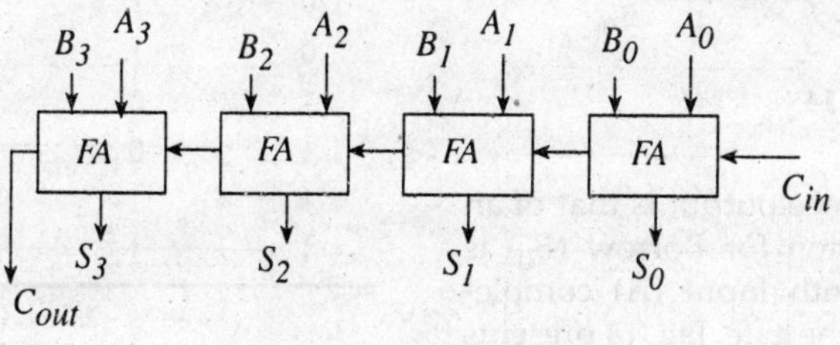

Fig. 12

3. What is a Half Subtractor?

Ans. : The subtraction of two given binary numbers can be obtained by adding 2's complement of the subtrahend to the minuend. With this, the

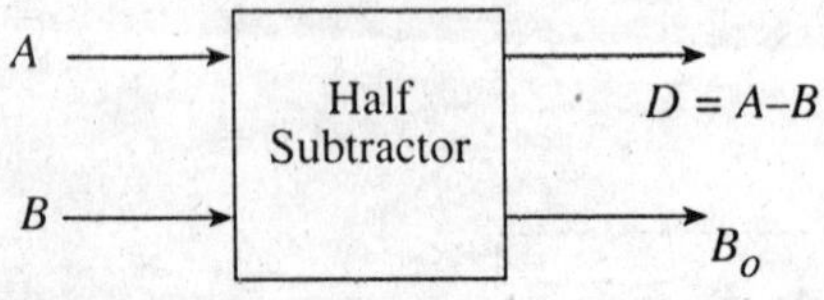

(a)

A	B	D	B_0
0	0	0	0
0	1	1	1
1	0	1	0
1	1	0	0

(b)

Fig. 13

subtraction operation can be hardware implemented with adder circuits. We have counterparts of half-adder and full adder in Half-subtractor and Full-subtractor for direct implementation of subtraction operation using logic gates.

Half subtractor is a combinational circuit that can be used to subtract one binary digit from another to produce a difference output and a borrow output. The borrow output specifies whether a '1' has been borrowed to perform the subtraction. The block diagram and the truth table of a half-subtractor is given in Figs. 13 (a) and (b) respectively.

The Boolean expressions for the two outputs can be written as,

$$D = \overline{A} . B + A . \overline{B}$$

$$B_0 = \overline{A} . B.$$

The two expressions indicate that there is no scope for any further simplification. While the

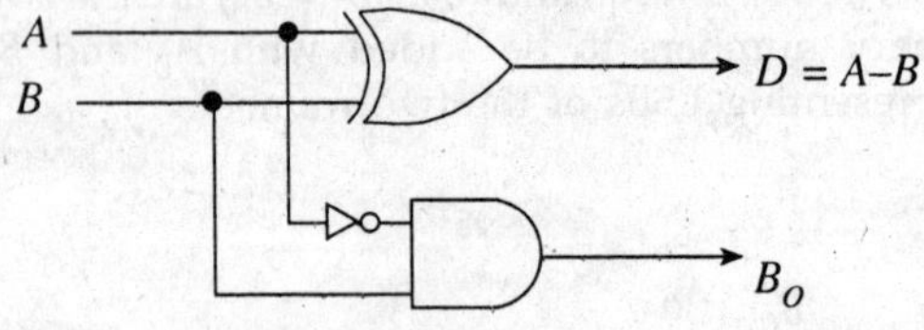

Fig. 14

expression for the difference output is that of an EX-OR gate, the expression for Borrow (B_0) is that of an AND gate with input (A) complemented before it is fed to the gate. Fig. 14 presents the logic complementation of a half-subtractor.

4. What is a Full-Subtractor ?

Ans. : Full-Subtractor is used to perform subtraction operation on two bits, a minuend and a subtrahend and also takes into consideration the practical fact whether a '1' had already been borrowed by the next adjacent lower minuend bit or not. There are three bits to be handled at the input of a full subtractor, viz., the two bits to be subtracted and a borrow bit designated as (B_{in}). There are two outputs, viz., the difference output (D) and the borrow output (B_0). Borrow output bit tells whether the minuend bit needs to borrow a '1' from the next possible higher minuend bit. Fig. 15(a) gives the block schematic representation and Fig. 15(b) shows the truth table of a full subtractor.

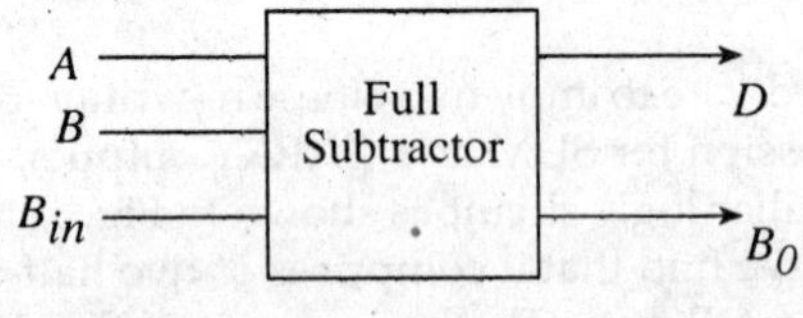

(a)

Minuend (A)	Subtrahend (B)	Borrow In (B_{in})	Difference (D)	Borrow Out (B_o)
0	0	0	0	0
0	0	1	1	1
0	1	0	1	1
0	1	1	0	1
1	0	0	1	0
1	0	1	0	0
1	1	0	0	0
1	1	1	1	1

(b)

Fig. 15

The Boolean expressions for the two output variables can be written as

$D = \overline{A} \cdot \overline{B} \cdot B_{in} + \overline{A} \cdot B \cdot \overline{B}_{in} + A \cdot \overline{B} \cdot \overline{B}_{in} + A \cdot B \cdot B_{in}$ and

$B_o = \overline{A} \cdot \overline{B} \cdot B_{in} + \overline{A} \cdot B \cdot \overline{B}_{in} + \overline{A}\ B + B_{in} + A \cdot B \cdot B_{in}$

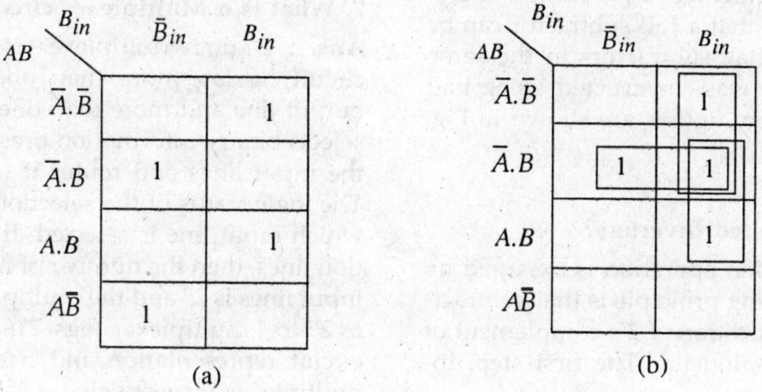

(a) (b)

Fig. 16

The Karnaugh maps for the two expressions are given in Fig. 16(*a*) for difference output (*D*) and (*b*) for borrow output (B_0).

The two Karnaugh maps indicate that no simplification possible for the difference output (*D*). The simplified expression for (B_o) is expressed as

$$B_o = \overline{A} \cdot B + \overline{A} \cdot B_{in} + BB_{in}$$

Also, $D = \overline{A} \cdot \overline{B} \cdot B_{in} + \overline{A} \cdot B \cdot \overline{B}_{in} + A \cdot \overline{B} \cdot \overline{B}_{in} + A \cdot B \cdot B_{in}$

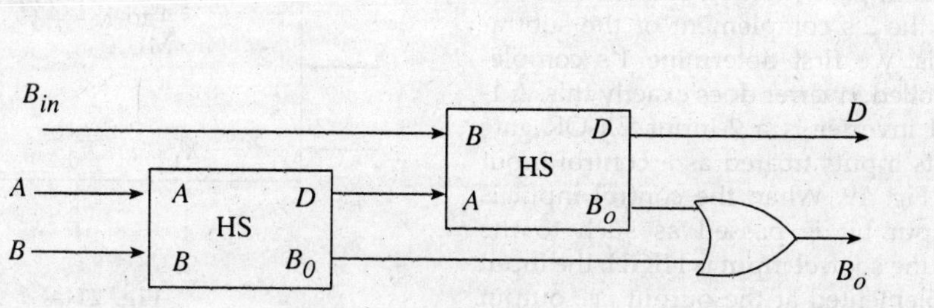

Fig. 17

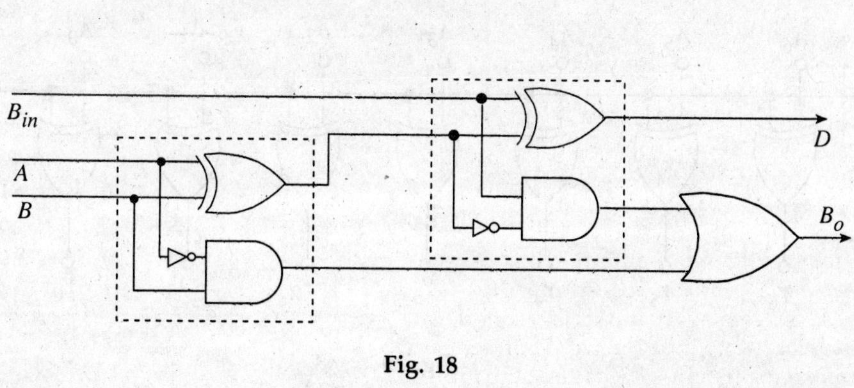

Fig. 18

Comparing these expressions with those of full-adder, we find that the expression for (*D*) is same as that for SUM output and expression for borrow out (B_o) is similar to the expression for carry out (C_o). They become same if (*A*) input for subtractor is complemented. By a similar analysis, it can be shown that a full subtractor can be implemented with half-subtractors in the same way as a full-adder was constructed using half adders. The necessary figures are shown in Fig. 17 and Fig. 18.

5. What are Controlled Inverters?

Ans. : The subtraction operation is executed by adders, the underlying principle is that subtracting is nothing but addition of 2's complement of subtrahend to the minuend. The first step, towards practical implementation of a subtractor is to determine the 2's complement of the subtrahend. For this, we first determine 1's complement. A controlled inverter does exactly this. A 1-bit controlled inverter is a 2-input EX-OR gate with one of its inputs treated as a control input as shown in Fig. 19. When the control input is LOW, the input bit is passed as such to the output. But if the control input is HIGH, the input bit gets complemented at the output *i.e.*, output is complement of the input. Fig. 20 gives an 8-bit controlled inverter of this type.

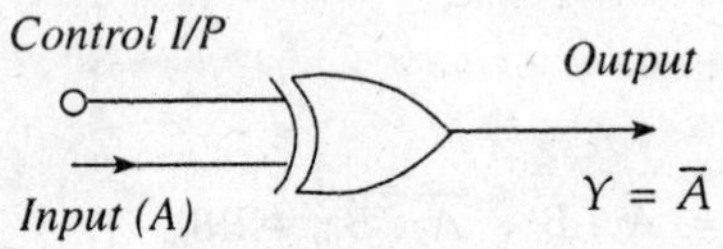

Fig. 19

6. What are Magnitude Comparators?

Ans. : Magnitude Comparator is a combinational circuit that compares two given numbers and finds whether one is equal to, less than or greater than the other.

7. What is a Multiplexer circuit?

Ans. : Digital Multiplexer is a combinational circuit having more than one input line, one output line and more than one selection lines. It selects binary information present on any one of the input lines and routes it to the output line. The logic status of the selection lines determines which input line is selected. If there are *n* selection lines, then the number of maximum possible input lines is 2^n and the multiplexer is referred to as 2^n-to-1 multiplexer. Figs. 21(*a*) and (*b*) show the circuit representation and truth table of 4-to-1 multiplexer respectively.

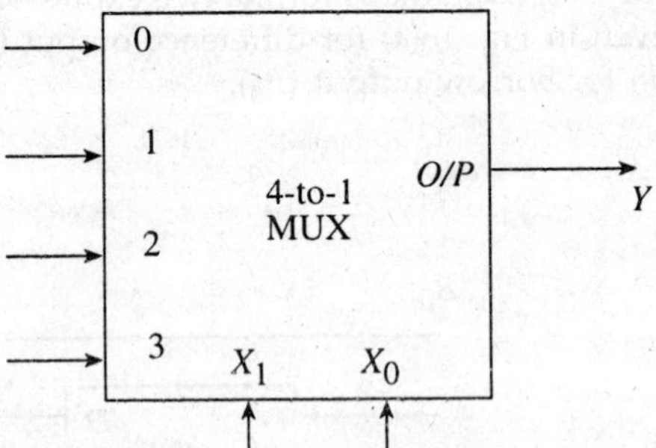

Fig. 21(a)

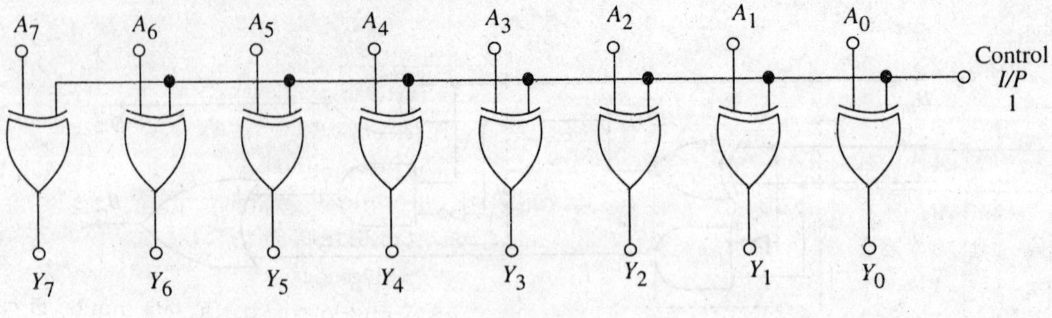

Fig. 20

X_1	X_0	Y
0	0	I_0
0	1	I_1
1	0	I_2
1	1	I_3

Fig. 21(b)

8. What is a De-multiplexer? How does a decoder differ from a de-multiplexer?

Ans. : De-multiplexer is a combinational logic circuit with an input line, 2^n output lines and n select lines. It routes the information present on the input line to any of the output lines. The output line that gets the information present on the input line is determined by the bit status of the selection lines. A decoder is a special case of a demultiplexer without the input line. Fig. 22(*a*) gives the circuit representation of a (1 × 4) demultiplexer. Fig. 22(*b*) gives the truth table of the demultiplexer when the input line is held HIGH.

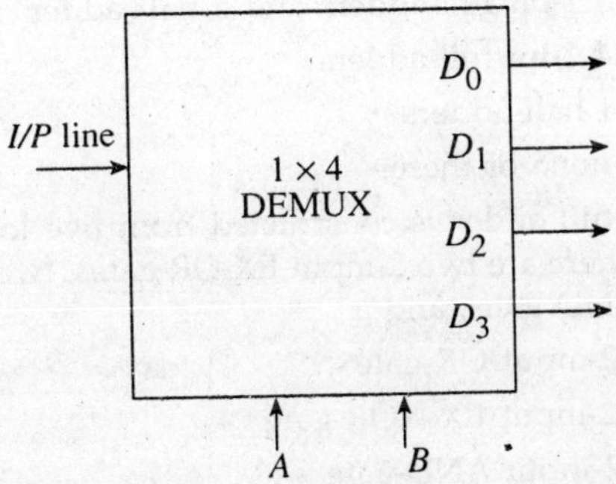

Fig. 22(a)

I/P	Select		O/P			
	A	B	D_0	D_1	D_2	D_3
1	0	0	1	0	0	0
1	0	1	0	1	0	0
1	1	0	0	0	1	0
1	1	1	0	0	0	1

Fig. 22(b)

9. What is an Encoder?

Ans. : Encoder is a multiplexer without the single output line of a multiplexer. It is a combinational logic function that has 2^n (or less) input lines and n output lines, which were n selection lines in a multiplexer. The n output lines generate the binary code for the possible 2^n input lines.

10. Outline the functional principle of a ROM. How is a ROM used to implement a Boolean expression?

Ans. : Read Only Memory (ROM) is a memory device that can be used to store a certain fixed set of binary information. These devices have certain inherent links which can be fused (or made) or broken to store any user specified binary information in the device. Once a given pattern is developed, it remains as such even if power is turned off and on. There are types of ROMs where the user can erase the data already stored on the ROM chip and load it with fresh data.

11. What is a Programmable Logic Array (PLA)?

Ans. : A Programmable Logic Array (PLA) enables logic functions expressed in sum-of-products form to be implemented directly.

12. Multiple Choice Questions

(i) A decoder is nothing but a demultiplexer without

(a) control inputs
(b) data input
(c) enable input
(d) none of these.

(ii) An arithmetic digital building block that has one data input, more than one data outputs and more than one control inputs is the

(a) Data Selector
(b) Multiplexer
(c) Decoder
(d) Demultiplexer.

(iii) A data selector is also called a

(a) De-multiplexer
(b) Priority encoder
(c) Multiplexer
(d) Decoder.

(iv) A multiplexer has n data inputs, m control inputs and one output, then

(a) $2^m = n$
(b) $2^n = m$

(c) $m^n = 2$
(d) $n^2 = m$.

(v) In a 1-to-8 demultiplexer, the control inputs are designated as *ABCD* with *A* as the LSB and *D* as the *MSB*, the ouputs are designated as 0 to 15. If the data input is steered to the output designated 7 when clocked, then the status of control inputs in the order *DCBA* just prior to clocking is
(a) 1110
(b) 0111
(c) 1000
(d) none of these.

(vi) In a 1-to-16 de-multiplexer, the number of control inputs will be
(a) 4
(b) 1
(c) 2
(d) 16.

(vii) If the arithmetic and logic operations are to be done on larger bit numbers and they are to be done with high speed, a digital building block that is invariably used is
(a) the lock ahead carry generator
(b) a RAM
(c) a ROM
(d) an ALU (Arithmetic Logic Unit).

(viii) A BCD-to-Decimal decoder has
(a) 10 inputs and 10 outputs
(b) 4 inputs and 10 outputs
(c) 10 inputs and 4 outputs
(d) none of these.

(ix) A de-multiplexer has one data input, *m* control inputs and *n* outputs, then
(a) $2^n = m$
(b) $2^m = n$
(c) $n^m = 2$
(d) $m^n = 2$.

(x) A full adder circuit has
(a) two inputs and one output
(b) two inputs and two outputs
(c) two inputs and three outputs
(d) three inputs and two outputs.

(xi) A half-adder circuit has two inputs and
(a) one output
(b) two outputs
(c) three outputs
(d) none of these

(xii) There is 10-line decimal designated 0, 1, 2, 3, 4, 5, 7, 8, 9 to 4-line *BCD* priority encoder with a LOW deciding the priority. The encoder outputs are active LOW and are designated as D (MSB) C B A (LSB). If the input bit stream is 1010111011 with the bit on extreme left representing 0-line and that on extreme right representing 9-line, then the encoded output is
(a) 0111
(b) 1000
(c) 1001
(d) 0000.

(xiii) A 4-bit binary adder could be constructed with a minimum of
(a) 3 1-bit full-adders and a half adder
(b) 4 1-bit full-adders
(c) 4 half-adders
(d) none of these.

(xiv) A full adder is constructed from five logic gates. There are two 2-input EX-OR gates, two 2-input AND gates and a
(a) 2-input OR-gate
(b) 2-input EX-NOR gate
(c) 2-input AND-gate
(d) 2-input NAND-gate.

(xv) A half-adder is constructed from two-input logic gates. One of them is a 2-input EX-OR gate. The other is a
(a) 2-input AND-gate
(b) 2-input NAND-gate
(c) 2-input OR-gate
(d) 2-input NOR-gate.

(xvi) A full subtractor can be constructed from two half subtractors and a
(a) 2-input NAND
(b) 2-input NOR
(c) 2-input OR
(d) 2-input AND.

(xvii) The half-subtractor can be constructed from a 2-input EX-OR, a 2-input AND and
(a) a 2-input OR
(b) a 2-input EX-NOR

(c) an inverter

(d) none of these.

(xviii) The half-subtractor has

(a) 2-inputs and 2-outputs

(b) 3-inputs and 2-outputs

(c) 2-inputs and 1-output

(d) none of these.

(xix) A ROM is usually not preferred to implement those Boolean functions which

(a) are very complex

(b) can otherwise be implemented using a PLA

(c) have a large number of 'Don't care conditions'

(d) have large number of outputs

(xx) It is desired to implement a binary multiplier capable of multiplying two 3-bit binary numbers. Size of ROM would be

(a) 64 × 4

(b) 1024 × 6

(c) 64 × 8

(d) 128 × 6.

(xxi) A masked programmed version of a PAL device that is pin compatible to its PAL counterpart is known as

(a) a PLA device

(b) HAL device

(c) a PLS device

(d) PGA device.

(xxii) A Programmable Array Logic (PAL) device has a

(a) programmable AND-array at the input and a fixed OR-array at the output

(b) fixed AND-array at the input and a programmable OR-array at the output

(c) fixed AND-array at the input and a fixed OR-array at the output

(d) none of these.

(xxiii) The programmable logic device (PLD) having a programmable AND-array at the input and a programmable OR-array at the output is called a

(a) Programmable Logic Array (PLA)

(b) Programmable Array Logic (PAL)

(c) Programmable Gae Array (PGA)

(d) Application Specific Integrated Circuit (ASIC).

(xxiv) It is said that a full-adder can be used as a subtractor. Let us say, you want to build a 4-bit parallel subtractor. All that you need is one 7483 (which is a 4-bit full-adder) and

(a) four inverters

(b) four 2-input AND-gates

(c) sixteen 2-input NAND-gates

(d) eight 2-input NAND-gates.

(xxv) Given that 7483 is a 4-bit parallel adder chip, you can build a 16-bit parallel adder circuit

(a) by cascaded arrangement of 16 numbers of 7483's

(b) by a cascaded arrangement of 4 numbers of 7483's

(c) 16-bit parallel adder cannot be built using 7483's

(d) none of these.

ANSWERS

(i) (b), (ii) (d), (iii) (c), (iv) (a), (v) (b), (vi) (a), (vii) (a), (viii) (b), (ix) (b), (x) (d), (xi) (b), (xii) (a), (xiii) (a), (xiv) (a), (xv) (a), (xvi) (c), (xvii) (c), (xviii) (a), (xix) (c), (xx) (a), (xxi) (b), (xxii) (a), (xxiii) (a), (xxiv) (a), (xxv) (b).